DEGENERATE DIFFERENTIAL EQUATIONS IN BANACH SPACES

PURE AND APPLIED MATHEMATICS

A Program of Monographs, Textbooks, and Lecture Notes

EXECUTIVE EDITORS

MONOGRAPHS AND TEXTBOOKS IN
PURE AND APPLIED MATHEMATICS

55. *C. W. Groetsch,* Elements of Applicable Functional Analysis (1980)
56. *I. Vaisman,* Foundations of Three-Dimensional Euclidean Geometry (1980)
57. *H. I. Freedan,* Deterministic Mathematical Models in Population Ecology (1980)
58. *S. B. Chae,* Lebesgue Integration (1980)
59. *C. S. Rees et al.,* Theory and Applications of Fourier Analysis (1981)
60. *L. Nachbin,* Introduction to Functional Analysis (R. M. Aron, trans.) (1981)
61. *G. Orzech and M. Orzech,* Plane Algebraic Curves (1981)
62. *R. Johnsonbaugh and W. E. Pfaffenberger,* Foundations of Mathematical Analysis (1981)
63. *W. L. Voxman and R. H. Goetschel,* Advanced Calculus (1981)
64. *L. J. Corwin and R. H. Szczarba,* Multivariable Calculus (1982)
65. *V. I. Istrăţescu,* Introduction to Linear Operator Theory (1981)
66. *R. D. Järvinen,* Finite and Infinite Dimensional Linear Spaces (1981)
67. *J. K. Beem and P. E. Ehrlich,* Global Lorentzian Geometry (1981)
68. *D. L. Armacost,* The Structure of Locally Compact Abelian Groups (1981)
69. *J. W. Brewer and M. K. Smith, eds.,* Emmy Noether: A Tribute (1981)
70. *K. H. Kim,* Boolean Matrix Theory and Applications (1982)
71. *T. W. Wieting,* The Mathematical Theory of Chromatic Plane Ornaments (1982)
72. *D. B. Gauld,* Differential Topology (1982)
73. *R. L. Faber,* Foundations of Euclidean and Non-Euclidean Geometry (1983)
74. *M. Carmeli,* Statistical Theory and Random Matrices (1983)
75. *J. H. Carruth et al.,* The Theory of Topological Semigroups (1983)
76. *R. L. Faber,* Differential Geometry and Relativity Theory (1983)
77. *S. Barnett,* Polynomials and Linear Control Systems (1983)
78. *G. Karpilovsky,* Commutative Group Algebras (1983)
79. *F. Van Oystaeyen and A. Verschoren,* Relative Invariants of Rings (1983)
80. *I. Vaisman,* A First Course in Differential Geometry (1984)
81. *G. W. Swan,* Applications of Optimal Control Theory in Biomedicine (1984)
82. *T. Petrie and J. D. Randall,* Transformation Groups on Manifolds (1984)
83. *K. Goebel and S. Reich,* Uniform Convexity, Hyperbolic Geometry, and Nonexpansive Mappings (1984)
84. *T. Albu and C. Năstăsescu,* Relative Finiteness in Module Theory (1984)
85. *K. Hrbacek and T. Jech,* Introduction to Set Theory: Second Edition (1984)
86. *F. Van Oystaeyen and A. Verschoren,* Relative Invariants of Rings (1984)
87. *B. R. McDonald,* Linear Algebra Over Commutative Rings (1984)
88. *M. Namba,* Geometry of Projective Algebraic Curves (1984)
89. *G. F. Webb,* Theory of Nonlinear Age-Dependent Population Dynamics (1985)
90. *M. R. Bremner et al.,* Tables of Dominant Weight Multiplicities for Representations of Simple Lie Algebras (1985)
91. *A. E. Fekete,* Real Linear Algebra (1985)
92. *S. B. Chae,* Holomorphy and Calculus in Normed Spaces (1985)
93. *A. J. Jerri,* Introduction to Integral Equations with Applications (1985)
94. *G. Karpilovsky,* Projective Representations of Finite Groups (1985)
95. *L. Narici and E. Beckenstein,* Topological Vector Spaces (1985)
96. *J. Weeks,* The Shape of Space (1985)
97. *P. R. Gribik and K. O. Kortanek,* Extremal Methods of Operations Research (1985)
98. *J.-A. Chao and W. A. Woyczynski, eds.,* Probability Theory and Harmonic Analysis (1986)
99. *G. D. Crown et al.,* Abstract Algebra (1986)
100. *J. H. Carruth et al.,* The Theory of Topological Semigroups, Volume 2 (1986)
101. *R. S. Doran and V. A. Belfi,* Characterizations of C*-Algebras (1986)
102. *M. W. Jeter,* Mathematical Programming (1986)
103. *M. Altman,* A Unified Theory of Nonlinear Operator and Evolution Equations with Applications (1986)
104. *A. Verschoren,* Relative Invariants of Sheaves (1987)
105. *R. A. Usmani,* Applied Linear Algebra (1987)
106. *P. Blass and J. Lang,* Zariski Surfaces and Differential Equations in Characteristic $p > 0$ (1987)
107. *J. A. Reneke et al.,* Structured Hereditary Systems (1987)
108. *H. Busemann and B. B. Phadke,* Spaces with Distinguished Geodesics (1987)
109. *R. Harte,* Invertibility and Singularity for Bounded Linear Operators (1988)
110. *G. S. Ladde et al.,* Oscillation Theory of Differential Equations with Deviating Arguments (1987)

111. L. Dudkin et al., Iterative Aggregation Theory (1987)
112. T. Okubo, Differential Geometry (1987)
113. D. L. Stancl and M. L. Stancl, Real Analysis with Point-Set Topology (1987)
114. T. C. Gard, Introduction to Stochastic Differential Equations (1988)
115. S. S. Abhyankar, Enumerative Combinatorics of Young Tableaux (1988)
116. H. Strade and R. Farnsteiner, Modular Lie Algebras and Their Representations (1988)
117. J. A. Huckaba, Commutative Rings with Zero Divisors (1988)
118. W. D. Wallis, Combinatorial Designs (1988)
119. W. Wiesław, Topological Fields (1988)
120. G. Karpilovsky, Field Theory (1988)
121. S. Caenepeel and F. Van Oystaeyen, Brauer Groups and the Cohomology of Graded Rings (1989)
122. W. Kozlowski, Modular Function Spaces (1988)
123. E. Lowen-Colebunders, Function Classes of Cauchy Continuous Maps (1989)
124. M. Pavel, Fundamentals of Pattern Recognition (1989)
125. V. Lakshmikantham et al., Stability Analysis of Nonlinear Systems (1989)
126. R. Sivaramakrishnan, The Classical Theory of Arithmetic Functions (1989)
127. N. A. Watson, Parabolic Equations on an Infinite Strip (1989)
128. K. J. Hastings, Introduction to the Mathematics of Operations Research (1989)
129. B. Fine, Algebraic Theory of the Bianchi Groups (1989)
130. D. N. Dikranjan et al., Topological Groups (1989)
131. J. C. Morgan II, Point Set Theory (1990)
132. P. Biler and A. Witkowski, Problems in Mathematical Analysis (1990)
133. H. J. Sussmann, Nonlinear Controllability and Optimal Control (1990)
134. J.-P. Florens et al., Elements of Bayesian Statistics (1990)
135. N. Shell, Topological Fields and Near Valuations (1990)
136. B. F. Doolin and C. F. Martin, Introduction to Differential Geometry for Engineers (1990)
137. S. S. Holland, Jr., Applied Analysis by the Hilbert Space Method (1990)
138. J. Okniński, Semigroup Algebras (1990)
139. K. Zhu, Operator Theory in Function Spaces (1990)
140. G. B. Price, An Introduction to Multicomplex Spaces and Functions (1991)
141. R. B. Darst, Introduction to Linear Programming (1991)
142. P. L. Sachdev, Nonlinear Ordinary Differential Equations and Their Applications (1991)
143. T. Husain, Orthogonal Schauder Bases (1991)
144. J. Foran, Fundamentals of Real Analysis (1991)
145. W. C. Brown, Matrices and Vector Spaces (1991)
146. M. M. Rao and Z. D. Ren, Theory of Orlicz Spaces (1991)
147. J. S. Golan and T. Head, Modules and the Structures of Rings (1991)
148. C. Small, Arithmetic of Finite Fields (1991)
149. K. Yang, Complex Algebraic Geometry (1991)
150. D. G. Hoffman et al., Coding Theory (1991)
151. M. O. González, Classical Complex Analysis (1992)
152. M. O. González, Complex Analysis (1992)
153. L. W. Baggett, Functional Analysis (1992)
154. M. Sniedovich, Dynamic Programming (1992)
155. R. P. Agarwal, Difference Equations and Inequalities (1992)
156. C. Brezinski, Biorthogonality and Its Applications to Numerical Analysis (1992)
157. C. Swartz, An Introduction to Functional Analysis (1992)
158. S. B. Nadler, Jr., Continuum Theory (1992)
159. M. A. Al-Gwaiz, Theory of Distributions (1992)
160. E. Perry, Geometry: Axiomatic Developments with Problem Solving (1992)
161. E. Castillo and M. R. Ruiz-Cobo, Functional Equations and Modelling in Science and Engineering (1992)
162. A. J. Jerri, Integral and Discrete Transforms with Applications and Error Analysis (1992)
163. A. Charlier et al., Tensors and the Clifford Algebra (1992)
164. P. Biler and T. Nadzieja, Problems and Examples in Differential Equations (1992)
165. E. Hansen, Global Optimization Using Interval Analysis (1992)
166. S. Guerre-Delabrière, Classical Sequences in Banach Spaces (1992)
167. Y. C. Wong, Introductory Theory of Topological Vector Spaces (1992)

168. *S. H. Kulkarni and B. V. Limaye*, Real Function Algebras (1992)
169. *W. C. Brown*, Matrices Over Commutative Rings (1993)
170. *J. Loustau and M. Dillon*, Linear Geometry with Computer Graphics (1993)
171. *W. V. Petryshyn*, Approximation-Solvability of Nonlinear Functional and Differential Equations (1993)
172. *E. C. Young*, Vector and Tensor Analysis: Second Edition (1993)
173. *T. A. Bick*, Elementary Boundary Value Problems (1993)
174. *M. Pavel*, Fundamentals of Pattern Recognition: Second Edition (1993)
175. *S. A. Albeverio et al.*, Noncommutative Distributions (1993)
176. *W. Fulks*, Complex Variables (1993)
177. *M. M. Rao*, Conditional Measures and Applications (1993)
178. *A. Janicki and A. Weron*, Simulation and Chaotic Behavior of α-Stable Stochastic Processes (1994)
179. *P. Neittaanmäki and D. Tiba*, Optimal Control of Nonlinear Parabolic Systems (1994)
180. *J. Cronin*, Differential Equations: Introduction and Qualitative Theory, Second Edition (1994)
181. *S. Heikkilä and V. Lakshmikantham*, Monotone Iterative Techniques for Discontinuous Nonlinear Differential Equations (1994)
182. *X. Mao*, Exponential Stability of Stochastic Differential Equations (1994)
183. *B. S. Thomson*, Symmetric Properties of Real Functions (1994)
184. *J. E. Rubio*, Optimization and Nonstandard Analysis (1994)
185. *J. L. Bueso et al.*, Compatibility, Stability, and Sheaves (1995)
186. *A. N. Michel and K. Wang*, Qualitative Theory of Dynamical Systems (1995)
187. *M. R. Darnel*, Theory of Lattice-Ordered Groups (1995)
188. *Z. Naniewicz and P. D. Panagiotopoulos*, Mathematical Theory of Hemivariational Inequalities and Applications (1995)
189. *L. J. Corwin and R. H. Szczarba*, Calculus in Vector Spaces: Second Edition (1995)
190. *L. H. Erbe et al.*, Oscillation Theory for Functional Differential Equations (1995)
191. *S. Agaian et al.*, Binary Polynomial Transforms and Nonlinear Digital Filters (1995)
192. *M. I. Gil'*, Norm Estimations for Operation-Valued Functions and Applications (1995)
193. *P. A. Grillet*, Semigroups: An Introduction to the Structure Theory (1995)
194. *S. Kichenassamy*, Nonlinear Wave Equations (1996)
195. *V. F. Krotov*, Global Methods in Optimal Control Theory (1996)
196. *K. I. Beidar et al.*, Rings with Generalized Identities (1996)
197. *V. I. Arnautov et al.*, Introduction to the Theory of Topological Rings and Modules (1996)
198. *G. Sierksma*, Linear and Integer Programming (1996)
199. *R. Lasser*, Introduction to Fourier Series (1996)
200. *V. Sima*, Algorithms for Linear-Quadratic Optimization (1996)
201. *D. Redmond*, Number Theory (1996)
202. *J. K. Beem et al.*, Global Lorentzian Geometry: Second Edition (1996)
203. *M. Fontana et al.*, Prüfer Domains (1997)
204. *H. Tanabe*, Functional Analytic Methods for Partial Differential Equations (1997)
205. *C. Q. Zhang*, Integer Flows and Cycle Covers of Graphs (1997)
206. *E. Spiegel and C. J. O'Donnell*, Incidence Algebras (1997)
207. *B. Jakubczyk and W. Respondek*, Geometry of Feedback and Optimal Control (1998)
208. *T. W. Haynes et al.*, Fundamentals of Domination in Graphs (1998)
209. *T. W. Haynes et al.*, Domination in Graphs: Advanced Topics (1998)
210. *L. A. D'Alotto et al.*, A Unified Signal Algebra Approach to Two-Dimensional Parallel Digital Signal Processing (1998)
211. *F. Halter-Koch*, Ideal Systems (1998)
212. *N. K. Govil et al.*, Approximation Theory (1998)
213. *R. Cross*, Multivalued Linear Operators (1998)
214. *A. A. Martynyuk*, Stability by Liapunov's Matrix Function Method with Applications (1998)
215. *A. Favini and A. Yagi*, Degenerate Differential Equations in Banach Spaces (1999)

Additional Volumes in Preparation

DEGENERATE DIFFERENTIAL EQUATIONS IN BANACH SPACES

Angelo Favini
University of Bologna
Bologna, Italy

Atsushi Yagi
Osaka University
Osaka, Japan

MARCEL DEKKER, INC.　　　NEW YORK · BASEL · HONG KONG

ISBN: 0-8247-1677-9

This book is printed on acid-free paper.

Headquarters
Marcel Dekker, Inc.
270 Madison Avenue, New York, NY 10016
tel: 212-696-9000; fax: 212-685-4540

Eastern Hemisphere Distribution
Marcel Dekker AG
Hutgasse 4, Postfach 812, CH-4001 Basel, Switzerland
tel: 44-61-261-8482; fax: 44-61-261-8896

World Wide Web
http://www.dekker.com

The publisher offers discounts on this book when ordered in bulk quantities. For more information, write to Special Sales/Professional Marketing at the headquarters address above.

Current printing (last digit)
10 9 8 7 6 5 4 3 2 1

PRINTED IN THE UNITED STATES OF AMERICA

To Maurizia and Gian Piero
with love

PREFACE

This book is concerned mainly with the initial value problem

$$
(1) \qquad \begin{cases} \dfrac{d}{dt} Mv = Lv + f(t), & 0 < t \le T, \\ Mv(0) = v_0, \end{cases}
$$

in a Banach space X, where M and L are closed linear operators on X, $f(\cdot)$ is an X-valued continuous function on $[0, T]$, and v_0 is a given element in X.

Since M^{-1} is not continuous in general, the classical theory of C_0 semigroups is not obviously applicable. On the other hand, a large number of partial differential equations arising in physics and in applied sciences can be written in this form; among them there are some famous examples, such as the pseudo-parabolic equations and the Sobolev equation.

Perhaps the first reference in this area is the monograph of R. W. Carroll and R. E. Showalter, *Singular and Degenerate Cauchy Problems* [48] published in 1976, but a large number of important theoretical results and applications have since enriched the field.

In general we look for regular (classical) solutions to equation (1) and treat weak solutions only occasionally. Roughly speaking there are two major approaches we use to treat equations of the form (1): the semigroup approach and the operational method developed by P. Grisvard and G. Da Prato [62,63,64,65] in a number of papers devoted to parabolic equations and elliptic equations.

The first approach reduces the problem to the multivalued differential equation

$$
(2) \qquad \begin{cases} \dfrac{du}{dt} \in Au + f(t), & 0 < t \le T, \\ u(0) = v_0, \end{cases}
$$

where $A = LM^{-1}$ and $u(t) = Mv(t)$. It has shown all its power when A is dissipative, because then one is able to treat even nonlinear operators A via the Crandall-Liggett theorem (see M. G. Crandall and T. Liggett [59]). However, maximal dissipativity properties of A are in general achieved only with respect to weak norms, for example, norms of distribution spaces H^{-m}, and this fails to imply regularity of the solutions. Here we exploit the linearity of the operator involved to develop a theory for (2) that seems much more satisfactory from this point of view.

The second method, although very abstract, just in virtue of its generality, permits us to handle various types of equations, from Cauchy problems to boundary value problems.

The main part of this book represents the original contributions of the authors. To explain simply the types of equations we can handle and the main directions of our theory, we begin the exposition with an Introduction, devoted to describing a few well known concrete partial differential equations of interest in various areas of applied mathematics, for which our results furnish more information (e.g., on regularity of their solutions) than previous treatments in the literature.

The reader is assumed to be familiar with some basic functional analysis, real and complex analysis and partial differential equations. To make the book reasonably self-contained, some classical results pertaining to linear operators, evolution equations and interpolation theory are included. The style of the book is in the Bourbaki spirit, which is typical of European pure mathematics. We think that this style is appropriate for the theory we are presenting, since it points out just the salient aspects.

In Chapter I, the main properties of multivalued linear operators which parallel the usual ones for single-valued linear operators are given.

Chapters II and III are devoted to existence theory for the multivalued linear Cauchy problem (2) in the hyperbolic case and in the parabolic case, respectively. Chapter III includes, in particular, extensions of the maximal regularity property for parabolic equations due to P. Grisvard. This should allow the solution of nonlinear problems of interest, but that falls outside the scope of our book.

Chapter IV considers equations of parabolic type in the nonautonomous case, where the coefficient operator $A = A(t)$ depends on t. In this way we can handle singular equations as defined in the Carroll and Showalter monograph.

In Chapter V, equation (2) is studied with resolvent properties of $A = LM^{-1}$ analogous to generators of integrated semigroups and to this end the operational method is suitably modified.

Chapter VI discusses degenerate equations of higher order in time, in particular, second order equations. It depends heavily on the results in Chapters II and III.

The last chapter contains very recent results by one of the authors concerning regularity of semigroups generated by second order degenerate parabolic operators in various function spaces, which occur in genetics, probability theory and approximation theory. To this end, we apply the methods of Chapter III, and we also describe some pertinent tools for each particular case.

We express our profound gratitude to Professor Viorel Barbu of the University of Iasi, Rumania, and Professor Jerry Goldstein of the University of Memphis for their useful advice and encouragement during the preparation of the book. We also thank Professor Enrico Obrecht of the University of Bologna, Italy, for his help with the final version. Thanks also to Professor Bruno Pini of the University of Bologna, who initiated the first author in this area of research. The second author expresses his sincere thanks to the Mathematical Analysis staff in the Department of Mathematics of the University of Bologna for their hearty hospitality during his several stays at the Department. We express our gratitude to the staff of Marcel Dekker, Inc., for their assistance and, in particular, to Ms Maria Allegra for her special efficiency and kindness.

Angelo Favini
Atsushi Yagi

CONTENTS

INTRODUCTION

The work described here began about 1980, but it had a decisive impetus after the visit of the second author at the University of Bologna in 1991. We saw the opportunity to unify semigroup methods and operational approaches to treat the solvability in a strong sense to very interesting degenerate differential problems, having always in mind their applications to partial differential equations.

Our model was the monograph of S. G. Krein "*Linear Differential Equations in Banach Space*", published in English translation by Amer. Math. Soc. in 1970, and we asked if well known results in this book and some others in the preceding litterratue on degenerate equations that ensured only their weak solution on the ground of energy norms, could be extended to "*better*" function spaces with respect to regularity.

Here we shall give a few examples relative to each chapter in the monograph, highlighting the application and novelty of the results.

EXAMPLE 1. Let us consider the symmetric system

$$C(x)\frac{\partial u}{\partial t} + \sum_{i=1}^{n} A_i(x)\frac{\partial u}{\partial x_i} + A_0(x)u = f(x,t) \quad \text{in } \Omega \times [0,T],$$

where Ω is a region in $\mathbb{R}^n$, $C(x)$, $A_0(x), \ldots, A_n(x)$ are $m \times m$ smoothly varying matrices, $A_1(x), \ldots, A_n(x)$ being hermitian, $C(x) \geq 0$, and f is a given function in $\Omega \times [0,T]$.

The case $C(x) \equiv I$, the identity, is well known after the works by P. D. Lax and R. S. Phillips [126], K. O. Friedrichs [106] and the more general case $C(x) > 0$ investigated by C. H. Wilcox [179]. More recently, M. Povoas studied in [146] the degenerate case $C(x) \geq 0$, and this allowed her to deduce existence and uniqueness of a (strong or strict) solution of Maxwell's equations in a nonhomogeneous anisotropic medium in the ambient space $L^2(0,T; L^2(\Omega))$.

One of our results permits to treat the same problem where $\Omega = \mathbb{R}^n$ in the space $\mathcal{C}([0,T]; L^2(\mathbb{R}^n))$ of all $L^2(\mathbb{R}^n)$ valued continuous functions on $[0,T]$.

The key step consists in representing the equation under consideration in the form

$$M^*\frac{dMv}{dt} = Lv + f(t), \quad 0 \leq t \leq T,$$

1

in a Hilbert space X, where M is a bounded operator from X into itself and L is a single valued closed linear operator in X. Though $L(M^*M)^{-1}$ is not maximal dissipative in X, it can be shown that $(M^*)^{-1}LM^{-1}$ does have this property and this suffices, in virtue of Theorem 2.10 below, to conclude that under suitable regularity conditions on f and on the initial datum, there exists a unique $v \in \mathcal{C}([0,T]; \mathcal{D}(L))$ with $Mv \in \mathcal{C}^1([0,T]; X)$ satisfying the equation and the condition $Mv(0) = Mv_0$.

By using again Theorem 2.10, we are able to treat in $X = L^2(\Omega)$ the degenerate Poisson-wave equation

$$\left(m(x)\frac{\partial}{\partial t} \right)^2 u = \Delta u + f(x,t) \quad \text{in } \Omega \times [0,T],$$

where Ω is a region in $\mathbb{R}^n$ that may be bounded or unbounded, $m \in L^\infty(\Omega)$, $m(x) \geq 0$. See Example 2.3.

EXAMPLE 2. Let us consider the famous degenerate heat equation

$$\frac{\partial m(x)v}{\partial t} = \Delta v + f(x,t) \quad \text{in } \Omega \times (0,T),$$

with boundary-initial conditions $v = 0$ on $\partial\Omega \times (0,T)$ and $m(x)v(x,0) = u_0(x)$ in Ω, where in this case Ω is a bounded region in $\mathbb{R}^n$, $n \geq 1$, with a smooth boundary, and $m \in L^\infty(\Omega)$, $m(x) \geq 0$ in Ω.

It is known that the theory of maximal dissipative operators works (see R. W. Carroll and R. E. Showalter [48], R. E. Showalter [155]) in the space $X = H^{-1}(\Omega)$, where $L = \Delta$ and M is the operator of multiplication by $m(x)$ which acts from $H_0^1(\Omega)$ into $H^{-1}(\Omega)$.

The choice of the space $H^{-1}(\Omega)$ is crucial in order to conclude that LM^{-1} is maximal dissipative. Indeed, as it is shown in Chapter III, such a property for LM^{-1} in the space $L^2(\Omega)$ holds no longer in general, because one has really the estimate

$$\|M(\lambda M - L)^{-1}f\|_{L^2} \leq C|\lambda|^{-\frac{1}{2}}\|f\|_{L^2} \quad \text{for all } f \in L^2(\Omega)$$

for any complex number λ such that $\Re e\lambda \geq -c(1 + |\Im m\lambda|)$, where $c > 0$. See Example 3.3. This is precisely the situation discussed by S. G. Krein [121] and K. Taira [165,166] when $M = I$.

Our first approach allows us to extend the results by Krein and Taira to this situation.

On the other hand, much work has been done in the last decades on maximal regularity of solutions v to the Cauchy problem of parabolic type

(1) $$\frac{dv}{dt} = Lv + f(t), \quad 0 \leq t \leq T,$$

(2) $$v(0) = v_0 \in X.$$

More precisely, what type of hypotheses on f and v_0 ensures that if f is Hölder continuous of exponent $\theta \in (0,1)$ from $[0,T]$ into X, then the solution v exists in the strict sense $v \in C^1([0,T];X)$ and, in addition, its derivative $\frac{dv}{dt}$ has the same time regularity as f?

The answer to this question was chiefly given by P. Grisvard in his pioneering work [113], but substantial applications of it have been made by G. Da Prato , A. Lunardi, E. Sinestrari, B. Terreni, P. Acquistapace and refer to local and global solutions to nonlinear parabolic partial differential equations of large interest.

The key role in the matter is played by the real interpolation spaces of J. L. Lions and J. Peetre $(X,Y)_{\theta,p}$, $0 < \theta < 1$, $1 \leq p \leq \infty$, between two Banach spaces X and Y, also known as "mean spaces" in the literature. For a detailed discussion of them we refer to the recent monograph by H. Amann [5] and to the even later text by H. Tanabe [168].

The main assumptions for this maximal regularity concern the analyticity of the semigroup generated by L, which may have a non dense domain $\mathcal{D}(L)$, too, and a condition like $Lv_0 + f(0) \in (X, \mathcal{D}(L))_{\theta,\infty}$, $0 < \theta < 1$.

Since the natural operator associated with the problem

$$(3) \qquad \frac{dMv}{dt} = Lv + f(t), \quad 0 < t \leq T,$$

$$(4) \qquad Mv(0) = u_0,$$

is the multivalued linear operator $A = LM^{-1}$ (after the change of unknown $Mv = u$), we are then led to investigate and possibly to characterize the space $(X, \mathcal{D}(A))_{\theta,\infty}$, $0 < \theta < 1$, following the lines of J. L. Lions and J. Peetre [130] and P. Grisvard [114], under the weaker assumption

$$(5) \qquad \|(\lambda - A)^{-1}f\|_X \leq C|\lambda|^{-\beta}\|f\|_X, \quad f \in X,$$

where $0 < \beta \leq 1$ and λ varies in a suitable region of the complex plane containing $\Re e\lambda \geq 0$.

Of course, we must expect that, unless $\beta = 1$, the regularity of the solution decreases with respect to the data. This in fact happens, but we are able to find a precise amount of this loss of regularity. See Theorem 3.24 below.

At the same time, in reaching our goal we do not confine ourselves to study a differential problem (3), (4), but we study the maximal regularity of solutions v to the operator equation

$$(6) \qquad \mathcal{T}v \in \mathcal{A}v + f,$$

where $\mathcal{T}$ is a single valued linear operator in X and $\mathcal{A}$ satisfies (5). In this manner we extend some results of C. Wild [180] to the multivalued case.

When applied to the Poisson-heat equation above, our results say that if $f \in \mathcal{C}^\theta([0,T]; L^2(\Omega))$, $\frac{1}{2} < \theta < 1$, and $u_0 = mv_0$, where $v_0 \in H^2(\Omega) \cap H_0^1(\Omega)$ and f are connected by the relation

$$\sup_{\xi>0} \xi^\theta \|\Delta(m\xi - \Delta)^{-1}\{\Delta v_0 + f(\cdot, 0)\}\|_{L^2} < \infty,$$

then such a problem has a unique strict (classical) solution v satisfying $\frac{\partial}{\partial t} mv \in \mathcal{C}^{\theta - \frac{1}{2}}([0,T]; L^2(\Omega))$.

EXAMPLE 3. Consider the degenerate parabolic equation

$$\frac{\partial u}{\partial t} = a(x)\Delta u + f(x,t) \quad \text{in } \Omega \times (0,T),$$

with boundary-initial conditions $u = 0$ on $\partial\Omega \times (0,T)$ and $u(x,0) = u_0(x)$ in Ω, where Ω is a bounded region in $\mathbb{R}^n$ with a smooth boundary $\partial\Omega$ and $a(x) > 0$ a.e. is a given function in $L^\infty(\Omega)$.

In Example 3.9 we prove that if

$$a^{-1} \in L^r(\Omega) \left\{ \begin{array}{l} \text{with some } r \geq 2 \text{ when } n = 1, \\ \text{with some } r > 2 \text{ when } n = 2, \\ \text{with some } r \geq n \text{ when } n \geq 3, \end{array} \right.$$

then, for all $f \in \mathcal{C}^\theta([0,T]; L^2(\Omega))$, $\frac{n}{2r} < \theta < 1$, and any $u_0 \in H_0^1(\Omega)$ such that $a(x)\Delta u_0 + f(x,0)$ belongs to a well characterised interpolation space, the problem has a unique solution u with regularity $\frac{\partial u}{\partial t} \in \mathcal{C}^{\theta - \frac{n}{2r}}([0,T]; L^2(\Omega))$.

EXAMPLE 4. Consider the problem

$$m(t)\frac{\partial u}{\partial t} = \Delta u + f(x,t) \quad \text{in } \Omega \times (0,T)$$

with boundary-initial conditions $u = 0$ on $\partial\Omega \times (0,T)$ and $u(x,0) = u_0(x)$ in Ω, where Ω is a bounded region in $\mathbb{R}^n$ with smooth $\partial\Omega$, $m(t) \geq 0$ is a function in $\mathcal{C}^{1+\mu}([0,T]; \mathbb{R})$, $0 < \mu \leq 1$, $m(0) = 0$, satisfying

$$\left|\frac{dm}{dt}(t)\right| \leq Cm(t)^\nu, \quad 0 \leq t \leq T,$$

with some $0 < \nu \leq 1$. We know (see e.g. A. Lunardi [131; p. 97]) that Δ with Dirichlet boundary conditions generates an analytic semigroup both in $X = L^p(\Omega)$, $1 < p < \infty$, and in $X = \mathcal{C}(\overline{\Omega})$ (here it has a non dense domain), provided that

$$\mathcal{D}(\Delta) = W_p^2(\Omega) \cap \overset{\circ}{W}{}_p^1(\Omega) \text{ and}$$

$$\mathcal{D}(\Delta) = \{u \in \bigcap_{p\geq 1} W_{p,\ell oc}^2(\Omega); \, u, \, \Delta u \in \mathcal{C}(\overline{\Omega}), \, u_{|\partial\Omega} = 0\},$$

respectively.

This is then a special case of the equation

$$M(t)\frac{du}{dt} = L(t)u + f(t), \quad 0 < t \leq T,$$

that is treated in Chapter IV, where various extensions of papers by A. Yagi [183,184], A. Favini [81,82,83,84,85], A. Favini and P. Plazzi [92,93,94], A. Favini and A. Yagi [96,97] are described.

Here Proposition 4.18 is applicable, entailing that if $\Delta u_0(x) + f(x, 0) = 0$ and $f \in \mathcal{C}^{1+\sigma}([0, T]; X)$, $\sigma > 0$, then the problem has a unique solution u with $u \in \mathcal{C}^1([0, T]; X) \cap \mathcal{C}([0, T]; \mathcal{D}(\Delta))$. Moreover, $\Delta\frac{\partial u}{\partial t}(x, 0) = \frac{\partial f}{\partial t}(x, 0)$. We refer to Example 4.2.

EXAMPLE 5. Differential-algebraic equations like $F(t, y(t), y'(t)) = 0$, where F and y are vector valued and the jacobian of F with respect to y' is singular, have been the subject of much attention by scientists and engineers in several areas for over twenty five years.

More recently many researchers focused their attention to the numerical solution of this type of equations, on the ground of their applications to circuits, constrained variable problems, robotics, optimal control.

We refer to K. E. Brenan, S. L. Campbell and L. R. Petzold [34] and to E. Griepentrog and R. März [112]. Very useful tools for this type of treatment can be found in the monograph by S. L. Campbell and C. D. Meyer Jr. [44] on generalized inverses of linear operators.

Since we confine ourselves to linear operators, even if we will discuss later the case where $\lambda = 0$ is a pole of $(\lambda + LM^{-1})^{-1} = M(\lambda M + L)^{-1}$ of arbitrary order, we need to develop a theory permitting $\|(\lambda - A)^{-1}\|$ to have a polymomial growth on the half plane $\Re e\lambda \geq -\lambda_0$, $\lambda_0 > 0$. The degree m of the polynomial $p(\lambda)$ such that $\|L(\lambda M + L)^{-1}\|_{\mathcal{L}(X)} \leq p(\lambda)$ has obviously a key role. See Theorem 5.7 below. In particular, we see, much more generally, that under these hypotheses for all $f \in \mathcal{C}^{m+2}([0, T]; X)$ with $f^{(j)}(0) = 0$, $j = 0, 1, \ldots, m + 1$, and $u_0 = Mv_0$, when Lv_0 belongs to the range of $(ML^{-1})^{m+2}$, then (3), (4) has a unique solution v such that $Mv \in \mathcal{C}^1([0, T]; X)$.

Here we want to single out two important examples. Consider the equation

$$\frac{\partial m(x)v}{\partial t} = \Delta v + f(x, t) \quad \text{in } \Omega \times (0, T)$$

of Example 2 with the Dirichlet boundary condition, but where this time we seek for $\mathcal{C}(\overline{\Omega})$ valued solutions. In other words, the domain of Δ is precisely the one indicated by H. B. Stewart [161] and described in Example 4.

In Example 5.10 below we shall prove that the involved operators satisfy our conditions with $m = 1$.

As a second example, we take the preceding equation but $X = L^1(\Omega)$. Of course, the domain of Δ must be suitably chosen according to H. Brezis and W. Strauss [37]; again our operational method works with $m = 1$.

In fact one proves that the related operators M and L satisfy the more accurate estimate

$$\|M(\lambda M - L)^{-1}f\|_{L^1} \le (\Re\lambda)^{-1}\|f\|_{L^1}, \quad \Re\lambda > 0,$$

that is the Hille-Yosida condition of Chapter II. Therefore we can apply Theorem 2.6, too, which says that if $f \in \mathcal{C}^1([0,T]; L^1(\Omega))$ and $\Delta v_0 + f(x,0) \in$ the closure in $L^1(\Omega)$ of the range $M(\mathcal{D}(L))$, then the present Cauchy problem

$$\frac{\partial m(x)v}{\partial t} = \Delta v + f(x,t) \quad \text{in } \Omega \times (0,T)$$
$$\|m(\cdot)[v(x,t) - v_0(\cdot)]\|_{L^1} \to 0 \quad \text{as } t \to 0$$

with Dirichlet boundary conditions has a unique solution v with $m(x)v \in \mathcal{C}^1([0,T]; L^1(\Omega))$.

EXAMPLE 6. The case where $\lambda = 0$ is a polar singularity of $(\lambda - A)^{-1}$ has some special features that allow to weaken somewhat the regularity assumptions required by Theorem 5.7. They rely on the representation of the space X as a direct sum of $\mathcal{R}(A^{-k})$ and $\mathcal{K}(A^{-k})$ with a suitable positive integer k. A typical application is to the Sobolev type equations. To this purpose, we are able to handle these equations in the form (3) even when $\mathcal{D}(L) \subset \mathcal{D}(M)$, which is a rather delicate one. It has recently received further attention, see G. A. Sviridyuk [162,163]. Optimal control for this type of equations is to be found in G. A. Sviridyuk and A. A. Efremov [164].

In Example 5.14 it is shown how our method works with success for the (parabolic) equation

$$\frac{\partial}{\partial t}(s - \Delta)v = -b\Delta^2 v + a\Delta v \quad \text{in } \Omega \times (0,T),$$

where a, $b > 0$ and s is an eigenvalue of Δ either in $L^2(\Omega)$ or in $\mathcal{C}(\overline{\Omega})$. It models the evolution of a free surface of the filtered fluid, G. A. Sviridyuk [163].

Section 5.6 can be viewed as the natural extension to infinite dimensional spaces of the classical theory of F. R. Gantmacher [108] and S. L. Campbell [40,41].

EXAMPLE 7. Consider the abstract control system in a Hilbert space X

$$(7) \qquad \frac{du}{dt} = Lu + Kf(t), \quad 0 < t < \infty,$$

with $u(0) = u_0$, where L generates a C_0 semigroup in X and the control operator K is continuous for the Hilbert space U to $\mathcal{D}(L^*)'$, the dual space of $\mathcal{D}(L^*)$ with respect to the X topology. Here L^* is the adjoint of L in X.

If $R \in \mathcal{L}(X)$, $R^*R \geq cI$, $c > 0$, it is known from I. Lasiecka and R. Triggiani [125] that under mild assumptions there exists the unique solution $\{f^0(t; u_0), u^0(t; u_0)\}$ of the optimal control problem associated to (7) which minimizes over all $u \in L^2((0, \infty); U)$ the quadratic cost

$$J(f, u) = \int_0^\infty \{\|Ru(t)\|_X^2 + \|f(t)\|_U^2\}dt,$$

and it is given by

$$f^0(t; u_0) = -KK^*Pu^0(t; u_0) \in L^2((0, \infty); U),$$

where P is the unique solution of the algebraic Riccati equation

$$L^*P + PL + R^*R = PKK^*P$$

in the sense explained in [125].

The optimal solution is in general described by a one-time integrated semigroup $\Phi(t)$ on X with feedback generator $L_F = L - KK^*P$. In some special cases, $\Phi(t)$ has additional properties such as analyticity or belonging to the Gevrey class. For example, if $L = iS$, $K = iS^\theta$, $0 \leq \theta \leq 1$, $R = I$, then $L_F = -S^\theta + iS$. If $\theta = 1$, then e^{tL_F} is analytic, but if $0 < \theta < 1$, e^{tL_F} is in the Gevrey class $> \frac{1}{\theta}$ for all $t > 0$. This follows from Theorem 5.14. Gevrey class regularity of solutions to possibly degenerate equations is investigated in Section 5.4. However, major application concerns the generation property of the sum $-A + iB$, $A = A^* > 0$, $B = B^*$ with $\mathcal{D}(A) \not\subset \mathcal{D}(B)$.

EXAMPLE 8. Consider the hyperbolic-parabolic problem

$$m_1(x)\frac{\partial^2 u}{\partial t^2} + m_2(x)\frac{\partial u}{\partial t} - \Delta u = f(x, t) \quad \text{in } \Omega \times [0, T],$$

with boundary-initial conditions $u = 0$ on $\partial\Omega \times [0, T]$ and $u(x, 0) = u_0(x)$, $\frac{\partial u}{\partial t}(x, 0) = u_1(x)$ in Ω, where $m_1(x)$, $m_2(x) \geq 0$ are two continuous non negative functions on $\overline{\Omega}$. The problem is strongly degenerate since it may change type from parabolic to hyperbolic to elliptic. It has been investigated in recent years by A. Bensoussan, J. L. Lions and G. C. Papanicolau

[27] and others, but in any case the usual approach of Hilbert triplets $V \subset H \subset V'$ is used for. Hence the solution to the problem satisfies $\Delta u \in L^2((0,T); H^{-1}(\Omega))$.

The case of nonlinear operators acting on the first derivative $\frac{\partial u}{\partial t}$ is of large interest. See the recent paper by K. T. Andrews, K. L. Kuttler and M. Shillor [6].

The linear character of multiplication by $m_2(x)$ allows us to treat the indicated problem in the space $X = L^2(\Omega)$ under compatibility relations involving $f(x,0)$, u_0, u_1 and the derivative $\frac{\partial f}{\partial t}(x,0)$. See Example 6.18. Notice that in this way one can treat a much more general case of the Poisson wave equation described in Example 2.

EXAMPLE 9. We recall from D. Henry [115; p. 43], that if A_1, A_2 are the two alleles in a population at a single gene locus for $0 \leq p$, $x \leq 1$, $t > 0$, and $\phi(x,t;p)dx$ denotes the probability that the frequency at A_1 at time t lies in $(x, x + dx)$, under the condition that the A_1-frequency is p at time $t = 0$, then ϕ satisfies the parabolic equation

$$\frac{\partial \phi}{\partial t} = \tfrac{1}{2} \frac{\partial^2}{\partial x^2}(V(x,t)\phi) - \frac{\partial}{\partial x}(M(x,t)\phi).$$

Here M, V are the mean and variance of the change in gene frequency for generation. For example, $V(x,t) = \frac{1}{2N}x(1-x)$, and M may then describe the action of selection, mutation, etc. on the gene frequencies, when random mating among a population of size N is assumed. Let us call W the elliptic operator on the right hand side of the equation. W is singular at $x = 0$ and $x = 1$, even if the singularity is not too surprising, since $x = 0$ and $x = 1$ are absorbing states in the absence of mutation.

After the pioneering papers by W. Feller [101,102], a large amount of work was done in studying generation properties of W in different function spaces with and without boundary conditions. In particular, the space of continuous functions is relevant for the most pertinent applications.

One of the main results in this subject is due to Ph. Clément and C. A. Timmermans [53], where necessary and sufficient conditions are given in order that $Wu = \alpha u'' + \beta u'$ with the so called Wentzell boundary conditions generates a C_0 semigroup in $\mathcal{C}([0,1])$.

After this result and related ones by J. A. Goldstein and others (see a. e., J. A. Goldstein and C.-Y. Lin [111]), an important question for its applications is the regularity of the semigroup generated by W.

Here, in particular, we prove that under suitable hypotheses satisfied by $\alpha(x) = x(1-x)$, $\beta(x) \equiv 0$, the most important case, the operator W generates either an analytic semigroup or a real analytic semigroup in different function spaces, including $\mathcal{C}([0,1])$.

For some other interesting examples of application to one dimensional Markov processes and diffusion models in genetics, we refer to P. Mandl [133; pp. 47-49].

PRELIMINARIES AND NOTATIONS

This preliminary chapter is devoted to fixing the notations and recalling some basic notions of interpolation theory.

0.1 LINEAR OPERATORS

If X is a complex normed space we shall write $\| \cdot \|$ for its norm. If several normed spaces are considered, we shall provide their norms with suitable subscripts such as $\| \cdot \|_X$. Throughout this monograph, if it is not mentioned, the normed spaces are always over the complex field $\mathbb{C}$ and therefore complex normed spaces.

If X is a real or complex Banach space with the norm $\| \cdot \|_X$ and $I \subset \mathbb{R}$ is an interval, $\mathcal{B}(I; X)$, $\mathcal{C}(I; X)$, $\mathcal{C}^m(I; X)$, $m \in \mathbb{N}$, $\mathcal{C}^\infty(I; X)$ consist respectively of all bounded, continuous, m-times continuously differentiable, infinitely many times differentiable functions $f \colon I \to X$. $\mathcal{B}(I; X)$ is endowed with the supremum norm

$$\|f\|_{\mathcal{B}(I;X)} = \sup_{t \in I} \|f(t)\|_X.$$

Sometimes, if no confusion may arise, we shall use $\| \cdot \|_\infty$ for $\| \cdot \|_{\mathcal{B}(I;X)}$. Moreover, if $X = \mathbb{R}$ or $\mathbb{C}$, the shorter notations $\mathcal{B}(I)$, $\mathcal{C}(I)$, $\ldots$ shall be used instead of $\mathcal{B}(I; X)$, $\mathcal{C}(I; X)$, $\ldots$.

The space of Hölder continuous functions $\mathcal{C}^\alpha(I; X)$, $\mathcal{C}^{m+\alpha}(I; X)$, $m \in \mathbb{N}$, $\alpha \in]0, 1[$, are defined by means of

$$\mathcal{C}^\alpha(I; X) = \{f \in \mathcal{B}(I; X) \cap \mathcal{C}(I; X); \sup_{t \neq s} \frac{\|f(t) - f(s)\|_X}{|t - s|^\alpha} < \infty\},$$

$$\mathcal{C}^{m+\alpha}(I; X) = \{f \in \mathcal{C}^m(I; X); \ f^{(k)} \in \mathcal{B}(I; X) \cap \mathcal{C}(I; X), \ k = 0, \ldots, m,$$
$$f^{(m)} \in \mathcal{C}^\alpha(I; X)\}.$$

If Ω is a nonempty open subset of $\mathbb{R}^n$, m is a nonnegative integer, D^m denotes m-th order derivatives, that is, $D^m = D^\alpha = \frac{\partial^m}{\partial x_1^{\alpha_1} \ldots \partial x_n^{\alpha_n}}$, where $\alpha = (\alpha_1, \ldots, \alpha_n)$ is a multi-index, $\alpha_j \in \mathbb{N} \cup \{0\}$, $|\alpha| = m = \alpha_1 + \cdots + \alpha_n$, then $\mathcal{C}^m(\Omega)$ denotes the set of all functions whose derivatives of order up to

m are continuous on Ω. One puts $\mathcal{C}^0(\Omega) = \mathcal{C}(\Omega)$. $\mathcal{C}_0^m(\Omega)$ is then the totality of the function belonging to $\mathcal{C}^m(\Omega)$ with compact support in Ω. One puts $\mathcal{C}_0^0(\Omega) = \mathcal{C}_0(\Omega)$. $\mathcal{C}^\infty(\Omega)$ and $\mathcal{C}_0^\infty(\Omega)$ are correspondingly defined. $\mathcal{C}(\overline{\Omega})$ denotes space of all continuous and bounded functions in $\overline{\Omega}$. Analogously, $\mathcal{C}^m(\overline{\Omega})$ denotes the set of all m-times continuously differentiable functions in Ω whose derivatives up to the order m are bounded and continuously extendable up to the boundary, $m \geq 1$. A corresponding definition holds for $m = \infty$.

If Ω is a measurable subset in $\mathbb{R}^n$, for $1 \leq p \leq \infty$, $L^p(\Omega)$ denotes the set of all scalar functions u in Ω such that u is (Lebesgue) measurable in Ω and $|u(x)|^p$ is summable on Ω if $1 \leq p < \infty$ and essentially bounded in Ω if $p = \infty$. The norm of $L^p(\Omega)$ is defined by

$$\|u\|_p = \begin{cases} \left(\int_\Omega |u(x)|^p dx \right)^{\frac{1}{p}}, & 1 \leq p < \infty, \\ \operatorname{esssup}_{x \in \Omega} |u(x)|, & p = \infty. \end{cases}$$

If X is a Banach space, then $L^p(0, T; X)$, $1 \leq p < \infty$, $0 < T < \infty$, is the set of all X-valued functions u that are strongly measurable in $(0, T)$ and $\|u(t)\|_X^p$ is summable on $(0, T)$ if $p < \infty$ and essentially bounded in $(0, T)$ if $p = \infty$. The norms in $L^p(0, T; X)$ are correspondingly defined.

If $1 \leq p \leq \infty$ and Ω is an open subset in $\mathbb{R}^n$, the Sobolev space $W_p^m(\Omega)$, $m \in \mathbb{N}$, is the totality of the functions whose distributional derivatives of order up to m belong to $L^p(\Omega)$. Its norm is usually given by

$$\|u\|_{m,p} = \left(\sum_{k=0}^m |u|_{k,p}^p \right)^{\frac{1}{p}}, \quad 1 \leq p < \infty,$$

$$\|u\|_{m,\infty} = \max_{k=0,\dots,m} |u|_{k,\infty},$$

where

$$|u|_{k,p} = \begin{cases} \left(\sum_{|\alpha|=k} \int_\Omega |D^\alpha u|^p dx \right)^{\frac{1}{p}}, & 1 \leq p < \infty, \\ \max_{|\alpha|=k} \operatorname{ess\,sup}_{x \in \Omega} |D^\alpha u(x)|, & p = \infty. \end{cases}$$

The closure of $\mathcal{C}_0^m(\Omega)$ in $W_p^m(\Omega)$, $1 \leq p < \infty$, $m \in \mathbb{N}$, is denoted by $\mathring{W}_p^m(\Omega)$. If $p = 2$ we shall use to write $H^m(\Omega)$ and $H_0^m(\Omega)$ instead of $W_2^m(\Omega)$ and $\mathring{W}_2^m(\Omega)$, respectively. $H^{-m}(\Omega)$ is the dual space of $H_0^m(\Omega)$, $m \in \mathbb{N}$. It is well known that if Ω is uniformly regular of class $\mathcal{C}^m$, $m \in \mathbb{N}$ (see H. Tanabe [168; p. 102]), $1 \leq p < \infty$, then the boundary values $(\gamma_j u)(x) = (\frac{\partial}{\partial \nu})^j u_{|\partial\Omega}$, $j = 0, \dots, m - 1$, are defined for all functions $u \in W_p^m(\Omega)$, where ν is the outward normal vector of $\partial\Omega$, so that $u \in W_p^m(\Omega)$ belongs to $\mathring{W}_p^m(\Omega)$ if and only if the trace $\gamma_j u$ of u on the boundary vanish for $j = 0, \dots, m - 1$. Of course $\gamma_0 u$ is the trace of u on $\partial\Omega$. At last, $W^m_{p,\ell oc}(\Omega)$ denotes the totality of functions u on Ω such that $u \in W^m_p(U)$ for all open set U with closure $\overline{U} \subset \Omega$.

If X, Y are two complex Banach spaces, the linear space of all continuous linear operators from X into Y shall be denoted by $\mathcal{L}(X,Y)$ and by $\mathcal{L}(X)$ if $X = Y$. For $B \in \mathcal{L}(X,Y)$, the norm of B is given by

$$\|B\| = \|B\|_{\mathcal{L}(X,Y)} = \sup_{\|x\|_X \leq 1} \|Bx\|_Y$$
$$= \inf\{c \geq 0;\ \|Bx\|_Y \leq c\|x\|_X\ \text{ for all }\ x \in X\}.$$

The topology on $\mathcal{L}(X,Y)$ induced by this norm is the *uniform operator topology*.

$\mathcal{K}(X,Y)$ (resp. $\mathcal{K}(X)$) shall denote the space of all compact operators from X into Y (resp. from X into itself).

If A is a linear operator from its domain $\mathcal{D}(A)$, a linear subspace of X, into Y, then we shall write

$$A{:}X \supset \mathcal{D}(A) \to Y.$$

By $\text{graph}(A) \subset X \times Y$, we denote the set

$$\text{graph}(A) = \{(x,y) \in X \times Y;\ x \in \mathcal{D}(A),\ y = Ax\}.$$

Such an operator is said to be *closed* if the $\text{graph}(A)$ is a closed subset of $X \times Y$. with respect to the product topology.

If $A{:}\mathcal{D}(A) \to Y$ is a linear operator, a norm is introduced on $\mathcal{D}(A)$ by setting
$$\|x\|_{\mathcal{D}(A)} = \|x\| + \|Ax\|$$
for all $x \in \mathcal{D}(A)$ and we always think of $\mathcal{D}(A)$ as being equipped with this norm, the socalled *graph norm* on $\mathcal{D}(A)$. It is seen that A is a closed operator if and only if $\mathcal{D}(A)$ is a Banach space. Moreover, if A is a bijection from $\mathcal{D}(A)$ onto Y and its inverse A^{-1} is a bounded operator from Y into X, then
$$\|x\|_{\mathcal{D}(A)} = \|Ax\|_X, \quad u \in \mathcal{D}(A),$$
defines a norm equivalent to the graph norm on $\mathcal{D}(A)$.

Finally, if $A{:}\mathcal{D}(A) \to Y$ is a linear operator, we set

$$\mathcal{N}(A) = \{x \in \mathcal{D}(A);\ Ax = 0\}, \quad \mathcal{R}(A) = \{Ax \in Y;\ x \in \mathcal{D}(A)\}.$$

$\mathcal{N}(A)$ and $\mathcal{R}(A)$ are linear subspaces of X and Y respectively, and are called *the kernel* (or *null space*) and *the range* (or *image*) of A, respectively.

If X and Y are Banach spaces such that $X \subset Y$ as sets and the inclusion (identity) mapping $i\colon X \to Y$ is continuous, X is said to be *continuously* *imbedded in Y*. If two normed spaces X and Y are equal as sets and have equivalent norms, we write $X = Y$.

If $X \subset Y$ is continuously imbedded and $A: \mathcal{D}(A) \subset Y \to Y$ is a closed linear operator, the operator $A_X: \mathcal{D}(A_X) \to X$ defined by $\mathcal{D}(A_X) = \{x \in \mathcal{D}(A) \cap X;\ Ax \in X\}$ and $A_X x = Ax$ is called *the X-realization of A*. It is easily recognized that A_X is a closed operator too.

If X is a Banach space over $\mathbb{C}$, a linear functional on X is a linear operator from X into $\mathbb{C}$. $X' = \mathcal{L}(X, \mathbb{C})$ is the topological dual space of X. If $x' \in X'$, $x \in X$, we write:

$$\langle x', x \rangle = \langle x', x \rangle_{X',X} = x'(x),$$

the duality between X' and X.

If $B \in \mathcal{L}(X, Y)$, when Y is another Banach space, *the dual operator B' of B* is the uniquely determined operator $B' \in \mathcal{L}(Y', X')$ such that

$$\langle B'y', x \rangle_{X',X} = \langle y', Bx \rangle_{Y',Y}$$

for all $y' \in Y'$ and $x \in X$. If X and Y are two Hilbert spaces, *the adjoint B^* of B* is defined by

$$\langle B^*y, x \rangle_X = \langle y, Bx \rangle_Y$$

for all $y \in Y$ and $x \in X$, where $\langle \cdot, \cdot \rangle_X$ and $\langle \cdot, \cdot \rangle_Y$ denote the inner product in X and in Y, respectively.

The dual (resp. *adjoint*) operator of a closed linear operator from X into Y is defined provided that B is densely defined:

$$\mathcal{D}(B') = \{y' \in Y';\ x \to \langle y', Bx \rangle \ \text{is continuous on } \mathcal{D}(B)$$

$$\text{endowed with the } X\text{-norm}\},$$

$$\langle B'y', x \rangle_{X',X} = \langle y', Bx \rangle_{Y',Y}$$

for all $y' \in \mathcal{D}(B')$ and all $x \in \mathcal{D}(B)$. An analogous definition holds for the adjoint of a densely defined linear operator from a Hilbert space to another Hilbert space.

If X is a Banach space over $\mathbb{C}$ and A is a closed linear operator from $\mathcal{D}(A) \subset X$ into X, *the resolvent set $\rho(A)$ of A* is the open set of $\mathbb{C}$ given by

$$\rho(A) = \{\lambda \in \mathbb{C};\ R(\lambda, A) = (\lambda - A)^{-1} \ \text{exists and lies in } \mathcal{L}(X)\},$$

where $\lambda - A = \lambda I - A$, I being the identity operator of X. *The spectrum $\sigma(A)$ of A* is the complement of $\rho(A)$ in $\mathbb{C}$: $\sigma(A) = \mathbb{C} \setminus \rho(A)$. Hence, $\sigma(A)$ is a closed set. *An eigenvalue of A* is an element of the point spectrum $\sigma_p(A)$ of A, i.e.

$$\sigma_p(A) = \{\lambda \in \mathbb{C};\ \lambda - A \ \text{is not injective}\}.$$

If λ is an eigenvalue of A, $\mathcal{N}(\lambda - A)$ is *its eigenspace* and any non zero element of it is said to be *an eigenvector* for the eigenvalue λ. If $A \in \mathcal{L}(X)$, then $\sigma(A)$ is a compact set and *the spectral radius* of A defined by $r(A) = \sup\{|\lambda|;\ \lambda \in \sigma(A)\}$ coincides with $\lim_{n \to \infty} \|A^n\|^{\frac{1}{n}}$.

0.2 INTERPOLATION SPACES

In this section we only recall these concepts of interpolation theory that are needed for our purposes. For the details we refer the reader to H. Triebel [172] and H. Amann [5].

The pair (E_0, E_1) of Banach spaces E_0 and E_1 is said to be *an interpolation couple* if there exists a locally convex topological space X such that $E_i \subset X$, $i = 0, 1$, continuously. Then, $E_0 \cap E_1$ and $E_0 + E_1$ are well defined Banach spaces. Notice that, if $E_1 \subset E_0$ continuously, then $E_0 \cap E_1 = E_1$ and $E_0 + E_1 = E_0$ (in the sense that they coincide with equivalence of norms), so that one can then take $X = E_0$.

If (E_0, E_1) is an interpolation couple and

$$E_0 \cap E_1 \subset E \subset E_0 + E_1 \quad \text{continuously,}$$

then the Banach space E is said to be an intermediate space with respect to (E_0, E_1).

If $\mathbb{K}$ is either $\mathbb{R}$ or $\mathbb{C}$, let $\mathcal{B}$ be the category of ($\mathbb{K}$-)Banach spaces, that is, the objects of $\mathcal{B}$ are the Banach spaces over $\mathbb{K}$, its morphisms are the bounded linear operators, and the composition is the usual composition of mappings. We denote by $\mathcal{B}_1$ the category of interpolation couples, so that the objects of $\mathcal{B}_1$ are the interpolation couples, the morphisms of $\mathcal{B}_1$ are the elements A of $\mathcal{L}(E_0 + E_1, F_0 + F_1)$ satisfying $A \in \mathcal{L}(E_i, F_i)$, $i = 0, 1$, where (E_0, E_1), (F_0, F_1) are interpolation couples and composition is the natural composition of mappings. By writing $A\colon (E_0, E_1) \to (F_0, F_1)$ one signifies that (E_0, E_1) and (F_0, F_1) are interpolation couples and A is a morphism of $\mathcal{B}_1$.

If (E_0, E_1) and (F_0, F_1) are interpolation couples, the Banach spaces E, F are said to be *interpolation spaces* with respect to (E_0, E_1) and (F_0, F_1) if E and F are intermediate spaces with respect to (E_0, E_1) and (F_0, F_1) respectively, and $A \in \mathcal{L}(E, F)$ whenever $A\colon (E_0, E_1) \to (F_0, F_1)$. Moreover, E and F are said to be *interpolation spaces of exponent* θ, where $0 < \theta < 1$, (with respect to (E_0, E_1) and (F_0, F_1)) if there exists $c = c_\theta > 0$ such that

$$\|A\|_{\mathcal{L}(E,F)} \le c_\theta \|A\|_{\mathcal{L}(E_0, F_0)}^{1-\theta} \|A\|_{\mathcal{L}(E_1, F_1)}^{\theta}$$

for $A\colon (E_0, E_1) \to (F_0, F_1)$. If $c_\theta = 1$, then E and F are *exact* interpolation spaces of exponent θ with respect to (E_0, E_1) and (F_0, F_1).

At last, a covariant functor $\mathcal{F}$ from $\mathcal{B}_1$ into $\mathcal{B}$ is said to be *an interpolation functor*, an *exact* interpolation functor, an interpolation functor

*of exponent θ if, given (E_0, E_1) and (F_0, F_1), $\mathcal{F}(E_0, E_1)$ and $\mathcal{F}(F_0, F_1)$
are interpolation spaces, exact interpolation spaces, interpolation spaces
of exponent θ with respect to (E_0, E_1) and (F_0, F_1) and if $\mathcal{F}(A) = A \in
\mathcal{L}(\mathcal{F}(E_0, E_1), \mathcal{F}(F_0, F_1))$, where $A: (E_0, E_1) \to (F_0, F_1)$.*

We observe that, if $\mathcal{F}_\theta$ is an interpolation functor of exponent θ, for all
interpolation couples (E_0, E_1), then setting $E_\theta = \mathcal{F}_\theta(E_0, E_1)$ gives that

$$\|x\|_{E_\theta} \leq c_\theta \|x\|_{E_0}^{1-\theta} \|x\|_{E_1}^{\theta}, \quad x \in E_0 \cap E_1,$$

see [5; p. 25]. Moreover, we have the following result.

PROPOSITION 0.1. *Let (E_0, E_1) and (F_0, F_1) be interpolation couples
with $E_1 \subset E_0$ and $F_1 \subset F_0$ continuously. If $\mathcal{F}$ is an arbitrary interpolation
functor from $\mathcal{B}_1$ into $\mathcal{B}$, then $(E_0 \times F_0, E_1 \times F_1)$ is an interpolation couple,
too, and*

$$E_1 \times F_1 \subset \mathcal{F}(E_0, E_1) \times \mathcal{F}(F_0, F_1) = \mathcal{F}(E_0 \times F_0, E_1 \times F_1) \subset E_0 \times F_0$$

continuously.

For the proof see [5; p. 27].

We also recall the main interpolation functors which enter in concrete
applications.

THE REAL INTERPOLATION FUNCTORS.

Let X be a Banach space. For an X-valued strongly measurable function
$v = v(t)$, $t \in (0, \infty)$, we set

$$\|v\|_{L_p^*(X)} = \left(\int_0^\infty \|v(t)\|_X^p \frac{dt}{t} \right)^{\frac{1}{p}}, \quad 1 \leq p < \infty,$$

and

$$\|v\|_{L_\infty^*(X)} = \sup_{0 < t < \infty} \|v(t)\|_X.$$

Given an interpolation couple (E_0, E_1), let either $1 \leq p_0, p_1 < \infty$ or $p_0 =
p_1 = \infty$. Define for $\theta \in (0, 1)$, $\frac{1}{p} = (1 - \theta)\frac{1}{p_0} + \theta\frac{1}{p_1}$ if $1 \leq p_0, p_1 <
\infty$, $p = \infty$ if $p_0 = p_1 = \infty$. Then $(E_0, E_1)_{\theta,p}$ denotes the set of all
elements $a \in E_0 + E_1$ represented by $a = v_0(t) + v_1(t), 0 < t < \infty$, $v_j(\cdot)$
is E_j-continuous (equivalently, E_j-measurable, equivalently, infinitely E_j-
differentiable) with

$$(0.1) \qquad \|t^{-\theta} v_0(t)\|_{L_{p_0}^*(E_0)} + \|t^{1-\theta} v_1(t)\|_{L_{p_1}^*(E_1)} < \infty.$$

$(E_0, E_1)_{\theta,p}$ endowed with the norm

$$\|a\|_{(E_0, E_1)_{\theta,p}} = \inf\{\|t^{-\theta} v_0(t)\|_{L_{p_0}^*(E_0)} + \|t^{1-\theta} v_1(t)\|_{L_{p_1}^*(E_1)}\},$$

where the infimum is taken over all the representations of a satisfying (0.1), is a Banach space. Let

$$\mathcal{F}_{\theta,p}(E_0, E_1) = (E_0, E_1)_{\theta,p} \text{ and } \mathcal{F}_{\theta,p}(A) = A$$

for $A\colon (E_0, E_1) \to (F_0, F_1)$. Then, given $p \in [1, \infty]$ and $\theta \in (0, 1)$, $\mathcal{F}_{\theta,p}$ is an exact interpolation functor of exponent θ. It is usually denoted by $(\cdot, \cdot)_{\theta,p}$ and is called the real interpolation functor of exponent θ and parameter q. The real interpolation spaces were introduced by J. L. Lions and J. Peetre [130].

THE COMPLEX INTERPOLATION FUNCTORS.

Let $\mathbb{K} = \mathbb{C}$ and let (E_0, E_1) be an interpolation couple. Let $\mathcal{F}(E_0, E_1)$ be the set of all bounded and continuous functions f from $\overline{S}$ into $E_0 + E_1$, where $S = \{\lambda \in \mathbb{C};\ 0 < \Re\lambda < 1\}$ such that $f_{|S}$ is holomorphic and $f_{|S_j} \in \mathcal{C}_0(S_j, E_j)$, $j = 0,\ 1$, where $S_j = \{\lambda \in \mathbb{C};\ \Re\lambda = j\}$ and $\mathcal{C}_0(S_j, E_j)$ is the space of all continuous E_j-valued functions on S_j vanishing at infinity. $\mathcal{F}(E_0, E_1)$ is a Banach space with the norm

$$\|f\|_{\mathcal{F}(E_0, E_1)} = \max\{\sup_{\eta} \|f(i\eta)\|_{E_0},\ \sup_{\eta} \|f(1 + i\eta)\|_{E_1}\}.$$

Given $\theta \in (0, 1)$, put

$$[E_0, E_1]_\theta = \{x \in E_0 + E_1;\ f(\theta) = x \text{ for some } f \in \mathcal{F}(E_0, E_1)\}$$

and endow this space with the norm

$$\|x\|_\theta = \inf\{\|f\|_{\mathcal{F}(E_0, E_1)};\ f(\theta) = x\}.$$

Letting $\mathcal{F}_\theta(E_0, E_1) = [E_0, E_1]_\theta$ and $\mathcal{F}_\theta(A) = A$ for $A\colon (E_0, E_1) \to (F_0, F_1)$, $\mathcal{F}_\theta$ is an exact interpolation functor of exponent θ (see [5; p. 28]).

If $\mathbb{K} = \mathbb{R}$ and (E_0, E_1) is an interpolation couple, then we put

$$\mathcal{F}(E_0, E_1) = ([(E_0)_{\mathbb{C}}, (E_1)_{\mathbb{C}}]_\theta \cap (E_0 + E_1),\ \|\cdot\|_\theta),$$

where $E_{\mathbb{C}}$ denotes the complexification of the real space E. In both the cases that $\mathbb{K} = \mathbb{R}$ or $\mathbb{C}$, the interpolation functor introduced above is denoted by $[\cdot, \cdot]_\theta$ and is called *the complex interpolation functor*. The complex interpolation method was introduced by J. L. Lions and A. P. Calderón.

We here present some fundamental results on real and complex interpolation spaces. For the proofs see [172] or [5].

PROPOSITION 0.2. *Let (E_0, E_1) be an interpolation couple, $\theta \in (0,1)$. If we define*

$$(E_0, E_1)_{\theta,\infty}^\circ = \text{the closure of } E_0 \cap E_1 \text{ in } (E_0, E_1)_{\theta,\infty},$$

then the following continuous injections

$$E_0 \cap E_1 \subset_d (E_0, E_1)_{\zeta,q} \subset_d (E_0, E_1)_{\eta,1} \subset_d [E_0, E_1]_\eta$$
$$\subset_d (E_0, E_1)_{\eta,\infty}^\circ \subset (E_0, E_1)_{\eta,\infty} \subset_d (E_0, E_1)_{\xi,q} \subset E_0 + E_1$$

are true for $1 \leq q < \infty$, $0 < \xi < \eta < \zeta < 1$, here $\subset_d$ denoting continuous and dense imbedding. In addition,

$$(E_0, E_1)_{\theta,q} \subset_d (E_0, E_1)_{\theta,r} \subset_d (E_0, E_1)_{\theta,\infty}^\circ$$

for $1 \leq q < r < \infty$, $0 < \theta < 1$.

PROPOSITION 0.3. *If (E_0, E_1) is an interpolation couple, then the same is true for (E_1, E_0) with the relations*

$$(E_0, E_1)_{\theta,q} = (E_1, E_0)_{1-\theta,q}, \quad [E_0, E_1]_\theta = [E_1, E_0]_{1-\theta},$$

for $0 < \theta < 1$, $1 \leq q \leq \infty$.

PROPOSITION 0.4. *Let $1 \leq q < \infty$, $1 < q' \leq \infty$ such that $\frac{1}{q} + \frac{1}{q'} = 1$ and let $E_0 \cap E_1 \subset_d E_j$, $j = 0, 1$. Then*

$$(E_0, E_1)_{\theta,q}' = (E_0', E_1')_{\theta,q'}, \quad \{(E_0, E_1)_{\theta,\infty}^\circ\}' = (E_0', E_1')_{\theta,1},$$

for all $0 < \theta < 1$ with respect to the duality pairing naturally induced by the pairing $\langle \cdot, \cdot \rangle_{(E_0 \cap E_1)', E_0 \cap E_1}$.

Let next either E_0 or E_1 be reflexive and let $E_0 \cap E_1 \subset_d E_j$, $j = 0, 1$. Then

$$[E_0, E_1]_\theta' = [E_0', E_1']_\theta$$

for all $0 < \theta < 1$ with respect to the duality pairing naturally induced by $\langle \cdot, \cdot \rangle$; moreover, $[E_0, E_1]_\theta$ is reflexive for $0 < \theta < 1$.

PROPOSITION 0.5. *Let*

$$(E_0, E_1)_{\theta_j,1} \subset F_j \subset (E_0, E_1)_{\theta_j,\infty} \text{ continuously },$$

where $0 \leq \theta_j \leq 1$, $\theta_0 \neq \theta_1$ $((E_0, E_1)_{0,q} = E_0, (E_0, E_1)_{1,q} = E_1)$. Then,

$$(F_0, F_1)_{\eta,q} = (E_0, E_1)_{(1-\eta)\theta_0 + \eta\theta_1, q}, \quad 0 < \eta < 1, 1 \leq q \leq \infty.$$

If, similarly, we put $[E_0, E_1]_j = E_j$, $j = 0, 1$, *and* $E_0 \cap E_1 \subset_d E_j$, $j = 0, 1$, *then*

$$[[E_0, E_1]_{\theta_0}, [E_0, E_1]_{\theta_1}]_\eta = [E_0, E_1]_{(1-\eta)\theta_0 + \eta\theta_1}, \quad 0 \le \theta_0, \theta_1, \eta \le 1,$$

and hence

$$([E_0, E_1]_{\theta_0}, [E_0, E_1]_{\theta_1})_{\eta,q} = (E_0, E_1)_{(1-\eta)\theta_0 + \eta\theta_1, q},$$
$$\theta_0 \ne \theta_1, 0 < \theta_0, \theta_1, \eta < 1, 1 \le q \le \infty.$$

It can be shown that

$$[(E_0, E_1)_{\theta_0, q_0}, (E_0, E_1)_{\theta_1, q_1}]_\eta = (E_0, E_1)_{(1-\eta)\theta_0 + \eta\theta_1, q},$$
$$\theta_0 \ne \theta_1, 0 < \theta_0, \theta_1, \eta < 1, 1 \le q_j \le \infty, \tfrac{1}{q} = \tfrac{1-\eta}{q_0} + \tfrac{\eta}{q_1}$$

with the exclusion of $q_0 = q_1 = \infty$.

COROLLARY 0.6. *If*

$$(E_0, E_1)_{\theta_j, 1} \subset F_j \subset (E_0, E_1)_{\theta_j, \infty} \ \text{continuously}, \ \theta_0 \ne \theta_1, 0 \le \theta_j \le 1,$$

then

$$(F_0, F_1)^\circ_{\eta, \infty} = (E_0, E_1)^\circ_{(1-\eta)\theta_0 + \eta\theta_1, \infty}, \quad 0 < \eta < 1.$$

0.3 FRACTIONAL POWERS AND INTERPOLATION

Let A be a closed linear operator from $\mathcal{D}(A) \subset X \to X$ such that $(-\infty, 0) \subset \rho(A)$ and there exists $M > 0$:

(0.2) $$(1 + \xi)\|(\xi + A)^{-1}\|_{\mathcal{L}(X)} \le M, \quad \xi > 0.$$

Then, there exist $c > 0$, $\varphi \in (0, \pi)$ such that

$$R_\varphi = \{\lambda; \ |\arg \lambda| \le \varphi\} \cup \{|\lambda| \le \varphi\} \subset \rho(-A),$$
$$|\lambda|\|(\lambda + A)^{-1}\|_{\mathcal{L}(X)} \le c, \quad \lambda \in R_\varphi.$$

Let Γ be any piecewise smooth curve in R_φ running from $\infty e^{-i\varphi}$ to $\infty e^{i\varphi}$ and avoiding $(0, \infty)$. Then for any $z \in \mathbb{C}$, $\Re z < 0$, we define the fractional powers of A by

$$A^z = \tfrac{1}{2\pi i} \int_\Gamma (-\lambda)^z (\lambda + A)^{-1} d\lambda.$$

In view of (0.2) it is verified that $A^z \in \mathcal{L}(X)$.

If $\mathcal{D}(A)$ is dense in X, the definition of A^z can be extended to all $z \in \mathbb{C}$ so that A^z is closed and densely defined for all z, $A^{z_1} A^{z_2} u = A^{z_1 + z_2} u$, $u \in \mathcal{D}(A^{2m})$ when $m \in \mathbb{N}$, $m \geq 2$, $\max\{\Re e z_1, \Re e z_2\} < m$. Further,

$$(X, \mathcal{D}(A^m))_{\Re e \frac{z}{m}, 1} \subset_d \mathcal{D}(A^z) \subset_d (X, \mathcal{D}(A^m))_{\Re e \frac{z}{m}, \infty}^{\circ},$$

for $0 < \Re e z < m$, $m \in \mathbb{N}$.

Moreover, if the purely imaginary powers of A are locally uniformly bounded, that is, $A^{iy} \in \mathcal{L}(X)$, $-1 \leq y \leq 1$, and there exists $K > 0$ such that $\|A^{iy}\|_{\mathcal{L}(X)} \leq K$, then

$$[\mathcal{D}(A^\alpha), \mathcal{D}(A^\beta)]_\theta = \mathcal{D}(A^{(1-\theta)\alpha + \theta\beta}),$$

$0 \leq \Re e \alpha < \Re e \beta$, $0 < \theta < 1$. See [154] or [172; p. 101].

We also have the following characterizations of the spaces $(X, \mathcal{D}(A^m))_{\theta, p}$, $0 < \theta < 1$, $1 \leq p \leq \infty$. For the proofs, see [172; p. 78].

PROPOSITION 0.7. *Let $m \in \mathbb{N}$, $1 \leq p \leq \infty$, $0 < \theta < 1$ and let A be the generator of the bounded semigroup e^{tA}, $0 \leq t < \infty$, i.e. $\sup_{t \geq 0} \|e^{tA}\|_{\mathcal{L}(X)} < \infty$. Then,*

$$(X, \mathcal{D}(A^m))_{\theta, p} = \{u \in X; \ \|t^{-\theta m}(e^{tA} - 1)^m u\|_{L_p^*(X)} < \infty\}$$

and $\|u\|_{\theta, p}$ is equivalent to $\|u\|_X + \|t^{-tm}(e^{tA} - 1)^m u\|_{L_p^*(X)}$.

This result will be applied in the sequel by taking $X = \mathcal{C}_0([0, T]; E)$, the space of all E-valued continuous functions on $[0, T]$ vanishing at $t = 0$ with the supremum norm, $A = \frac{d}{dt}$, $\mathcal{D}(A) = \{u \in \mathcal{C}^1([0, T]; E); \ u(0) = u'(0) = 0\}$, so that it is readily seen, according to Proposition 0.7, that $(X, \mathcal{D}(A))_{\theta, \infty}$, $0 < \theta < 1$, coincides with the space $\mathcal{C}_0^\theta([0, T]; E)$ of E-valued Hölder continuous functions on $[0, T]$, with exponent θ, and vanishing at $t = 0$.

If A is a positive operator acting in the Banach X, that is, $(-\infty, 0] \subset \rho(A)$, A being densely defined and closed with (0.2), then an important characterization of the spaces $(X, \mathcal{D}(A^m))_{\theta, p}$, $m \in \mathbb{N}$, $0 < \theta < 1$, $1 \leq p \leq \infty$, is given by the following result.

PROPOSITION 0.8. *Let A be a positive operator in X and let $m \in \mathbb{N}$, $0 < \theta < 1$, $1 \leq p \leq \infty$. Then,*

$$(X, \mathcal{D}(A^m))_{\theta, p} = \{u \in X; \ \|u\|^* = \|\xi^{\theta m}\{A(\xi + A)^{-1}\}^m u\|_{L_p^*(X)} < \infty\}$$

and $\|u\|^*$ is an equivalent norm in the space $(X, \mathcal{D}(A^m))_{\theta, p}$.

In the case of A generator of an analytic semigroup, one has the far-reaching result as follows.

PROPOSITION 0.9. *Let A be the generator of an analytic semigroup e^{tA} in the Banach space X and let $m \in \mathbb{N}$, $0 < \theta < 1$, $1 \leq p \leq \infty$. Moreover, let*

$$(0.3) \qquad \|e^{tA}\|_{\mathcal{L}(X)} \leq C e^{\beta t}, \quad 0 \leq t < \infty.$$

Then,

$$(X, \mathcal{D}(A^m))_{\theta,p} = \{u \in X; \ \|a\|^{**} = \|t^{(1-\theta)m} A^m e^{tA} u\|_{L_p^*(X)} + \|u\|_X < \infty\}$$

and $\|u\|^{**}$ is an equivalent norm in the space $(X, \mathcal{D}(A^m))_{\theta,p}$.
If $\beta < 0$ in (0.3), then $\|u\|^{**}$ is equivalent to $\|t^{(1-\theta)m} A^m e^{tA} u\|_{L_p^*(X)}$.

CHAPTER I

MULTIVALUED LINEAR OPERATORS

We shall generalize the notion of linear operators to the multivalued operators in a natural way. For example, the inverse of a linear operator is always a multivalued linear operator, and the dual operator of a linear operator is always defined as a multivalued linear operator even if the domain is not dense.

1.1 DEFINITIONS

Let X be a Banach space over the complex numbers $\mathbb{C}$. For two subsets F, G of X, we define: $F + G = \{f + g; f \in F, g \in G\}$ and, for a number $\lambda \in \mathbb{C}, \lambda F = \{\lambda f; f \in F\}$.

DEFINITION. A mapping A from X into 2^X is called *a multivalued linear operator in* X if the domain $\mathcal{D}(A) = \{u \in X; Au \neq \emptyset\}$ is a linear subspace of X and A satisfies:

$$\begin{cases} Au + Av \subset A(u + v) & \text{for } u, v \in \mathcal{D}(A), \\ \lambda Au \subset A(\lambda u) & \text{for } \lambda \in \mathbb{C} \text{ and } u \in \mathcal{D}(A). \end{cases}$$

$\mathcal{R}(A) = \bigcup_{u \in \mathcal{D}(A)} Au$ is called the range of A.

In this chapter we shall abbreviate multivalued linear to m. l. As a special case, a linear operator is a m. l. operator. The following properties of m. l. operators are verified as an immediate consequence of the definition.

THEOREM 1.1. $Au + Av = A(u + v)$ *for* $u, v \in \mathcal{D}(A)$. $\lambda Au = A(\lambda u)$ *for* $u \in \mathcal{D}(A)$ *if* $\lambda \neq 0$.

PROOF. By definition, $A(u + v) - Av \subset A(u + v) + A(-v) \subset Au$; hence, $A(u + v) \subset Au + Av$. Similarly, $A(\lambda u) = \lambda \lambda^{-1} A(\lambda u) \subset \lambda Au$ if $\lambda \neq 0$.

THEOREM 1.2. $A0$ *is a linear subspace of* X. $Au = f + A0$ *with any* $f \in Au$. *In particular, A is single valued if and only if $A0 = 0$.*

PROOF. From the definition, $A0 + A0 = A0$ and $\lambda A0 = A0$; hence the first assertion follows. Clearly, $f + A0 \subset Au$ for any $f \in Au$; on the other hand, if $g \in Au, g - f \in Au - Au = A0$; so that, $Au \subset f + A0$.

The inverse A^{-1} of a m. l. operator A is defined as

$$\begin{cases} \mathcal{D}(A^{-1}) = \mathcal{R}(A), \\ A^{-1}f = \{u \in \mathcal{D}(A); Au \ni f\}. \end{cases}$$

THEOREM 1.3. A^{-1} is also a m. l. operator in X. $u \in A^{-1}f$ if and only if $f \in Au$; in particular, $(A^{-1})^{-1} = A$.

PROOF. If $u \in A^{-1}f$ and $v \in A^{-1}g$, then $f \in Au$ and $g \in Av$; so that, $f + g \in A(u + v)$; hence, $u + v \in A^{-1}(f + g)$. Similarly, $u \in A^{-1}f$ implies $\lambda f \in A(\lambda u)$ and $\lambda u \in A^{-1}(\lambda f)$; hence, $\lambda A^{-1}f \subset A^{-1}(\lambda f)$. The second assertion is then obvious.

When two m. l. operators A, B satisfy $\mathcal{D}(A) \subset \mathcal{D}(B)$ with the inclusion $Au \subset Bu$ for all $u \in \mathcal{D}(A)$, B is called an extension of A and we put $A \subset B$. Obviously, $A = B$ if and only if $A \subset B$ and $A \supset B$.

When there exists a single valued linear operator A° which satisfies $\mathcal{D}(A^\circ) = \mathcal{D}(A)$ and $A^\circ \subset A$, A° is called a linear section of A.

1.2 SUM AND PRODUCT OF MULTIVALUED LINEAR OPERATORS

Let A and B be two m. l. operators in X. The sum and the product of A, B are defined respectively as follows:

$$\begin{cases} \mathcal{D}(A + B) = \mathcal{D}(A) \cap \mathcal{D}(B), \\ (A + B)u = Au + Bu; \end{cases}$$

$$\begin{cases} \mathcal{D}(AB) = \{u \in \mathcal{D}(B); \mathcal{D}(A) \cap Bu \neq \emptyset\}, \\ ABu = \displaystyle\bigcup_{f \in \mathcal{D}(A) \cap Bu} Af. \end{cases}$$

Then we easily verify the following theorems.

THEOREM 1.4. Let A, B be m. l. operators. Then, $A + B$ is also a m. l. operator in X.

PROOF. Let $u, v \in \mathcal{D}(A + B)$; if $f \in (A + B)u$ and $g \in (A + B)v$, then $f = f_1 + f_2$ and $g = g_1 + g_2$ with $f_1 \in Au, f_2 \in Bu, g_1 \in Av$ and $g_2 \in Bv$; so that, $f + g = f_1 + g_1 + f_2 + g_2 \in (A + B)(u + v)$; this shows that $(A + B)u + (A + B)v \subset (A + B)(u + v)$. Similarly, $\lambda(A + B)u \subset (A + B)(\lambda u)$ for $\lambda \in \mathbb{C}$ and $u \in \mathcal{D}(A + B)$.

THEOREM 1.5. *Let A, B be m. l. operators. Then, AB is also a m. l. operator in X. $\varphi \in ABu$ if and only if there exists $f \in \mathcal{D}(A) \cap \mathcal{R}(B)$ such that $\varphi \in Af$ and $f \in Bu$.*

PROOF. The second assertion is nothing more than the definition. Let $\varphi \in ABu$ and $\psi \in ABv$; then, $\varphi \in Af, f \in Bu$ and $\psi \in Ag, g \in Bv$; so that, $\varphi + \psi \in A(f + g)$ and $f + g \in B(u + v)$; since these mean that $\varphi + \psi \in AB(u + v)$, it follows that $ABu + ABv \subset AB(u + v)$. Similarly, $\lambda ABu \subset AB(\lambda u)$ for $\lambda \in \mathbb{C}$ and $u \in \mathcal{D}(AB)$.

Combining the last theorem with Theorem 1.3 we can observe that

$$(AB)^{-1} = B^{-1}A^{-1}.$$

1.3 RESOLVENT OF MULTIVALUED LINEAR OPERATORS

For a m. l. operator A, the set of all numbers $\lambda \in \mathbb{C}$ such that $\mathcal{R}(\lambda - A) = \mathcal{D}((\lambda - A)^{-1}) = X$ and $(\lambda - A)^{-1}$ is a *single valued* bounded operator on X is called the resolvent set of A and is denoted by $\rho(A)$. The bounded linear operator $(\lambda - A)^{-1}, \lambda \in \rho(A)$, is called the resolvent of A.

THEOREM 1.6. *$\rho(A)$ is an open set of $\mathbb{C}$. The resolvent $(\lambda - A)^{-1}$ is a holomorphic function in $\rho(A)$ with values in $\mathcal{L}(X)$.*

PROOF. I) We first consider the case where $(\lambda - A)^{-1} \neq 0$ for every $\lambda \in \rho(A)$. Let $\lambda_0 \in \rho(A)$ and let $|\lambda - \lambda_0| < 1/\|(\lambda_0 - A)^{-1}\|_{\mathcal{L}(X)}$. For any $f \in X$, put

$$g = \{1 + (\lambda - \lambda_0)(\lambda_0 - A)^{-1}\}^{-1}f \quad \text{and} \quad u = (\lambda_0 - A)^{-1}g;$$

since $g + (\lambda - \lambda_0)u = f$ and $g \in (\lambda_0 - A)u$, we have: $f \in (\lambda - A)u$; hence, $\mathcal{R}(\lambda - A) = X$. Conversely, let $(\lambda - A)^{-1}0 \ni u$ or $0 \in (\lambda - A)u$; then there is an element $g \in (\lambda_0 - A)u$ such that $g + (\lambda - \lambda_0)u = 0$; thus, $\{1 + (\lambda - \lambda_0)(\lambda_0 - A)^{-1}\}u = 0$; this shows that $u = 0$. Thus we have proved that $\lambda \in \rho(A)$ and

$$(\lambda - A)^{-1} = (\lambda_0 - A)^{-1}\{1 + (\lambda - \lambda_0)(\lambda_0 - A)^{-1}\}^{-1}.$$

II) Consider now the case where $(\lambda_0 - A)^{-1} = 0$ for some $\lambda_0 \in \rho(A)$. Then, $\lambda_0 - A = O_\infty$ is the inverse of the zero operator and it is seen that $\mathcal{D}(O_\infty) = \{0\}$ and $O_\infty 0 = X$; so that, $A = O_\infty$. It is also easily observed that $\rho(O_\infty) = \mathbb{C}$ and $(\lambda - O_\infty)^{-1} = 0$ identically; hence the desired result is proved.

THEOREM 1.7. $(\lambda-A)^{-1}A \subset \lambda(\lambda-A)^{-1}-1 \subset A(\lambda-A)^{-1}$ for $\lambda \in \rho(A)$. In particular, $(\lambda - A)^{-1}A$ is single valued on $\mathcal{D}(A)$ and $(\lambda - A)^{-1}Au = (\lambda - A)^{-1}f$ with any $f \in Au$.

PROOF. Let $f \in Au$; then, $\lambda u - f \in (\lambda - A)u$; so that, $(\lambda - A)^{-1}f = \lambda(\lambda-A)^{-1}u-u$; therefore, $(\lambda-A)^{-1}A \subset \lambda(\lambda-A)^{-1}-1$. On the other hand, let $v = (\lambda - A)^{-1}f$ or $\lambda v - f \in Av$; then, $\lambda(\lambda - A)^{-1}f - f \in A(\lambda - A)^{-1}f$; hence, $\lambda(\lambda - A)^{-1} - 1 \subset A(\lambda - A)^{-1}$.

DEFINITION. By Theorem 1.7, $\lambda(\lambda-A)^{-1}-1$ is a bounded linear section of the operator $A(\lambda - A)^{-1}, \lambda \in \rho(A)$. This linear section will be denoted (with some abuse of notation) by $A^\circ(\lambda - A)^{-1}$. Note here that A° does not necessarily denote a linear section of A itself.

We shall next establish the resolvent equation of m. l. operators.

THEOREM 1.8. For $\lambda, \mu \in \rho(A)$,

$$(\lambda - A)^{-1} - (\mu - A)^{-1} = -(\lambda - \mu)(\lambda - A)^{-1}(\mu - A)^{-1}.$$

PROOF. Putting $\lambda_\mu = \lambda - \mu$ and $A_\mu = A - \mu$, we apply Theorem 1.7 to A_μ. Then, since $0, \lambda_\mu \in \rho(A_\mu)$, we obtain that

$$(\lambda_\mu - A_\mu)^{-1}A_\mu \subset \lambda_\mu(\lambda_\mu - A_\mu)^{-1} - 1,$$
$$(\lambda_\mu - A_\mu)^{-1}A_\mu(A_\mu)^{-1} \subset \lambda_\mu(\lambda_\mu - A_\mu)^{-1}(A_\mu)^{-1} - (A_\mu)^{-1}.$$

While, since the inclusion $1 \subset A_\mu(A_\mu)^{-1}$ holds, Theorem 1.7 yields again that $(\lambda_\mu - A_\mu)^{-1}A_\mu(A_\mu)^{-1} = (\lambda_\mu - A_\mu)^{-1}$. Therefore, the desired equation is verified.

THEOREM 1.9. Let A, B be two m. l. operators the resolvent sets of which contain 0. Then, for $\lambda \in \rho(A) \cap \rho(B)$,

$$(\lambda - A)^{-1} - (\lambda - B)^{-1} = -A^\circ(\lambda - A)^{-1}(A^{-1} - B^{-1})B^\circ(\lambda - B)^{-1}.$$

PROOF. Writing:

$$(\lambda - A)^{-1} - (\lambda - B)^{-1} = \{\lambda(\lambda - A)^{-1} - 1\}(\lambda - B)^{-1}$$
$$- (\lambda - A)^{-1}\{\lambda(\lambda - B)^{-1} - 1\},$$

we replace $(\lambda - A)^{-1}$ and $(\lambda - B)^{-1}$ by $\{\lambda(\lambda - A)^{-1} - 1\}A^{-1}$ and by $B^{-1}\{\lambda(\lambda - B)^{-1} - 1\}$ respectively. Then it turns out that

$$(\lambda - A)^{-1} - (\lambda - B)^{-1} = -\{\lambda(\lambda - A)^{-1} - 1\}(A^{-1} - B^{-1})\{\lambda(\lambda - B)^{-1} - 1\},$$

which is the desired equation.

1.4 FRACTIONAL POWERS
AND INTERPOLATION SPACES

Let A be a m. l. operator in X such that the resolvent set contains the non positive real axis $(-\infty, 0]$. We assume that A satisfies the estimate

$$\|(\lambda - A)^{-1}\|_{\mathcal{L}(X)} \leq \frac{M}{(|\lambda| + 1)^{\beta}}, \quad \lambda \leq 0,$$

with some exponent $0 < \beta \leq 1$ and constant M. Then, by the same argument as in the proof of Theorem 1.6, it is obtained that

$$\rho(A) \supset \Sigma = \{\lambda \in \mathbb{C}; |\Im m\lambda| \leq (2M)^{-1}(-\Re e\lambda + c)^{\beta}, \Re e\lambda \leq c\},$$

with a bound

$$\|(\lambda - A)^{-1}\|_{\mathcal{L}(X)} \leq \frac{M'}{(|\lambda| + 1)^{\beta}}, \quad \lambda \in \Sigma,$$

where c, M' are some constants. For $\theta > 1 - \beta$, the fractional powers $A^{-\theta}$ of A are defined by the Dunford integrals

$$A^{-\theta} = \frac{1}{2\pi i} \int_{\Gamma} \lambda^{-\theta}(\lambda - A)^{-1} d\lambda$$

in $\mathcal{L}(X)$, where Γ is the integral contour: $\lambda = \xi \pm i(2M)^{-1}(-\xi + c)^{\beta}$, $-\infty < \xi \leq c$, lying in $\Sigma - (-\infty, 0]$. If we use Theorem 1.8, we verify the semigroup property

(1.1) $$A^{-\theta}A^{-\theta'} = A^{-(\theta + \theta')}, \quad \theta, \theta' > 1 - \beta.$$

For $\theta > 1 - \beta$, A^{θ} is defined as the inverse of $A^{-\theta}$; by Theorem 1.3, A^{θ} are m. l. operators in X.

THEOREM 1.10. $A^{\theta}A^{\theta'} = A^{\theta + \theta'}$ for $\theta, \theta' > 1 - \beta$.

PROOF. The result is an immediate consequence of (1.1) and Theorem 1.5.

In analogy to the single valued linear operators, we now consider real interpolation spaces between the domain $\mathcal{D}(A)$ and X. First of all, we need to specify the topology on $\mathcal{D}(A)$. In fact, we equip $\mathcal{D}(A)$ with the norm

$$\|u\|_{\mathcal{D}(A)} = \inf_{f \in Au} \|f\|_X.$$

Since $A^{-1} \in \mathcal{L}(X)$ is assumed, this norm is equivalent to the graph norm

$$\inf_{f \in Au} \|f\|_X \leq \|u\|_X + \inf_{f \in Au} \|f\|_X \leq (\|A^{-1}\|_{\mathcal{L}(X)} + 1) \inf_{f \in Au} \|f\|_X.$$

PROPOSITION 1.11. $\mathcal{D}(A)$ *is a Banach space with the norm* $\|\cdot\|_{\mathcal{D}(A)}$.

PROOF. Let $\mathcal{G}(A) = \{(u, f) \in X \times X; \ u \in \mathcal{D}(A), f \in Au\}$ be the graph of A. Since $A^{-1} \in \mathcal{L}(X)$, $\mathcal{G}(A)$ is a closed subspace of $X \times X$. In addition, $\{0\} \times A0$ is a closed subspace of $\mathcal{G}(A)$. Therefore, the quotient space $\mathcal{G}(A)/\{0\} \times A0$ is a Banach space with the quotient norm. Here it is easily seen that

$$\|[(u, f)]\|_{\mathcal{G}(A)/\{0\} \times A0} = \|u\|_X + \inf_{g \in Au} \|g\|_X$$

for each $(u, f) \in \mathcal{G}(A)$, where $[(u, f)]$ is the quotient class of (u, f). This then means that $\pi : \mathcal{G}(A) \to \mathcal{D}(A)$, $\pi(u, f) = u$ is an isomorphism of normed space, and hence the result is verified.

Note that A is a m. l. operator if its graph is a linear subspace of $X \times X$. For each $0 < \theta < 1$ we introduce the (intermediate) space

$$X_A^\theta = \{u \in X; \sup_{\xi > 0} \xi^\theta \|A^\circ(\xi + A)^{-1} u\|_X < \infty\}$$

equipped with the norm

$$\|u\|_{X_A^\theta} = \|u\|_X + \sup_{\xi > 0} \xi^\theta \|A^\circ(\xi + A)^{-1} u\|_X.$$

Obviously, X_A^θ becomes a Banach space.

We then prove, by the similar argument as for the single valued linear operator case (cf. H. Triebel [172]), the following relation.

THEOREM 1.12. *Let* $(X, \mathcal{D}(A))_{\theta,\infty}$, $0 < \theta < 1$, *be the real interpolation spaces. Then,*

$$X_A^\theta \subset (X, \mathcal{D}(A))_{\theta,\infty}, \quad 0 < \theta < 1,$$
$$(X, \mathcal{D}(A))_{\theta,\infty} \subset X_A^{\theta+\beta-1}, \quad 1 - \beta < \theta < 1.$$

In particular, when $\beta = 1$, $X_A^\theta = (X, \mathcal{D}(A))_{\theta,\infty}$.

PROOF. According to [172: Theorem 1.5.2], (see Chapter 0, too), it is known that

$$(X, \mathcal{D}(A))_{\theta,\infty} = \{u \in X; \ u = u_0(t) + u_1(t), 0 < t < \infty,$$
$$\text{with } u_0 \in \mathcal{C}((0, \infty); X), \ u_1 \in \mathcal{C}((0, \infty); \mathcal{D}(A))$$
$$\text{and with } t^{-\theta} u_0 \in L^\infty((0, \infty); X), \ t^{1-\theta} u_1 \in L^\infty((0, \infty); \mathcal{D}(A))\}.$$

Then the proof is quite similar as for the case that A is single valued if we notice from the definition that

$$u = A^\circ(t^{-1} + A)^{-1} u + t^{-1}(t^{-1} + A)^{-1} u = u_0(t) + u_1(t), \quad 0 < t < \infty.$$

1.5 DUAL OPERATORS

For a m. l. operator A in a Banach space X, we shall define its dual A' which acts in the dual space X'. In fact, for $\eta, \xi \in X'$, the relation $\eta \in A'\xi$ is defined by means of

$$\xi(f) = \eta(u) \text{ for all pairs } (u, f) \text{ such that } f \in Au.$$

THEOREM 1.13. A' defined above is a m. l. operator in X'. If $\xi \in \mathcal{D}(A')$, then ξ vanishes on $A0$.

PROOF. Let $\eta \in A'\xi$ and $\mu \in A'\zeta$; then, $\xi(f) = \eta(u)$ and $\zeta(f) = \mu(u)$ for all (u, f) such that $f \in Au$; so that, $(\xi + \zeta)(f) = (\eta + \mu)(u)$ for all $f \in Au$; therefore, $\eta + \mu \in A'(\xi + \zeta)$ that is $A'\xi + A'\zeta \subset A'(\xi + \zeta)$. Similarly, $\lambda A'\xi \subset A'(\lambda \xi)$ for $\lambda \in \mathbb{C}$ and $\xi \in \mathcal{D}(A')$. The second assertion is obvious from the definition.

From the definition and from Theorems 1.3 and 1.5, it is also immediate to verify the following properties:

(1.2) $$(A')^{-1} = (A^{-1})';$$

(1.3) $$A \subset A'', \text{ viewing } X \text{ as } X \subset X'';$$

(1.4)
$$A' + B' \subset (A + B)' \text{ and } A'B' \subset (BA)' \text{ for two m. l. operators } A, B;$$

(1.5) $$\text{If } B \in \mathcal{L}(X), \text{then } A' + B' = (A + B)'.$$

When X is a Hilbert space, the adjoint operator A^* of A is defined in an analogous way. Indeed, the relation $\eta \in A^*\xi$ means that

$$(\xi, f) = (\eta, u) \text{ for all pairs } (u, f) \text{ such that } f \in Au,$$

where $(\cdot, \cdot)$ denotes the inner product of X. Then, A^* is proved as above to be a m. l. operator in X. Analogous properties to (1.2,3,4,5) are verified in the present case, too.

1.6 MODIFIED RESOLVENTS

Let L and M be two single valued, closed linear operators in a Banach space X with $\mathcal{D}(L) \subset \mathcal{D}(M)$. We are concerned with the resolvent of the m. l. operator LM^{-1}.

In order to represent $(\lambda - LM^{-1})^{-1}$ by L, M, we introduce the notion of M modified resolvent of L.

DEFINITION. The set $\{\lambda \in \mathbb{C}; \lambda M - L \text{ has a single valued and bounded inverse on } X\}$ is called the M modified resolvent set of L (or simply the M resolvent set of L) and is denoted by $\rho_M(L)$. The bounded operator $(\lambda M - L)^{-1}$ is called the M modified resolvent of L (or simply the M resolvent of L).

Then the following result holds.

THEOREM 1.14. $\rho_M(L) \subset \rho(LM^{-1})$; in addition, $M(\lambda M - L)^{-1} = (\lambda - LM^{-1})^{-1}$ for $\lambda \in \rho_M(L)$.

PROOF. By Theorems 1.3 and 1.5, it is verified that $M(\lambda M - L)^{-1} = (\lambda - LM^{-1})^{-1}$ in the m. l. operator sense for every $\lambda \in \mathbb{C}$. Hence, the assertion is obvious.

Let now two single valued, closed linear operators L and M in X be densely defined with $\mathcal{D}(L') \subset \mathcal{D}(M')$ and consider the m. l. operator $M^{-1}L$. By (1.2), (1.3) and (1.4) it is verified that

$$(1.6) \qquad M^{-1}L \subset (M'')^{-1}L'' \subset \{L'(M')^{-1}\}',$$

viewing X as a subspace of X''. Since it is very difficult to handle the inverse $(\lambda - M^{-1}L)^{-1}$ directly, we shall consider $(\lambda - \{L'(M')^{-1}\}')^{-1}$.

THEOREM 1.15. $\rho_{M'}(L') \subset \rho(\{L'(M')^{-1}\}')$; in addition, we verify that $\{M'(\lambda M' - L')^{-1}\}' = (\lambda - \{L'(M')^{-1}\}')^{-1}$ for $\lambda \in \rho_{M'}(L')$.

PROOF. By (1.2) and (1.5),

$$(\lambda - \{L'(M')^{-1}\}')^{-1} = \{(\lambda - L'(M')^{-1})^{-1}\}'.$$

Moreover, by Theorem 1.14,

$$(\lambda - L'(M')^{-1})^{-1} = M'(\lambda M' - L')^{-1}.$$

Hence, the result is proved.

Even when X is a Hilbert space, it is similarly verified that

$$(1.7) \qquad M^{-1}L \subset \{L^*(M^*)^{-1}\}^*, \text{ viewing } X \text{ as } X = X^{**},$$

and an analogous result to Theorem 1.15 holds.

CHAPTER II

DEGENERATE EQUATIONS OF HYPERBOLIC TYPE

Under the condition of Hille-Yosida type, generation of semi-group will be studied for multivalued linear operators in a Banach space. As a class of operators satisfying this condition, the notion of maximal dissipative operator will be generalized for multivalued linear operators in a Hilbert space.

The main difficulty of the treatment is due to the fact that the domain of the involved multivalued linear operator A is not necessarily dense in the Banach space X. Existence and uniqueness results for the Cauchy problem $\frac{du}{dt} \in Au + f(t)$, $t \geq 0$, $u(0) = u_0$ are given, together with their application to $\frac{d}{dt}(Mv) - Lv = f(t)$, $Mv(0) = u_0$.

Various examples of application to partial differential equations illustrate the abstract theory.

2.1 GENERATION OF SEMIGROUP

Let A be a m. l. operator in a Banach space X. We assume that A satisfies the Hille-Yosida condition (cf. K. Yosida [187]):

(H-Y) The resolvent set $\rho(A)$ contains a real half line (β, ∞), $-\infty < \beta < \infty$, and the resolvent $(\lambda - A)^{-1}$ satisfies

$$\|(\lambda - A)^{-n}\|_{\mathcal{L}(X)} \leq \frac{M}{(\lambda - \beta)^n}, \quad \lambda > \beta, \ n = 1, 2, 3, \cdots,$$

with a constant M.

Condition (H-Y) yields first the following proposition.

PROPOSITION 2.1. *The sum* $X_0 = \overline{\mathcal{D}(A)} + A0$ *of the two closed subspaces* $\overline{\mathcal{D}(A)}$ *and* $A0$ *is a closed subspace of* X; *in addition, the sum is topologically direct.*

PROOF. Let us notice that $A0$ is closed; indeed, since $A0 = (\lambda_0 - A)0$, $A0$ coincides with the kernel of the bounded operator $(\lambda_0 - A)^{-1}$, $\lambda_0 > \beta$. To verify $\overline{\mathcal{D}(A)} \cap A0 = \{0\}$, we shall use the following lemma.

LEMMA 2.2. *For each integer* $n > \beta$, *set* $J_n = n(n - A)^{-1} = (1 - n^{-1}A)^{-1}$. *Then, as* $n \to \infty$, J_n *converges to the identity strongly on* $\overline{\mathcal{D}(A)}$.

PROOF OF THE LEMMA. Let $u \in \mathcal{D}(A)$; according to Theorem 1.7, $(J_n - 1)u = -(n - A)^{-1}f$ with any $f \in Au$; therefore, $J_n u \to u$ in X. The assertion of the lemma then follows from the uniform boundedness of the norms $\|J_n\|_{\mathcal{L}(X)}$, $n > \beta$.

COMPLETION OF PROOF OF PROPOSITION 2.1. Let $f \in \overline{\mathcal{D}(A)} \cap A0$; as $n \to \infty$, $J_n f \to f$; on the other hand, $f \in A0$ implies that $J_n f = 0$ for every n; so that, $f = 0$, i.e. $\overline{\mathcal{D}(A)} \cap A0 = \{0\}$. Consider now $g \in \overline{\mathcal{D}(A)}$ and $h \in A0$; since $J_n(g + h) = J_n g \to g$, $\|g\|_X \leq M\|g + h\|_X$; in addition, $\|h\|_X \leq (M + 1)\|g + h\|_X$; these mean that the projections from X_0 onto $\overline{\mathcal{D}(A)}$ and onto $A0$ are continuous. Finally the closedness of X_0 is obvious; indeed, consider a sequence $f_m = g_m + h_m \in \overline{\mathcal{D}(A)} + A0$ which converges to f in X; as proved above, g_m (resp. h_m) is a Cauchy sequence in $\overline{\mathcal{D}(A)}$ (resp. $A0$); therefore, $f = g + h \in \overline{\mathcal{D}(A)} + A0$.

REMARK. When X is reflexive, X coincides with X_0. In fact, let $f \in X$; since the norms $\|J_n f\|_X$ are uniformly bounded, there exists a subsequence $J_{n_k} f$ which is weakly convergent to an element $g \in \overline{\mathcal{D}(A)}$ (note that $\overline{\mathcal{D}(A)}$ is weakly closed); on the other hand, the weak limit h of the sequence $(1 - J_{n_k})f$ belongs to $A0$, since $(\lambda_0 - A)^{-1}h = \text{w-}\lim_{n_k \to \infty}(\lambda_0 - A)^{-1}(1 - J_{n_k})f = \lim_{n_k \to \infty}(1 - J_{n_k})(\lambda_0 - A)^{-1}f = 0$, $\lambda_0 > \beta$ (note that $(\lambda_0 - A)^{-1}$ is continuous even in the weak topology); therefore, $f = g + h \in \overline{\mathcal{D}(A)} + A0$.

For each $n > \beta$, we define the Yosida approximation A_n of A by the formula

$$A_n = n\{-1 + n(n - A)^{-1}\} = n(J_n - 1).$$

Obviously A_n is a bounded (single valued) operator on X. By direct calculation it is also verified that $\rho(A_n) \supset (\beta_n, \infty)$, where $\beta_n = \frac{n\beta}{n-\beta}$, with

$$(2.1) \quad (\lambda - A_n)^{-1} = \frac{1}{\lambda + n} + \left(\frac{n}{\lambda + n}\right)^2 \left(\frac{\lambda n}{\lambda + n} - A\right)^{-1}, \quad \lambda > \beta_n.$$

Under (H-Y), A_n converges to A in the following sense.

PROPOSITION 2.3. *For each* $\lambda > \beta$, $(\lambda - A_n)^{-1}$ *converges, as* $n \to \infty$, *to* $(\lambda - A)^{-1}$ *in* $\mathcal{L}(X)$. *Moreover, if* $u \in \mathcal{D}(A)$ *and* $\overline{\mathcal{D}(A)} \cap Au \neq \emptyset$, *then* $\overline{\mathcal{D}(A)} \cap Au$ *consists of a single element* g *and the Yosida approximation* $A_n u$ *converges to* g *in* X.

PROOF. The first assertion is verified directly from (2.1). Let $u \in \mathcal{D}(A)$ and let $g_1, g_2 \in \overline{\mathcal{D}(A)} \cap Au \neq \emptyset$; then, it follows from Proposition 2.1 that $g_1 - g_2 \in \overline{\mathcal{D}(A)} \cap A0 = \{0\}$; hence, $g_1 = g_2$. Now let $\overline{\mathcal{D}(A)} \cap Au = \{g\}$;

then, by Theorem 1.7 we have: $A_n u = J_n g$; hence, Lemma 2.2 yields that $A_n u \to g$ in X.

Consider now the semigroups

$$e^{tA_n} = e^{-nt} e^{n^2(n-A)^{-1}t} = e^{-nt} \sum_{i=0}^{\infty} \frac{\{n^2(n-A)^{-1}t\}^i}{i!}, \quad t \geq 0,$$

generated by A_n, $n > \beta$. Clearly, e^{tA_n} is a semigroup both on the whole space X and on the subspace X_0. Moreover it is immediately observed from (H-Y) that

$$\|e^{tA_n}\|_{\mathcal{L}(X)}, \ \|e^{tA_n}\|_{\mathcal{L}(X_0)} \leq M e^{-nt} \sum_{i=0}^{\infty} \left(\frac{n^2 t}{n-\beta}\right)^i \frac{1}{i!} \leq M e^{\beta n t}, \quad t \geq 0.$$

We shall prove two convergence theorems for e^{tA_n}.

THEOREM 2.4. *(Convergence on X_0.) Let A be a m. l. operator satisfying (H-Y). On the space X_0, e^{tA_n}, $t \geq 0$, converges strongly to a bounded operator $e^{tA} \in \mathcal{L}(X_0)$. e^{tA} defines a semigroup on X_0 with an estimate $\|e^{tA}\|_{\mathcal{L}(X_0)} \leq M e^{\beta t}$. Moreover, e^{tA} maps $\overline{\mathcal{D}(A)}$ into itself and defines a C_0 semigroup on $\overline{\mathcal{D}(A)}$, too; on the other hand, e^{tA} vanishes on $A0$ for every $t > 0$.*

PROOF. Define the subset $Y = \{u \in \mathcal{D}(A); \overline{\mathcal{D}(A)} \cap Au \neq \emptyset\}$. Since

$$\|(e^{tA_n} - e^{tA_m})u\|_X = \left\|\int_0^t e^{(t-\tau)A_m} e^{\tau A_n}(A_m - A_n)u\,d\tau\right\|_X$$

$$\leq M^2 \int_0^t e^{(t-\tau)\beta_m} e^{\tau \beta_n} \|(A_m - A_n)u\|_X \, d\tau,$$

Proposition 2.3 yields that, for $u \in Y$, $e^{tA_n}u$ is convergent in X uniformly on any finite interval $[0,T]$ of t. We next notice that Y is a dense subset of $\overline{\mathcal{D}(A)}$. Indeed, for $g \in \overline{\mathcal{D}(A)}$, consider the sequence $u_n = J_n^2 g$; since $nu_n - nJ_n g \in Au_n$, $u_n \in Y$; in addition, from Lemma 2.2 it follows that $u_n \to g$. This then yields that the same convergence as above is true for $e^{tA_n}g$, $g \in \overline{\mathcal{D}(A)}$, also. Moreover, since $e^{tA_n}g \in \overline{\mathcal{D}(A)}$, it is true for the limit. Let us now consider an element $h \in A0$. Since $(n-A)^{-1}h = 0$ for $n > \beta$, $e^{tA_n}h = e^{-tn}h$, $t \geq 0$; hence, $e^{tA_n}h \to 0$ if $t > 0$. In view of Proposition 2.1, we have thus verified the strong convergence of e^{tA_n} on the space X_0, and as the limit e^{tA} is defined. Obviously, e^{tA} is a semigroup and enjoys the desired properties.

On the whole space X, however, we shall only have weak convergence. For each $n > \beta$, let us consider the integral operator

$$(e^{tA_n} * f)(t) = \int_0^t e^{(t-\tau)A_n} f(\tau) d\tau, \quad 0 \le t < \infty,$$

defined for $f \in L^1((0, \infty); X)$. Clearly, $e^{tA_n} * f$ is a continuous function on $[0, \infty)$ with values in X and satisfies

(2.2) $\|(e^{tA_n} * f)(t)\|_X \le M \int_0^t e^{(t-\tau)\beta_n} \|f(\tau)\|_X d\tau, \quad 0 \le t < \infty.$

THEOREM 2.5. *(Convergence on X.) Let A be as in Theorem 2.4. For each $f \in L^1((0, \infty); X)$, $e^{tA_n} * f$ converges to a continuous function with values in X which is denoted by $e^{tA} * f$. The mapping: $f \mapsto e^{tA} * f$ is a linear operator from $L^1((0, \infty); X)$ to $\mathcal{C}([0, \infty); \overline{\mathcal{D}(A)})$ satisfying the estimate*

(2.3) $\|(e^{tA} * f)(t)\|_X \le M \int_0^t e^{(t-\tau)\beta} \|f(\tau)\|_X d\tau, \quad 0 \le t < \infty.$

PROOF. First let us consider the case where $f \in \mathcal{C}^1([0, \infty); X)$. Fix a number $\lambda_0 > \max\{\beta_n; n > \beta\}$. Then, since $\lambda_0 \in \cap_{n > \beta} \rho(A_n)$, we can write:

$$(e^{tA_n} * f)(t) = (\lambda_0 - A_n) \int_0^t e^{(t-\tau)A_n} (\lambda_0 - A_n)^{-1} f(\tau) d\tau$$

$$= \lambda_0 \int_0^t e^{(t-\tau)A_n} (\lambda_0 - A_n)^{-1} f(\tau) d\tau + \int_0^t \frac{\partial e^{(t-\tau)A_n}}{\partial \tau} (\lambda_0 - A_n)^{-1} f(\tau) d\tau.$$

Integration by parts yields

$$(e^{tA_n} * f)(t) = -e^{tA_n} (\lambda_0 - A_n)^{-1} f(0) + (\lambda_0 - A_n)^{-1} f(t)$$

$$+ \int_0^t e^{(t-\tau)A_n} (\lambda_0 - A_n)^{-1} \{\lambda_0 f(\tau) - f'(\tau)\} d\tau.$$

According to Proposition 2.3 and Theorem 2.4, we have: $e^{tA_n} (\lambda_0 - A_n)^{-1} \rightarrow e^{tA} (\lambda - A)^{-1}$ strongly on X. Therefore $e^{tA_n} * f$ is seen to converge to the function

$$(e^{tA} * f)(t) = -e^{tA} (\lambda_0 - A)^{-1} f(0) + (\lambda_0 - A)^{-1} f(t)$$

$$+ \int_0^t e^{(t-\tau)A} (\lambda_0 - A)^{-1} \{\lambda_0 f(\tau) - f'(\tau)\} d\tau$$

Obviously this function belongs to $C([0, \infty); \overline{\mathcal{D}(A)})$. The estimate (2.3) follows directly from (2.2).

Let us now consider the general case where $f \in L^1((0, \infty); X)$. The proof is however immediate from (2.3) if we notice that $C^1([0, T]; X)$ is a dense subspace of $L^1((0, T); X)$ for any finite T.

In the statement of Theorem 2.5 we used $e^{tA} *$ to denote a linear operator without considering what e^{tA} actually means on X. e^{tA} must in fact be viewed as a distribution semigroup on X generated by A. For all $\varphi \in C_0^\infty(\mathbb{R})$, we can define the products $\langle e^{tA}, \varphi \rangle$ by the formulas

$$\langle e^{tA}, \varphi \rangle f = \lim_{n \to \infty} \int_0^\infty e^{tA_n} \varphi(t) f \, dt, \quad f \in X.$$

Each $\langle e^{tA}, \varphi \rangle$ is a bounded operator on X with

$$\|\langle e^{tA}, \varphi \rangle f\|_X \leq M \int_0^\infty e^{\beta t} |\varphi(t)| dt \|f\|_X, \quad f \in X.$$

In addition it is easily seen that $\langle e^{tA}, \varphi * \psi \rangle = \langle e^{tA}, \varphi \rangle \langle e^{tA}, \psi \rangle$ for $\varphi, \psi \in C_0^\infty(\mathbb{R})$ such that $\varphi(t) = \psi(t) = 0$ for $t \leq 0$. Such an $\mathcal{L}(X)$ valued distribution e^{tA} was called a distribution semigroup by J. L. Lions [127] who first studied it. We also observe that

$$(e^{tA} * \varphi)(t) f = \lim_{n \to \infty} \int_0^t e^{(t-\tau)A_n} \varphi(\tau) f \, d\tau, \quad 0 \leq t < \infty, \ f \in X,$$

for all $\varphi \in C_0^\infty(\mathbb{R})$ such that $\varphi(t) = 0$ for $t \leq 0$, which then justifies our definition of $e^{tA} *$ in Theorem 2.5.

2.2 MULTIVALUED EQUATIONS $\frac{du}{dt} \in Au + f(t)$

Consider the multivalued evolution equation

(E) $$\begin{cases} \dfrac{du}{dt} \in Au + f(t), & 0 \leq t \leq T, \\ u(0) = u_0, \end{cases}$$

in a Banach space X. Here, A is a m. l. operator satisfying (H-Y), $f : [0, T] \to X$ is a given continuous function, $u_0 \in \mathcal{D}(A)$ is an initial value, $u : [0, T] \to \mathcal{D}(A)$ is the unknown function.

We shall seek an X-valued C^1 solution. Such a solution is often called a strict solution to (E).

DEFINITION. A function $u:[0,T] \to \mathcal{D}(A)$ is called *a strict solution to* (E) if $u \in \mathcal{C}^1([0,T];X)$ with $u(0) = u_0$ and if u satisfies the multivalued equation in (E) at each point $0 \le t \le T$.

If a strict solution u exists, then $u'(0) \in \overline{\mathcal{D}(A)} \cap \{f(0) + Au_0\}$. This then means that the condition

(C) $$\overline{\mathcal{D}(A)} \cap \{f(0) + Au_0\} \ne \emptyset$$

must always be assumed as a compatibility condition in seeking the strict solution. Note that, if (C) holds, then, by Proposition 2.1, $\overline{\mathcal{D}(A)} \cap \{f(0) + Au_0\}$ consists of a single point.

THEOREM 2.6. *Let A be a m. l. operator satisfying (H-Y). If $f \in \mathcal{C}^1([0,T];X)$ and if $u_0 \in \mathcal{D}(A)$ with the compatibility condition (C), then the function given by*

(2.4) $$u(t) = e^{tA}u_0 + (e^{tA} * f)(t), \quad 0 \le t \le T,$$

is a strict solution to (E). Conversely, if $f \in \mathcal{C}([0,T];X)$ and if $u_0 \in \mathcal{D}(A)$, then any strict solution to (E) is necessarily of the form (2.4) and hence is unique.

PROOF. Let us first prove that the function defined by (2.4) gives a strict solution. Consider the sequence of functions

$$u_n(t) = e^{tA_n}u_{0,n} + \int_0^t e^{(t-\tau)A_n} f(\tau)d\tau, \quad 0 \le t \le T, n > \beta.$$

Here, $u_{0,n} = u_0 + n^{-1}\{f(0) - g_0\}$, $n > \beta$, and g_0 is the unique element of $\overline{\mathcal{D}(A)} \cap \{f(0) + Au_0\}$. By Theorems 2.4 and 2.5, u_n converges to the function u pointwisely. In addition, we have:

$$A_n u_n(t) = e^{tA_n} A_n u_{0,n} - \int_0^t \frac{\partial e^{(t-\tau)A_n}}{\partial \tau} f(\tau)d\tau$$

$$= e^{tA_n} g_0 - f(t) + \int_0^t e^{(t-\tau)A_n} f'(\tau)d\tau$$

Here we used Theorem 1.7 to obtain that

$$A_n u_{0,n} = n(J_n - 1)\{u_0 + n^{-1}(f(0) - g_0)\} = g_0 - f(0).$$

Therefore, $A_n u_n$ also converges to the continuous function

$$-f(t) + e^{tA} g_0 + (e^{tA} * f')(t), \quad 0 \le t \le T,$$

pointwisely.

The proof is then immediate. Indeed, from $u_n(t) = (\lambda_0 - A_n)^{-1} \times (\lambda_0 - A_n)u_n(t)$ with some $\lambda_0 \geq \max\{\beta_n; n > \beta\}$, it is verified that $u(t) = (\lambda_0 - A)^{-1}\{\lambda_0 u(t) + f(t) - g(t)\}$, where $g(t) = e^{tA}g_0 + (e^{tA} * f')(t)$; that is, $u(t) \in \mathcal{D}(A)$ with $g(t) \in f(t) + Au(t)$ for every $0 \leq t \leq T$. On the other hand, letting $n \to \infty$ in

$$u_n(t) - u_{0,n} = \int_0^t \{f(\tau) + A_n u_n(\tau)\}d\tau,$$

it follows that $u(t) - u_0 = \int_0^t g(\tau)d\tau$; that is, $u \in C^1([0,T]; X)$ with $u' = g$.

Let us next prove that any strict solution u to (E) must be of the form (2.4). Consider the function $e^{(t-\tau)A_n}u(\tau)$ with respect to the variable τ, fixing $0 < t \leq T$. Then,

$$\frac{\partial e^{(t-\tau)A_n}u(\tau)}{\partial \tau} = -e^{(t-\tau)A_n}A_n u(\tau) + e^{(t-\tau)A_n}u'(\tau)$$

$$= e^{(t-\tau)A_n}\{J_n f(\tau) + (1 - J_n)u'(\tau)\}, \quad 0 \leq \tau \leq t.$$

Here we used again Theorem 1.7 to obtain that

$$A_n u(\tau) = J_n\{u'(\tau) - f(\tau)\}.$$

Hence,

$$u(t) - e^{tA_n}u_0 = J_n \int_0^t e^{(t-\tau)A_n}f(\tau)d\tau + (1 - J_n) \int_0^t e^{(t-\tau)A_n}u'(\tau)d\tau.$$

Let $n \to \infty$; then, (2.4) follows immediately from Lemma 2.2 and Theorem 2.5. (Recall that $e^{tA} * f$ and $e^{tA} * u'$ take their values in $\overline{\mathcal{D}(A)}$).

2.3 DISSIPATIVE MULTIVALUED LINEAR OPERATORS

In this section X denotes a Hilbert space with the inner product $(\cdot, \cdot)_X$. We shall consider dissipative m. l. operators acting in X in an analogous way as for the single valued operator case (cf. H. Tanabe [168; Chapter 7]).

DEFINITION. A m. l. operator A in X is called *dissipative* if

$$\Re e(f, u)_X \leq 0 \quad \text{for all pairs } f, u \text{ such that } f \in Au.$$

A dissipative m. l. operator A which has no strict extension as a dissipative operator is called *maximal dissipative*.

It is proved by a similar argument as for single valued operators (therefore we may omit the proof) that a dissipative m. l. operator A is maximal if and only if the range condition $\mathcal{R}(\lambda_0 - A) = X$ holds for some $\lambda_0 > 0$.

Noting this we prove the following result.

THEOREM 2.7. *Let A be a m. l. operator in X such that $A - \beta$ is maximal dissipative with some real number β. More precisely, let A satisfy*

$$(2.5) \qquad \Re e(f, u)_X \leq \beta \|u\|_X^2 \quad \text{for all} \quad f \in Au$$

with the range condition

$$(2.6) \qquad \mathcal{R}(\lambda_0 - A) = X \quad \text{for some} \quad \lambda_0 > \beta.$$

Then, $\rho(A) \supset (\beta, \infty)$ and A satisfies Condition (H-Y)

$$\|(\lambda - A)^{-1}\|_{\mathcal{L}(X)} \leq \frac{1}{\lambda - \beta}, \quad \lambda > \beta$$

with $M = 1$.

PROOF. Let $\lambda > \beta$. If $f \in (\lambda - A)u$, from (2.5) it follows that

$$(\lambda - \beta)\|u\|_X^2 \leq \Re e(f, u)_X \leq \|f\|_X \|u\|_X.$$

Therefore, $\|u\|_X \leq \frac{\|f\|_X}{\lambda - \beta}$. This shows that $(\lambda - A)^{-1}$ is a single valued operator satisfying

$$\|(\lambda - A)^{-1}f\|_X \leq \frac{\|f\|_X}{\lambda - \beta}, \quad f \in \mathcal{R}(\lambda - A).$$

To complete the proof it therefore suffices to verify that $\mathcal{R}(\lambda - A) = X$. But the proof is quite the same as for the single valued case.

Let A be a m. l. operator in X with a maximal dissipative $A - \beta$, $\beta \in \mathbb{R}$. By virtue of Theorem 2.4, a semigroup e^{tA} is generated on the whole space $X = \overline{\mathcal{D}(A)} + A0$ (see Remark to Proposition 2.1) with an estimate $\|e^{tA}\|_{\mathcal{L}(X)} \leq e^{\beta t}$, $0 \leq t < \infty$.

In addition, consider the multivalued evolution equation

$$(E) \qquad \begin{cases} \dfrac{du}{dt} \in Au + f(t), \ 0 \leq t \leq T, \\ u(0) = u_0. \end{cases}$$

In the present case the compatibility condition (C) is reduced to $u_0 \in \mathcal{D}(A)$. Indeed, if $u_0 \in \mathcal{D}(A)$ and $f_0 \in Au_0$, then $f(0) + f_0 = g_0 + h_0$ with $g_0 \in \overline{\mathcal{D}(A)}$, $h_0 \in A0$. Therefore,

$$f(0) + Au_0 = f(0) + f_0 + A0 = g_0 + h_0 + A0 = g_0 + A0 \ni g_0.$$

In virtue of Theorem 2.6 we then obtain the following theorem.

THEOREM 2.8. *Let A be a m. l. operator satisfying (2.5) and (2.6). For any $f \in C^1([0, T]; X)$ and any $u_0 \in \mathcal{D}(A)$, the function*

$$u(t) = e^{tA}u_0 + \int\limits_0^t e^{(t-\tau)A}f(\tau)d\tau, \quad 0 \leq t \leq T,$$

is the unique strict solution to (E).

2.4 DEGENERATE EVOLUTION EQUATIONS

We shall study three types of degenerate evolution equations in a Hilbert space X.

Consider first:

(D-E.1)
$$\begin{cases} \dfrac{dMv}{dt} = Lv + f(t), \quad 0 \le t \le T, \\ Mv(0) = u_0, \end{cases}$$

where M and L are single valued linear operators in X with $\mathcal{D}(L) \subset \mathcal{D}(M)$, $f{:}[0,T] \to X$ is a given function, u_0 is an initial value, and $v{:}[0,T] \to \mathcal{D}(L)$ is the unknown function.

We assume:

(2.7)
$$\Re(Lv, Mv)_X \le \beta \|Mv\|_X^2, \quad v \in \mathcal{D}(L);$$

(2.8) $\mathcal{R}(\lambda_0 M - L) = X$ and $(\lambda_0 M - L)^{-1}$ is single valued with some $\lambda_0 > \beta$.

Then, as a direct consequence of Theorem 2.8, we establish the following result.

THEOREM 2.9. *Let (2.7), (2.8) be satisfied. For any $f \in \mathcal{C}^1([0,T]; X)$ and any $u_0 \in M(\mathcal{D}(L))$, (D-E.1) possesses a unique strict solution v such that*

(2.9)
$$Mv \in \mathcal{C}^1([0,T]; X) \quad and \quad Lv \in \mathcal{C}([0,T]; X).$$

PROOF. By changing the unknown function to $u(t) = Mv(t)$, we rewrite (D-E.1) into a multivalued equation of the form (E) in Section 2 with the operator coefficient $A = LM^{-1}$. Then $A - \beta$ is shown to be maximal dissipative in X. Indeed, if $f \in Au$, then $f = Lv$, $Mv = u$ with some $v \in \mathcal{D}(L)$; so that, $(f, u)_X = (Lv, Mv)_X$, which shows that (2.5) follows from (2.7). On the other hand, for any $f \in X$, $f = (\lambda_0 M - L)v$ with some $v \in \mathcal{D}(L)$. If we put $u = Mv$, then $u \in M(\mathcal{D}(L)) = \mathcal{D}(A)$ and $f \in (\lambda_0 - A)u$, i.e. (2.6). Therefore, the multivalued equation (E) possesses a unique strict solution u. Clearly, u is a strict solution to (E) if and only if v is a strict solution to (D-E.1) in the sense (2.9). Similarly, $u_0 \in \mathcal{D}(A)$ if and only if $u_0 \in M(\mathcal{D}(L))$. Finally, the uniqueness of the solution v follows from (2.8).

Consider next:

(D-E.2)
$$\begin{cases} M^* \dfrac{dMv}{dt} = Lv + M^* f(t), \quad 0 \le t \le T, \\ Mv(0) = u_0, \end{cases}$$

where M is a bounded linear operator on X and L is a single valued linear operator in X, $f:[0,T] \to X$ is a given function, u_0 is an initial value, and $v:[0,T] \to \mathcal{D}(L)$ is the unknown function.

We assume:

(2.10) $\Re(Lv,v)_X \le \beta\|Mv\|_X^2, \quad v \in \mathcal{D}(L);$

(2.11) $\mathcal{R}(\lambda_0 M^*M - L) \supset \mathcal{R}(M^*)$ and $(\lambda_0 M^*M - L)^{-1}$ is single valued on $\mathcal{R}(M^*)$ with some $\lambda_0 > \beta$.

Then we have as above the following theorem.

THEOREM 2.10. *Let (2.10) and (2.11) be satisfied. Then, for any* $f \in C^1([0,T];X)$ *and any* u_0 *satisfying*

(2.12) $u_0 = Mv_0, \quad Lv_0 \in \mathcal{R}(M^*) \text{ with some } v_0 \in \mathcal{D}(L),$

(D-E.2) possesses a unique strict solution v *such that*

(2.13) $Mv \in C^1([0,T];X) \text{ and } Lv \in C([0,T];X).$

PROOF. By changing the unknown function to $u(t) = Mv(t)$, (D-E.2) is also rewritten into a multivalued equation (E) with the operator coefficient $A = (M^*)^{-1}LM^{-1}$. Then $A - \beta$ is shown to be maximal dissipative in X. Indeed, if $f \in Au$, then $M^*f = Lv$, $Mv = u$ with some $v \in \mathcal{D}(L)$; so that, $(f,u)_X = (f,Mv)_X = (Lv,v)_X$. On the other hand, for any $f \in X$, $M^*f = (\lambda_0 M^*M - L)v$ with some $v \in \mathcal{D}(L)$. If we put $u = Mv$, then $v \in M^{-1}u$ and $\lambda_0 u - f \in (M^*)^{-1}Lv$; that is, $f \in (\lambda_0 - A)u$. It is easy to verify that $u_0 \in \mathcal{D}(A)$ is equivalent to (2.12). Therefore, the reduced equation (E) possesses a unique strict u. It is also easy to verify that $u = Mv$ is a strict solution to (E) if and only if v is a strict solutio to (D-E.2) in the sense (2.13). The uniqueness of the solution v follows from (2.11).

It is also possible to handle a problem of the type

(D-E.2′) $\begin{cases} M^* \dfrac{dMv}{dt} = Lv + f(t), & 0 \le t \le T, \\ Mv(0) = u_0, \end{cases}$

provided that (2.11) is strengthened to

(2.14) $\mathcal{R}(\lambda_0 M^*M - L) = X$ and $(\lambda_0 M^*M - L)^{-1}$ is single valued and bounded on X with some $\lambda_0 > \beta$.

In fact, if we change the unknown function to

$$\tilde{v}(t) = v(t) - (\lambda_0 M^*M - L)^{-1}f(t),$$

then the reduced equation is of the form

$$M^* \frac{dM\tilde{v}}{dt} = L\tilde{v} + M^*M(\lambda_0 M^*M - L)^{-1}\{\lambda_0 f(t) - f'(t)\}.$$

COROLLARY 2.11. *If (2.10) and (2.14) are satisfied, then for any $f \in \mathcal{C}^2([0,T]; X)$ and any initial value u_0 satisfying $u_0 = Mv_0$, $Lv_0 + f(0) \in \mathcal{R}(M^*)$ with some $v_0 \in \mathcal{D}(L)$, the problem (D-E.2) possesses a unique strict solution v satisfying (2.13).*

At last let us consider the adjoint problem of (D-E.1):

(D-E.3)
$$\begin{cases} M\dfrac{du}{dt} = Lu + Mf(t), & 0 \le t \le T, \\ u(0) = u_0, \end{cases}$$

where M and L are densely defined single valued linear operators in X such that $\mathcal{D}(L^*) \subset \mathcal{D}(M^*)$, $f: [0,T] \to \mathcal{D}(M)$ is a given function, u_0 is an initial value, and $u = u(t)$ is the unknown function.

We assume:

(2.15)
$$\Re e(L^*\xi, M^*\xi)_X \le \beta \|\xi\|_X^2, \quad \xi \in \mathcal{D}(L^*);$$

(2.16)
$$\lambda_0 \in \rho_{M^*}(L^*) \quad \text{with some} \quad \lambda_0 > \beta.$$

Then, by the same argument as in the proof of Theorem 2.9, (2.15) and (2.16) yield that the m. l. operator $L^*(M^*)^{-1} - \beta$ is maximal dissipative in X. As a consequence, its adjoint $\{L^*(M^*)^{-1}\}^* - \beta$ is also maximal dissipative; so that, $\{L^*(M^*)^{-1}\}^*$ is the generator of a semigroup on X. On the other hand, clearly (D-E.3) is written in the form (E) with a coefficient operator $M^{-1}L$. Moreover, from (1.7), $M^{-1}L \subset \{L^*(M^*)^{-1}\}^*$. Therefore, we have verified the following theorem.

THEOREM 2.12. *Let (2.15) and (2.16) be satisfied. Then, for any $f \in \mathcal{C}^1([0,T]; X)$ and any $u_0 \in \mathcal{D}(L)$ such that $Lu_0 \in \mathcal{R}(M)$, an extended multivalued equation of (D-E.3) with the coefficient operator $\{L^*(M^*)^{-1}\}^*$ possesses a unique strict solution u.*

Of course, Theorem 2.12 obtains its major interest when $\{L^*(M^*)^{-1}\}^*$ and $M^{-1}L$ in fact coincide. The next proposition entails a condition sufficient to this end.

PROPOSITION 2.13. *If M is a bounded operator on X and if $\rho_M(L) \cap \rho_{M^*}(L^*) \ne \emptyset$, then $M^{-1}L = \{L^*(M^*)^{-1}\}^*$.*

PROOF. Let $\lambda_0 \in \rho_M(L) \cap \rho_{M^*}(L^*)$ be fixed. Since $(\lambda_0 - M^{-1}L)^{-1} = (\lambda_0 M - L)^{-1} M$, it is seen that $\lambda_0 \in \rho(M^{-1}L)$. On the other hand, according to Theorem 1.13, $\lambda_0 \in \rho(\{L^*(M^*)^{-1}\}^*)$. Therefore, noting that (1.7) implies the inclusion $(\lambda_0 - M^{-1}L)^{-1} \subset (\lambda_0 - \{L^*(M^*)^{-1}\}^*)^{-1}$, we obtain the desired result.

The more general equation of the form

(D-E.3′)
$$\begin{cases} M\dfrac{du}{dt} = Lu + f(t), & 0 \le t \le T, \\ u(0) = u_0, \end{cases}$$

can also be handled, provided that the additional conditions

(2.17) $\mathcal{D}(L) \subset \mathcal{D}(M)$ and $\rho_M(L) \neq \emptyset$

are assumed. In fact, change of the unkown function to $\tilde{u}(t) = u(t) - (\mu_0 M - L)^{-1} f(t)$, where $\mu_0 \in \rho_M(L)$, yields that

$$M\frac{d\tilde{u}}{dt} = L\tilde{u} + M(\mu_0 M - L)^{-1}\{\mu_0 f(t) - f'(t)\}.$$

COROLLARY 2.14. *Let (2.15), (2.16) and (2.17) be satisfied. Then, for any $f \in C^2([0,T]; X)$ and any $u_0 \in \mathcal{D}(L)$ such that $Lu_0 + f(0) \in \mathcal{R}(M)$, the extended multivalued equation of (D-E.3) with the coefficient operator $\{L^*(M^*)^{-1}\}^*$ possesses a unique strict solution u. If, in addition, M is bounded on X and $\rho_M(L) \cap \rho_{M^*}(L^*) \neq \emptyset$, then such a function u is the unique strict solution to (D-E.3).*

2.5 EXAMPLES

EXAMPLE 2.1. Consider:

(2.18)
$$\begin{cases} \dfrac{\partial m(x)v}{\partial t} = -\dfrac{\partial v}{\partial x} + f(x,t), & -\infty < x < \infty,\ 0 \le t \le T, \\ m(x)v(x,0) = u_0(x), & -\infty < x < \infty. \end{cases}$$

Here, $m(x)$ is the characteristic function of some measurable set $J \subset \mathbb{R}$ (i.e. $m(x) = 1$ for $x \in J$, $m(x) = 0$ for $x \notin J$), $f(x,t)$ is a given function, $u_0(x)$ is an initial function, and $v = v(x,t)$ is the unknown function.

We consider the problem in $X = L^2(\mathbb{R})$. Let M be the multiplication operator by $m(x)$ acting in X; clearly, M is bounded on X and enjoys the properties $M^* = M$ and $M^2 = M$. Therefore, (2.18) is formulated in X in the form

(2.19)
$$\begin{cases} M^*\dfrac{\partial Mv}{dt} = Lv + f(t), & 0 \le t \le T, \\ Mv(0) = u_0, \end{cases}$$

where $L = -\frac{d}{dx}$ with $\mathcal{D}(L) = H^1(\mathbb{R})$, L being a closed linear operator in X.

Since $\Re e(Lv, v)_X = 0$ for all $v \in \mathcal{D}(L)$, Condition (2.10) is satisfied with $\beta = 0$. So the question is whether the Conditions (2.11) and (2.14) are satisfied or not. Let us investigate this problem in some specific cases.

1) $J = (-\infty, a) \cup (b, \infty)$ $(a < b)$. In this case (2.14) is verified. Indeed, for $f \in X$, consider the problem

$$(2.20) \qquad m(x)v + \frac{dv}{dx} = f(x), \quad -\infty < x < \infty,$$

and seek a solution $v \in H^1(\mathbb{R})$. Obviously, (2.20) consists of three problems: $v_1 + \frac{dv_1}{dx} = f(x)$ for $x < a$ with $v_1(-\infty) = 0$; $\frac{dv_2}{dx} = f(x)$ for $a < x < b$ with $v_2(a) = v_1(a)$; and $v_3 + \frac{dv_3}{dx} = f(x)$ for $x > b$ with $v_3(b) = v_2(b)$, $v_3(\infty) = 0$. On the other hand, a trival calculation shows that the function v such that $v = v_1$ for $x < a$, $v = v_2$ for $a < x < b$, and $v = v_3$ for $x > b$, is the unique solution to (2.20) satisfying the estimate $\|v\|_{H^1} \leq Const.\|f\|_{L^2}$. Hence, (2.14) is verified.

2) $J = (a, \infty)$. Consider similarly the problem (2.20). Then we observe that the first problem: $\frac{dv_1}{dx} = f(x)$ for $x < a$ does not admit any solution $v_1 \in H^1((-\infty, a))$ in general; this means that (2.14) is no longer valid. On the contrary, (2.11) is verified. In fact, we are concerned with

$$(2.21) \qquad m(x)v + \frac{dv}{dx} = m(x)f(x), \quad -\infty < x < \infty.$$

In other words, $\frac{dv_1}{dx} = 0$ for $x < a$ with $v_1(-\infty) = 0$, and $v_2 + \frac{dv_2}{dx} = f(x)$ for $x > a$ with $v_2(a) = v_1(a) = 0$. This problem then admits a unique solution $v \in H^1(\mathbb{R})$.

3) $J = (-\infty, a)$. Neither (2.11) nor (2.14) are true. In fact, consider the problem (2.21). Since $\frac{dv_2}{dx} = 0$ for $x > a$, $v_2(a)$ must vanish; hence, $v_1 + \frac{dv_1}{dx} = f(x)$ for $x < a$ with $v_1(-\infty) = v_1(a) = 0$; but this does not happen in general.

EXAMPLE 2.2. Let

$$\mathrm{rot}\, E = -\frac{\partial B}{\partial t}, \quad \mathrm{rot}\, H = \frac{\partial D}{\partial t} + J$$

be Maxwell's equations in $\mathbb{R}^3$, where E (resp. H) denotes the electric (resp. magnetic) field intensity, B (resp. D) denotes the electric (resp. magnetic) flux density, and where J is the current density.

We assume that the medium which fills the space $\mathbb{R}^3$ is linear but it may be anisotropic and nonhomogeneous, that is,

$$D = \varepsilon E, \quad B = \mu H, \quad J = \sigma E + J'$$

with some 3×3 matrices $\varepsilon(x)$, $\mu(x)$ and $\sigma(x)$, $x \in \mathbb{R}^3$, where J' is a given forced current density. Then it is written as

$$\frac{\partial}{\partial t} \begin{pmatrix} \varepsilon(x) & 0 \\ 0 & \mu(x) \end{pmatrix} \begin{pmatrix} E \\ H \end{pmatrix} = \begin{pmatrix} 0 & \text{rot} \\ -\text{rot} & 0 \end{pmatrix} \begin{pmatrix} E \\ H \end{pmatrix}$$
$$- \begin{pmatrix} \sigma(x) & 0 \\ 0 & 0 \end{pmatrix} \begin{pmatrix} E \\ H \end{pmatrix} - \begin{pmatrix} J'(x,t) \\ 0 \end{pmatrix}$$

or

$$(2.22) \qquad \frac{\partial c(x)v}{\partial t} = \sum_{i=1}^{3} a_i \frac{\partial v}{\partial x_i} + b(x)v + f(x,t), \quad (x,t) \in \mathbb{R}^3 \times [0,T],$$

with

$$v = \begin{pmatrix} E \\ H \end{pmatrix}, \ c(x) = \begin{pmatrix} \varepsilon(x) & 0 \\ 0 & \mu(x) \end{pmatrix}, \ b(x) = - \begin{pmatrix} \sigma(x) & 0 \\ 0 & 0 \end{pmatrix},$$
$$f(x,t) = - \begin{pmatrix} J'(x,t) \\ 0 \end{pmatrix}$$

and with certain 6×6 symmetric matrices a_i, $i = 1, 2, 3$.

$\varepsilon(x)$, $\mu(x)$ and $\sigma(x)$ are assumed to be real matrices the components of which are bounded measurable functions in $\mathbb{R}^3$. In addition, we assume:

(2.23) $\varepsilon(x)$ are symmetric and $\varepsilon(x) \geq 0$ for all $x \in \mathbb{R}^3$;

(2.24) $\mu(x)$ are symmetric and $\mu(x) \geq \delta$ with some $\delta > 0$ uniformly in $x \in \mathbb{R}^3$; and

(2.25) $(\{\gamma\varepsilon(x) + \sigma(x)\}\xi, \xi) \geq \delta\|\xi\|^2$, $\xi \in \mathbb{R}^3$, with some $\delta > 0$ and $\gamma \geq 0$ uniformly in $x \in \mathbb{R}^3$.

Then we can formulate the problem (2.22) in the abstract form

$$(2.26) \qquad \begin{cases} M^* \dfrac{dMv}{dt} = Lv + f(t), \quad 0 \leq t \leq T, \\ Mv(0) = u_0, \end{cases}$$

in the space $X = \{L^2(\mathbb{R}^3)\}^6$, using the bounded operator M of multiplication by $\sqrt{c(x)}$ acting in X ($M = M^*$) and the closed linear operator L defined by

$$(2.27) \qquad \begin{cases} \mathcal{D}(L) = \{v \in X; \displaystyle\sum_{i=1}^{3} a_i \frac{\partial v}{\partial x_i} \in X\} \\ Lv = \displaystyle\sum_{i=1}^{3} a_i \frac{\partial v}{\partial x_i} + b(x)v. \end{cases}$$

Condition (2.10) is verified as follows. Let $v \in Y = \{H^1(\mathbb{R}^3)\}^6 \subset \mathcal{D}(L)$. Then,

$$(Lv, v)_X = -\left(v, \sum_{i=1}^{3} a_i \frac{\partial v}{\partial x_i} + b(x)v\right)_X + (b(x)v, v)_X + (v, b(x)v)_X;$$

so that,

$$\Re(Lv, v)_X = \Re(b(x)v, v)_X = -\Re(\sigma E, E)_{L^2},$$

where $v = \begin{pmatrix} E \\ H \end{pmatrix}$. (2.24) and (2.25) then yield that

$$\begin{aligned}
\Re(Lv, v)_X &\leq -\delta(\|E\|_{L^2}^2 + \|H\|_{L^2}^2) + \gamma(\varepsilon(x)E, E)_{L^2} + (\mu(x)H, H)_{L^2} \\
&\leq -\delta\|v\|_X^2 + \lambda\{(\varepsilon(x)E, E)_{L^2} + (\mu(x)H, H)_{L^2}\} \\
(2.28) \qquad &= -\delta\|v\|_X^2 + \lambda\|Mv\|_X^2,
\end{aligned}$$

where $\lambda \geq \max\{\gamma, 1\}$. This estimate is in fact verified for $v \in \mathcal{D}(L)$ also, because there exists a sequence $v_n \in Y$ such that $v_n \to v$ and $Lv_n \to Lv$ in X. Thus (2.10) holds with $\beta = \max\{\gamma, 1\}$.

Let us next verify (2.14). Since (2.28) yields that

$$\|(\lambda M^*M - L)v\|_X \geq \delta\|v\|_X, \quad v \in \mathcal{D}(L),$$

$(\lambda M^*M - L)$ is seen to be one-to-one and to have a closed range. Therefore it suffices to verify that $\mathcal{R}(\lambda M^*M - L)^\perp = \{0\}$. Let $w \in \mathcal{R}(\lambda M^*M - L)^\perp$; then, $w \in \mathcal{D}(L^*)$ and $(\lambda M^*M - L^*)w = 0$; on the other hand, since the principal part of L is symmetric, $w \in \mathcal{D}(L)$. Therefore, (2.28) yields

$$-\delta\|w\|_X^2 + \lambda\|Mw\|_X^2 \geq \Re(Lw, w)_X = \Re(w, L^*w)_X = \lambda\|Mw\|_X^2,$$

i.e. $w = 0$.

REMARK. Strong and weak solutions to the equation (2.22), even with some boundary conditions, have been studied by G. Duvaut and J. L. Lions [73], P. D. Lax and R. S. Phillips [126], M. Povoas [146], etc. Our assumptions are analogous to Povoas' ones in [146], where she studies strong solutions in a region in $\mathbb{R}^3$.

EXAMPLE 2.3. Let

$$(2.29) \qquad \begin{cases} \left(m(x)\dfrac{\partial}{\partial t}\right)^2 v = \Delta v + f(x, t), & (x, t) \in \Omega \times [0, T], \\ v = 0, & (x, t) \in \partial\Omega \times [0, T], \\ v(x, 0) = u_0(x),\ m(x)\dfrac{\partial v}{\partial t}(x, 0) = u_1(x), & x \in \Omega, \end{cases}$$

be the Poisson-wave equation in a (bounded or unbounded) region $\Omega \subset \mathbb{R}^n$ with a smooth boundary $\partial\Omega$. Here, $m \in L^\infty(\Omega)$, $m(x) \geq 0$ is a given function which is allowed to vanish in a bounded subset of Ω, $f(x, t)$ is also a given function, $u_0(x)$ and $u_1(x)$ are initial functions, and $v = v(x, t)$ is the unknown function.

Setting $v_0 = v$, $v_1 = \frac{\partial v}{\partial t}$, we rewrite the equation as a system

$$\begin{pmatrix} 1 & 0 \\ 0 & m(x) \end{pmatrix} \frac{\partial}{\partial t} \begin{pmatrix} 1 & 0 \\ 0 & m(x) \end{pmatrix} \begin{pmatrix} v_0 \\ v_1 \end{pmatrix} = \begin{pmatrix} 0 & 1 \\ \Delta & 0 \end{pmatrix} \begin{pmatrix} v_0 \\ v_1 \end{pmatrix} + \begin{pmatrix} 0 \\ f(x, t) \end{pmatrix}.$$

Then (2.29) is formulated in the form

$$\begin{cases} M^* \dfrac{dMV}{dt} = LV + F(t), & 0 \leq t \leq T, \\ MV(0) = U_0, \end{cases}$$

in the product space $X = H_0^1(\Omega) \times L^2(\Omega)$, where $M = \begin{pmatrix} 1 & 0 \\ 0 & m(x) \end{pmatrix}$ is a multiplication operator acting in X $(M = M^*)$, $L = \begin{pmatrix} 0 & 1 \\ \Delta & 0 \end{pmatrix}$ is a closed linear operator in X with $\mathcal{D}(L) = \{H^2(\Omega) \cap H_0^1(\Omega)\} \times H_0^1(\Omega)$, and where $F(t) = \begin{pmatrix} 0 \\ f(t) \end{pmatrix}$, $U_0 = \begin{pmatrix} u_0 \\ u_1 \end{pmatrix}$, $V = \begin{pmatrix} v_0 \\ v_1 \end{pmatrix}$.

When Ω is unbounded, we also assume:
(2.30) $m(x) \geq \delta$, $x \in \Omega - \Omega_R$, with some $\delta > 0$ and $R > 0$, where $\Omega_R = \{x \in \Omega; |x| \leq R\}$.

LEMMA 2.15. *The norm*

$$\|u\|_{\tilde{H}_0^1} = \sqrt{\|\nabla u\|_{L^2}^2 + \|mu\|_{L^2}^2}$$

is an equivalent norm of the space $H_0^1(\Omega)$.

PROOF OF THE LEMMA. Clearly, $\|u\|_{\tilde{H}_0^1} \leq \max\{1, \|m\|_{L^\infty}\}\|u\|_{H_0^1}$, $u \in H_0^1(\Omega)$. Let $\psi \in \mathcal{C}_0^\infty(\mathbb{R}^n)$, $0 \leq \psi \leq 1$, such that $\psi \equiv 0$ for $|x| \geq R + 1$ and $\psi \equiv 1$ for $|x| \leq R$. Then, from (2.30),

$$\|u\|_{L^2} \leq \|\psi u\|_{L^2} + \|(1 - \psi)u\|_{L^2} \leq \|\psi u\|_{L^2} + \delta^{-1}\|mu\|_{L^2}.$$

In addition, according to Poincaré inequality, we observe that

$$\|\psi u\|_{L^2} \leq C\|\nabla(\psi u)\|_{L^2} \leq C\{\|u\nabla\psi\|_{L^2} + \|\psi\nabla u\|_{L^2}\}$$
$$\leq C\{\|\nabla u\|_{L^2} + \delta^{-1}\|mu\|_{L^2}\}.$$

Hence, $\|u\|_{H_0^1} \leq Const.\|u\|_{\tilde{H}_0^1}$, $u \in H_0^1(\Omega)$.

Henceforth we shall endow $H_0^1(\Omega)$ with the new norm $\|\cdot\|_{\tilde{H}_0^1}$. Then, for

$$F = \begin{pmatrix} f_0 \\ f_1 \end{pmatrix} \in X, \|F\|_X = \sqrt{\|f_0\|_{\tilde{H}_0^1}^2 + \|f_1\|_{L^2}^2}.$$

Therefore, for $V = \begin{pmatrix} v_0 \\ v_1 \end{pmatrix} \in \mathcal{D}(L)$,

$$(LV, V)_X = (\nabla v_1, \nabla v_0)_{L^2} + (mv_1, mv_0)_{L^2} + (\Delta v_0, v_1)_{L^2}$$
$$= 2i\Im(\nabla v_1, \nabla v_0)_{L^2} + (mv_1, mv_0)_{L^2}.$$

Hence,

$$\Re(LV, V)_X \leq \{\|mv_0\|_{L^2}^2 + \|mv_1\|_{L^2}^2\} \leq \|MV\|_X^2,$$

i.e. (2.10) is valid with $\beta = 1$.

In order to verify (2.14), consider the sesquilinear form

$$a_\lambda(u, v) = (\nabla u, \nabla v)_{L^2} + \lambda^2(mu, mv)_{L^2}$$

on $H_0^1(\Omega)$ for each $\lambda > 1$. Since $a_\lambda(\cdot, \cdot)$ is coercive, the Lax-Milgram theorem (see [187; p. 92]) yields that, for any $F = \begin{pmatrix} f_0 \\ f_1 \end{pmatrix} \in X$, there exists $v_0 \in H_0^1(\Omega)$ such that $-\Delta v_0 + \lambda^2 m^2 v_0 = \lambda m^2 f_0 + f_1$; moreover, thanks to the a priori estimates, $v_0 \in H^2(\Omega)$. Set $v_1 = \lambda v_0 - f_0 \in H_0^1(\Omega)$, then obviously $V = \begin{pmatrix} v_0 \\ v_1 \end{pmatrix} \in \mathcal{D}(L)$ and $(\lambda M^* M - L)V = F$. Thus (2.14) is verified.

More general Poisson-wave equations, containing the first derivative $\frac{\partial v}{\partial t}$, shall be discussed later on. See Section 6.4.

CHAPTER III

DEGENERATE EQUATIONS OF PARABOLIC TYPE, I:
THE AUTONOMOUS CASE

Degenerate parabolic equations will be studied by using infinitely differentiable semigroups generated by multivalued linear operators. In addition, the maximal regularity property of solutions to degenerate equations will be also established.

More precisely, as a first step we shall extend the known results on the maximal regularity of solutions of $\frac{du}{dt} = Au + f(t)$ when A generates an analytic semigroup in the complex Banach space X to the multivalued equation $\frac{du}{dt} \in Au + f(t)$. See Section 3.4. Section 3.5 is devoted to handle, by means of an extension of the sum of operators' method, the general case of generators of infinitely differentiable semigroups. In both the cases, interpolation theory has a crucial role. Section 3.6 deals with convergence of solutions of approximating problems to a solution of a limit problem, according to the model of Trotter and Kato theorem.

Last paragraph contains various applications of the theory to important classes of degenerate partial differential equations of parabolic type.

3.1 GENERATION OF INFINITELY
DIFFERENTIABLE SEMIGROUPS

Let A be a m. l. operator in a Banach space X. A is assumed to satisfy the resolvent condition:

(P) The resolvent set $\rho(A)$ contains a region

$$\Sigma = \{\lambda \in \mathbb{C}; \Re e\lambda \geq -c(|\Im m\lambda| + 1)^\alpha\}$$

and the resolvent $(\lambda - A)^{-1}$ satisfies

(3.1) $$\|(\lambda - A)^{-1}\|_{\mathcal{L}(X)} \leq \frac{M}{(|\lambda| + 1)^\beta}, \quad \lambda \in \Sigma,$$

with some exponents $0 < \beta \leq \alpha \leq 1$ and constants c, $M > 0$.

Then, A generates an infinitely differentiable semigroup on X as precised below.

THEOREM 3.1. *Let A be a m. l. operator satisfying (P). Define the family of bounded linear operators e^{tA} by*

$$e^{tA} = \frac{1}{2\pi i} \int_\Gamma e^{t\lambda}(\lambda - A)^{-1}d\lambda, \quad t > 0,$$

where the Dunford integrals are viewed in $\mathcal{L}(X)$ and the contour Γ is parametrized by $\lambda = -c(|\eta| + 1)^\alpha + i\eta$, $-\infty < \eta < \infty$, lying in Σ. Define also $e^{0A} = 1$. Then, e^{tA}, $t \geq 0$, is a semigroup on X. e^{tA} is infinitely differentiable for $t > 0$ in the uniform operator norm with

(3.2) $$\frac{d}{dt}e^{tA} \subset Ae^{tA}, \quad t > 0.$$

PROOF. All the proof is quite similar to the single valued case. The semigroup property is verified from Theorem 1.8 directly. Differentiabity is checked by applying the Lebesgue dominated convergence theorem, as well as

$$\frac{d^k}{dt^k}e^{tA} = \frac{1}{2\pi i} \int_\Gamma \lambda^k e^{t\lambda}(\lambda - A)^{-1}d\lambda, \quad t > 0, \, k = 1, 2, 3, \ldots.$$

Finally, (3.2) is verified from $A^{-1}\frac{d}{dt}e^{tA} = e^{tA}$.

In the next proposition we list some estimates for e^{tA} and other properties generalizing the ones for the single valued case. In particular, we use the spaces X_A^θ introduced in Chapter I.

PROPOSITION 3.2. *For $\theta \geq 0$, let*

$$([-A]^\theta)^\circ e^{tA} = \frac{1}{2\pi i} \int_\Gamma (-\lambda)^\theta e^{t\lambda}(\lambda - A)^{-1}d\lambda, \quad t > 0.$$

Then,

(3.3) $$\|([-A]^\theta)^\circ e^{tA}\|_{\mathcal{L}(X)} \leq C_\theta t^{\frac{\beta-\theta-1}{\alpha}}, \quad t > 0.$$

For $0 < \theta < 1$,

$$\|e^{tA}\|_{\mathcal{L}(X, X_A^\theta)} \leq C_\theta t^{\frac{\beta-\theta-1}{\alpha}}, \quad t > 0.$$

Here, C_θ is a constant depending on the exponent θ.

PROOF. When $\frac{\theta+1-\beta}{\alpha} > 0$,

$$\|([-A]^\theta)^\circ e^{tA}\|_{\mathcal{L}(X)} \leq C \int_\Gamma |\lambda|^{\theta-\beta} e^{t\Re\lambda}|d\lambda|$$

$$\leq C \int_0^\infty \eta^{\theta-\beta} e^{-t\eta^\alpha} d\eta \leq Cc^{-\frac{\theta}{\alpha}} \Gamma\left(\frac{\theta+1-\beta}{\alpha}\right) t^{\frac{\beta-\theta-1}{\alpha}}.$$

When $\frac{\theta+1-\beta}{\alpha} = 0$ (that is when $\alpha = \beta = 1$ and $\theta = 0$), the estimate is verified by substituting an appropriate contour for Γ in a usual way (note that e^{tA} becomes an analytic semigroup).

Similarly, since

$$\xi^\theta A^\circ(\xi - A)^{-1}e^{tA} = \frac{1}{2\pi i}\int_\Gamma \frac{\xi^\theta \lambda}{\xi - \lambda}e^{t\lambda}(\lambda - A)^{-1}d\lambda, \quad \xi > 0,$$

it follows that

$$\xi^\theta \|A^\circ(\xi - A)^{-1}e^{tA}\|_{\mathcal{L}(X)} \leq C\int_\Gamma |\lambda|^{\theta - \beta}e^{t\Re\lambda}|d\lambda|.$$

As simple calculations show, this proves that the second estimate holds.

Since $[-A]^{-\theta}([-A]^\theta)^\circ e^{tA} = e^{tA}$ for $\theta > 1 - \beta$, $([-A]^\theta)^\circ e^{tA}$ is in fact a linear section of the m. l. operator $[-A]^\theta e^{tA}$ for each $t > 0$, $\theta > 1 - \beta$.

Let us now consider continuity of e^{tA} at $t = 0$.

THEOREM 3.3. e^{tA} is strongly continuous with respect to the seminorm $p_A(\cdot) = \|A^{-1}\cdot\|_X$. If $1 - \beta < \theta$, then e^{tA} is strongly continuous in the original norm of X on the subspaces $\mathcal{D}([-A]^\theta)$ and X_A^θ.

PROOF. Let $u \in \mathcal{D}([-A]^\theta)$, $\theta > 1 - \beta$ and let $f \in [-A]^\theta u$. Then, as $t \to 0$,

$$e^{tA}u = e^{tA}[-A]^{-\theta}f = \frac{1}{2\pi i}\int_\Gamma (-\lambda)^{-\theta}e^{t\lambda}(\lambda - A)^{-1}d\lambda$$

$$\to \frac{1}{2\pi i}\int_\Gamma (-\lambda)^{-\theta}(\lambda - A)^{-1}d\lambda = [-A]^{-\theta}f = u$$

in X.

In particular, since $A^{-1}e^{tA} \to A^{-1}$, the first assertion is also verified.

Let now $u \in X_A^\theta$, $\theta > 1 - \beta$, and set

$$w = \frac{1}{2\pi i}\int_\Gamma \lambda^{-1}A^\circ(\lambda - A)^{-1}u d\lambda.$$

Note here that, since

$$(\lambda - A)^{-1} = (|\lambda| - A)^{-1} + (|\lambda| - \lambda)(\lambda - A)^{-1}(|\lambda| - A)^{-1},$$

$u \in X_A^\theta$ implies

(3.4) $\|A^\circ(\lambda - A)^{-1}u\|_X \leq C|\lambda|^{1-\beta-\theta}\|u\|_{X_A^\theta}, \quad \lambda \in \Gamma.$

Clrealy, $A^{-1}w = A^{-1}u$; so that, $w - u \in A0$. On the other hand, for $\xi > 0$, it is observed that

$$(\xi - A)^{-1}w = (\xi - A)^{-1}u = \tfrac{1}{2\pi i} \int_\Gamma \frac{1}{\xi - \lambda}(\lambda - A)^{-1}u d\lambda$$

$$= \tfrac{1}{2\pi i} \int_\Gamma \frac{1}{\lambda(\xi - \lambda)} A^\circ(\lambda - A)^{-1}u d\lambda.$$

Therefore,

$$A^\circ(\xi - A)^{-1}w = -w + \xi(\xi - A)^{-1}w = \tfrac{1}{2\pi i} \int_\Gamma \frac{1}{\xi - \lambda} A^\circ(\lambda - A)^{-1}u d\lambda,$$

implies by (3.4) that

$$\|A^\circ(\xi - A)^{-1}w\|_X \le C \int_\Gamma \frac{|\lambda|^{1-\beta-\theta}}{|\xi - \lambda|}|d\lambda|\|u\|_{X_A^\theta} \le C \int_0^\infty \frac{\eta^{1-\beta-\theta}}{\xi + \eta} d\eta \|u\|_{X_A^\theta}.$$

This then shows that $w \in X_A^{\beta+\theta-1}$. We have therefore proved that $w - u \in A0 \cap X_A^{\beta+\theta-1} = \{0\}$.

Thus, as $t \to 0$,

$$e^{tA}u = \tfrac{1}{2\pi i} \int_\Gamma \lambda^{-1} e^{t\lambda} A^\circ(\lambda - A)^{-1}u d\lambda \to w = u$$

in X.

More strongly, Hölder exponent of $(e^{-tA} - 1)u$ can be calculated if u is assumed to belong to the subspaces $\mathcal{D}([-A]^\theta)$ and X_A^θ, respectively, with larger exponents θ. In fact, the norm of $A^\circ e^{tA}u$ is estimated as follows.

PROPOSITION 3.4. *For* $1 - \beta < \theta \le 1$,

$$\|A^\circ e^{tA}u\|_X \le C_\theta t^{\frac{\beta+\theta-2}{\alpha}}\|u\|_{\mathcal{D}([-A]^\theta)}, \quad u \in \mathcal{D}([-A]^\theta).$$

For $1 - \beta < \theta < 1$,

$$\|A^\circ e^{tA}u\|_X \le C_\theta t^{\frac{\beta+\theta-2}{\alpha}}\|u\|_{X_A^\theta}, \quad u \in X_A^\theta.$$

PROOF. Let $u \in \mathcal{D}([-A]^\theta)$, $1 - \beta < \theta \le 1$, and let $f \in [-A]^\theta u$. Since $-A^\circ e^{tA}u = -A^\circ e^{tA}[-A]^{-\theta}f = ([-A]^{1-\theta})^\circ e^{tA}f$, it follows from (3.3) that

$$\|A^\circ e^{tA}u\|_X \le C_\theta t^{\frac{\beta+\theta-2}{\alpha}}\|f\|_X.$$

So that the first estimate follows from the definition of $\|\cdot\|_{\mathcal{D}([-A]^\theta)}$.

Let next $u \in X_A^\theta$, $1 - \beta < \theta < 1$. Since

$$A^\circ e^{tA}u = \frac{1}{2\pi i} \int_\Gamma e^{t\lambda} A^\circ(\lambda - A)^{-1} d\lambda,$$

the second estimate is an immediate consequence of (3.4).

Proposition 3.4 combined with Theorem 3.3 yields the following result.

THEOREM 3.5. *Let* $2 - \alpha - \beta < \theta < 1$. *Then*

$$\|(e^{tA} - 1)u\|_X \le C_\theta t^{\frac{\alpha+\beta+\theta-2}{\alpha}} \|u\|_{\mathcal{D}([-A]^\theta)}, \quad u \in \mathcal{D}([-A]^\theta),$$

$$\|(e^{tA} - 1)u\|_X \le C_\theta t^{\frac{\alpha+\beta+\theta-2}{\alpha}} \|u\|_{X_A^\theta}, \quad u \in X_A^\theta.$$

We shall conclude this section by establishing further properties of the Yosida approximation A_n of the m. l. operator A and studying convergence of e^{tA_n}. Recall from Section 2.1 that

$$A_n = n\{n(n - A)^{-1} - 1\} = n(J_n - 1), \quad n = 1, 2, 3, \dots,$$

is the Yosida approximation of the m. l. operator A. Moreover, noting that

$$\frac{n\lambda}{n + \lambda} = \left|\frac{n}{n + \lambda}\right|^2 \lambda + \left|\frac{\lambda}{n + \lambda}\right|^2 n,$$

we deduce that there exist an integer n_0 and a number $0 < \tilde{c} < c$ such that, if $n \ge n_0$ and $\lambda \in \tilde{\Sigma} = \{\lambda \in \mathbb{C}; \Re\mathrm{e}\lambda \ge -\tilde{c}(|\Im\mathrm{m}\lambda| + 1)^\alpha\}$, then $\frac{n\lambda}{n-\lambda} \in \Sigma$. In other words, (P) implies that $\rho(A_n) \supset \tilde{\Sigma}$ for all $n \ge n_0$ as well as

$$(\lambda - A_n)^{-1} = \frac{1}{\lambda + n} + \left(\frac{n}{n + \lambda}\right)^2 \left(\frac{n\lambda}{n + \lambda} - A\right)^{-1}, \quad \lambda \in \tilde{\Sigma}.$$

Similarly, it is verified from (3.1) that

$$\|(\lambda - A_n)^{-1}\|_{\mathcal{L}(X)} \le \frac{\tilde{M}}{(|\lambda| + 1)^\beta}, \quad \lambda \in \tilde{\Sigma}, n \ge n_0,$$

with some constant $\tilde{M}$.

Therefore, it seen that, for each $\theta \ge 0$,

$$[-A_n]^\theta e^{tA_n} \to ([-A]^\theta)^\circ e^{tA}, \quad t > 0,$$

in $\mathcal{L}(X)$ with a uniform estimate

(3.5) $\|[-A_n]^\theta e^{tA_n}\|_{\mathcal{L}(X)} \le \tilde{C}_\theta t^{\frac{\beta-\theta-1}{\alpha}}, \quad n \ge n_0,$

$\tilde{C}_\theta$ being independent of n.

Next, we prove the following proposition.

PROPOSITION 3.6. *Let $J_n = (1 - n^{-1}A)^{-1}$. Then, for each $\theta \geq 0$,*

$$[-A_n]^\theta J_n e^{tA_n} \to ([-A]^\theta)^\circ e^{tA}, \quad t > 0,$$

in $\mathcal{L}(X)$ as $n \to \infty$ with a uniform estimate

$$\|[-A_n]^\theta J_n e^{tA_n}\|_{\mathcal{L}(X)} \leq \tilde{C}_\theta t^{\frac{\beta - \theta - 1}{\alpha}}, \quad n \geq n_0,$$

$\tilde{C}_\theta$ *being independent of n.*

PROOF. From Theorem 1.8 and (2.1), we have that

$$
\begin{aligned}
J_n(\lambda - A_n)^{-1} &= \frac{n}{\lambda + n}(n - A)^{-1} + \frac{n^3}{(n + \lambda)^2}(n - A)^{-1}\left(\frac{n\lambda}{n + \lambda} - A\right)^{-1} \\
&= -\frac{n}{n + \lambda}\left(\frac{n\lambda}{n + \lambda} - A\right)^{-1}.
\end{aligned}
$$

Therefore, as $n \to \infty$, $J_n(\lambda - A_n)^{-1}$ converges to $(\lambda - A)^{-1}$ with an estimate

$$\|J_n(\lambda - A_n)^{-1}\|_{\mathcal{L}(X)} \leq \frac{\tilde{M}}{(|\lambda| + 1)^\beta}, \quad \lambda \in \tilde{\Sigma},$$

$\tilde{M}$ being some constant independent of n. Then the proof is the same as for $[-A_n]^\theta e^{tA_n}$.

REMARK. If the resolvent set of a m. l. operator A contains a halfplane $\{\lambda \in \mathbb{C}; \Re e\lambda \geq 0\}$ and if the resolvent satisfies

$$\|(\lambda - A)^{-1}\|_{\mathcal{L}(X)} \leq \frac{M}{(|\lambda| + 1)^\beta}, \quad \Re e\lambda \geq 0,$$

with some $0 < \beta \leq 1$, then as in the proof of Theorem 1.6 it is verified that $\rho(A)$ contains a region $\Sigma = \{\lambda \in \mathbb{C}; \Re e\lambda \geq -c(|\Im m\lambda| + 1)^\beta\}$, $c > 0$, and that the estimate on $(\lambda - A)^{-1}$ is also extended over Σ. This fact then shows that, if the condition (P) holds with some exponents $0 < \alpha, \beta \leq 1$, then the relation $\beta \leq \alpha$ always takes place.

REMARK. When $\alpha = \beta = 1$, $\|e^{tA}\|_{\mathcal{L}(X)}$ is bounded as $t \to 0$ (Proposition 3.2). Therefore, e^{tA} is strongly continuous in the norm of X on the subspace $\overline{\mathcal{D}(A)}$, cf. Theorem 3.3.

3.2 MULTIVALUED LINEAR EQUATIONS
$\frac{du}{dt} \in Au + f(t)$ IN THE PARABOLIC CASE

Consider the multivalued evolution problem

(E)
$$\begin{cases} \dfrac{du}{dt} \in Au + f(t), & 0 < t \leq T, \\[2mm] u(0) = u_0, \end{cases}$$

in a Banach space X. Here, A is a m. l. operator satisfying (P), $f{:}[0,T] \to X$ is a given continuous function, $u_0 \in X$ is an initial value.

In the parabolic case, the strict solution to (E) is defined as follows.

DEFINITION. A function $u{:}[0,T] \to X$ is called *a strict solution to* (E) if $u \in \mathcal{C}^1((0,T];X)$ and $u(t) \in \mathcal{D}(A)$ for every $0 < t \leq T$, u satisfies the multivalued equation in (E) at each point $0 < t \leq T$, and if $u(0) = u_0$ holds with respect to the seminorm $p_A(\cdot) = \|A^{-1}\cdot\|_X$ (that is, $\lim_{t \to 0} p_A(u(t) - u_0) = 0$).

THEOREM 3.7. *Let A be a m. l. operator satisfying (P) with $2\alpha + \beta > 2$. For any Hölder continuous function $f \in \mathcal{C}^\sigma([0,T];X)$, $\frac{2-\alpha-\beta}{\alpha} < \sigma \ (\leq 1)$, and any initial value $u_0 \in X$, the function given by*

(3.6)
$$u(t) = e^{tA}u_0 + \int_0^t e^{(t-\tau)A} f(\tau)d\tau, \quad 0 \leq t \leq T,$$

is a strict solution to (E). Conversely, any strict solution to (E) with $f \in \mathcal{C}([0,T];X)$ and with $u_0 \in X$ is necessarily of the form (3.6).

PROOF. Let first $f \in \mathcal{C}^\sigma([0,T];X)$ with $\frac{2-\alpha-\beta}{\alpha} < \sigma$. We consider the sequence of functions

$$u_n(t) = e^{tA_n}u_0 + \int_0^t e^{(t-\tau)A_n} f(\tau)d\tau, \quad 0 \leq t \leq T, \ n \geq n_0.$$

Note that, since $2\alpha + \beta > 2$ implies $\frac{\beta-1}{\alpha} > -1$, the integral in this formula exists. Obviously, u_n converges, as $n \to \infty$, to u pointwise on $[0,T]$. Moreover, apply A_n to $u_n(t)$ and write:

$$A_n u_n(t) = A_n e^{tA_n}u_0 + \int_0^t A_n e^{(t-\tau)A_n} \{f(\tau) - f(t)\}d\tau + (e^{tA_n} - 1)f(t).$$

Then, in view of (3.5) (with $\theta = 0$), $A_n u_n$ is seen to converge to a continuous function $g \in \mathcal{C}((0,T]; X)$ pointwise on $(0,T]$. Since

$$u_n(t) - u_n(\varepsilon) = \int_\varepsilon^t \{A_n u_n(\tau) + f(\tau)\} d\tau, \quad \varepsilon \leq t \leq T,$$

with any $\varepsilon > 0$, it follows that $u(t) - u(\varepsilon) = \int_\varepsilon^t \{g(\tau) + f(\tau)\} d\tau$; in other words, $u \in \mathcal{C}^1((0,T]; X)$ with $u'(t) = f(t) + g(t), 0 < t \leq T$. On the other hand, since $u_n(t) = A_n^{-1} A_n u_n(t)$, it follows that $u(t) = A^{-1} g(t)$ or $g(t) \in Au(t), 0 < t \leq T$. Thus u satisfies the equation in (E). Since, as $t \to 0$, $\int_0^t e^{(t-\tau)A} f(\tau) d\tau \to 0$ in X, also the initial condition of (E) is verified.

Let now $f \in \mathcal{C}([0,T]; X)$ and assume that a strict solution u to (E) exists. We write:

$$u(t) - e^{(t-\varepsilon)A_n} u(\varepsilon) = \int_\varepsilon^t \frac{\partial}{\partial \tau} \{e^{(t-\tau)A_n} u(\tau)\} d\tau$$

$$= \int_\varepsilon^t e^{(t-\tau)A_n} \{-A_n u(\tau) + g(\tau)\} d\tau + \int_\varepsilon^t e^{(t-\tau)A_n} f(\tau) d\tau, \quad \varepsilon \leq t \leq T,$$

with any $\varepsilon > 0$, where $g(t) = u'(t) - f(t) \in Au(t), 0 < t \leq T$. Since $A_n u(t) = J_n g(t)$, Proposition 3.6 (with $\theta = 0$) yields that

$$u(t) - e^{(t-\varepsilon)A} u(\varepsilon) = \int_\varepsilon^t e^{(t-\tau)A} f(\tau) d\tau.$$

Let here ε tend to 0. The initial condition on u then implies that

$$e^{(t-\varepsilon)A} u(\varepsilon) = A^\circ e^{(t-\varepsilon)A} A^{-1} u(\varepsilon) \to A^\circ e^{tA} A^{-1} u_0 = e^{tA} u_0.$$

REMARK. By virtue of Theorem 3.3, if u_0 belongs to

$$u_0 \in \mathcal{D}([-A]^\theta) \cup X_A^\theta, \quad 1 - \beta < \theta < 1,$$

then the solution given by (3.6) is continuous at $t = 0$ in the norm of X, that is, the initial condition is satisfied in the strong sense that $\lim_{t \to 0} u(t) = u_0$ in X. Similarly, in the case when $\alpha = \beta = 1$, $u_0 \in \overline{\mathcal{D}(A)}$ implies that $\lim_{t \to 0} u(t) = u_0$ in X.

REMARK. As shown in the proof of Theorem 3.7, for $f \in C^\sigma([0,T];X)$, $\frac{2-\alpha-\beta}{\alpha} < \sigma$, the derivative of the solution is given by

$$\frac{du}{dt}(t) = e^{tA}f(t) + A^\circ e^{tA}u_0 + \int_0^t A^\circ e^{(t-\tau)A}\{f(\tau) - f(t)\}d\tau, \quad 0 < t \leq T.$$

Then, if u_0 satisfies the condition

$$\{f(0) + Au_0\} \cap \{\mathcal{D}([-A]^\theta) \cup X_A^\theta\} \neq \emptyset, \quad 1 - \beta < \theta < 1,$$

then we have:

$$\frac{du}{dt}(t) = e^{tA}\{f(0) + f_0\} + e^{tA}\{f(t) - f(0)\} + \int_0^t A^\circ e^{(t-\tau)A}\{f(\tau) - f(t)\}d\tau$$

with $f_0 \in Au_0$ such that $f(0)+f_0 \in \mathcal{D}([-A]^\theta)\cup X_A^\theta$. Therefore, $\lim_{t\to 0}\frac{du}{dt}(t)$ $= f(0) + f_0$. This shows that $u \in C^1([0,T];X)$ and that the equation in (E) holds even at $t = 0$. Note that $\{f(0) + Au_0\} \cap \{\mathcal{D}([-A]^\theta) \cup X_A^\theta\}$ always consists of a single point if it is nonempty, beause of the fact that $A0 \cap \{\mathcal{D}([-A]^\theta) \cup X_A^\theta\} = \{0\}$.

3.3 DEGENERATE EVOLUTION PROBLEMS OF PARABOLIC TYPE

Consider the degenerate evolution problem

(D-E.1)
$$\begin{cases} \dfrac{dMv}{dt} = Lv + f(t), & 0 < t \leq T, \\ Mv(0) = u_0, \end{cases}$$

in a Banach space X, where M and L are single valued closed linear operators in X with $\mathcal{D}(M) \supset \mathcal{D}(L)$, $f: [0,T] \to X$ is a given continuous function, $u_0 \in X$ is an initial value, and where $v: [0,T] \to X$ is the unknown function.

We assume:

(3.7) The M resolvent set $\rho_M(L)$ contains a region

$$\Sigma_\gamma = \{\lambda \in \mathbb{C}; \Re\mathfrak{e}(\lambda - \gamma) \geq -c(|\Im\mathfrak{m}\lambda| + 1)^\alpha\}, \quad \gamma \in \mathbb{R},$$

and the M resolvent satisfies

$$\|M(\lambda M - L)^{-1}\|_{\mathcal{L}(X)} \leq \frac{C}{(|\lambda - \gamma| + 1)^\beta}, \quad \lambda \in \Sigma_\gamma,$$

with some exponents $0 < \beta \leq \alpha \leq 1$ and constants $c, C > 0$.

Then Theorem 3.7 yields the following result.

THEOREM 3.8. *Let (3.7) be satisfied with* $2\alpha + \beta > 2$. *For any* $f \in$ $\mathcal{C}^\sigma([0,T]; X)$, $\frac{2-\alpha-\beta}{\alpha} < \sigma \ (\leq 1)$, *and any* $u_0 \in X$, *(D-E.1) possesses a unique strict solution* v *such that*

(3.8) $Mv \in \mathcal{C}^1((0,T]; X) \ and \ Lv \in \mathcal{C}((0,T]; X),$

provided that $Mv(0) = u_0$ *is understood in the seminorm sense that*

(3.9) $\|M(\gamma M - L)^{-1}\{Mv(t) - u_0\}\|_X \to 0 \ as \ t \to 0.$

PROOF. By changing the unknown function to $u(t) = Mv(t)$, we rewrite (D-E.1) into the form

(3.10)
$$\begin{cases} \dfrac{du}{dt} \in LM^{-1}u + f(t), & 0 < t \leq T, \\ u(0) = u_0. \end{cases}$$

In addition, change of unknown function to $u_\gamma(t) = e^{-\gamma t}u(t)$ yields that (3.10) is regarded as a multivalued equation of the form (E) in Section 3.2 with a coefficient operator

(3.11) $A = LM^{-1} - \gamma.$

From Theorem 1.14 it follows that $(\lambda - A)^{-1} = M((\lambda + \gamma)M - L)^{-1}$ for $\lambda + \gamma \in \rho_M(L)$. This then means that (3.7) yields directly Condition (P) for the m. l. operator A defined by (3.11). Therefore, the reduced multivalued equation (E) possesses a unique strict solution u. Clearly, u_γ is a strict solution to (E) if and only if v is a strict solution to (D-E.1) in the sense (3.8) and (3.9). Finally, the uniqueness of the solution v follows from the invertibility of $\gamma M - L$.

Continuity of $Mv(t)$ at $t = 0$ in the topology of X is obtained as follows.

THEOREM 3.9. *Let* $u_0 \in \overline{M(\mathcal{D}(L))}$ *if* $\alpha = \beta = 1$ *and let* $u_0 \in M(\mathcal{D}(L))$ *in the other case. Then, for the solution* v *obtained in Theorem 3.8,* $Mv(t)$ *is continuous at* $t = 0$ *in the norm of* X, *i.e.* $Mv \in \mathcal{C}([0,T]; X)$, *with* $Mv(0) = u_0$ *if* $u_0 \in M(\mathcal{D}(L))$.

PROOF. In the present case we have: $\mathcal{D}(A) = \mathcal{D}(LM^{-1}) = M(\mathcal{D}(L))$. Therefore, if u_0 is as assumed, then $e^{tA}u_0$ is continuous in the norm of X; see Theorem 3.3.

Consider next:

(D-E.2)
$$\begin{cases} M\dfrac{du}{dt} = Lu + Mf(t), & 0 < t \leq T, \\ u(0) = u_0, \end{cases}$$

in a Banach space Y. Here M is a bounded linear operator from X to Y and L is a single valued closed linear operator in Y with $\mathcal{D}(L) \subset X$, X being another Banach space such that $X \subset Y$ continuously. $f : [0,T] \to X$ is a given function, $u_0 \in X$ is an initial value, and $u = u(t)$ is the unknown function.

(D-E.2) is then written as a multivalued equation

$$(3.12) \qquad \begin{cases} \dfrac{du}{dt} \in M^{-1}Lu + f(t), & 0 < t \le T, \\ u(0) = u_0, \end{cases}$$

in the space X; notice that $M^{-1}L$ acts in X.

Let us assume:

(3.13) The M resolvent set $\rho_M(L)$ contains a region

$$\Sigma_\gamma = \{\lambda \in \mathbb{C};\ \Re e(\lambda - \gamma) \ge -c(|\Im m\lambda| + 1)^\alpha\}, \quad \gamma \in \mathbb{R},$$

and the M resolvent satisfies

$$\|(\lambda M - L)^{-1}M\|_{\mathcal{L}(X)} \le \frac{C}{(|\lambda - \gamma| + 1)^\beta}, \quad \lambda \in \Sigma_\gamma,$$

with some exponents $0 < \beta \le \alpha \le 1$ and constants c, $C > 0$.

Then we prove the following theorem.

THEOREM 3.10. *Let (3.13) be satisfied with* $2\alpha + \beta > 2$. *For any* $f \in C^\sigma([0,T];X)$, $\frac{2-\alpha-\beta}{\alpha} < \sigma\ (\le 1)$, *and any* $u_0 \in X$, *(D-E.2) possesses a unique strict solution u such that*

$$(3.14) \qquad u \in C^1((0,T];X) \quad and \quad Lu \in C((0,T];Y)$$

and $u(0) = u_0$ in the seminorm sense that

$$(3.15) \qquad \|(\gamma M - L)^{-1}M\{u(t) - u_0\}\|_X \to 0 \ as \ t \to 0.$$

PROOF. Change of the unkown function to $u_\gamma(t) = e^{-\gamma t}u(t)$ yields that (3.12) is regarded as an equation of the form (E) with the operator coefficient $A = M^{-1}L - \gamma$. Since

$$(\lambda - A)^{-1} = ((\lambda + \gamma)M - L)^{-1}M,$$

(3.13) yields that A satisfies Condition (P). Therefore, (3.12) possesses a unique strict solution $u \in C^1((0,T];X)$. It is immediate to observe that this u is the unique strict solution to (D-E.2) in the sense (3.14) and (3.15).

Continuity of the solution $u(t)$ at $t = 0$ in the topology of X is achieved as follows.

THEOREM 3.11. *Let $u_0 \in L^{-1}(\mathcal{R}(M))$. In the case when $\alpha = \beta = 1$, let $u_0 \in \overline{L^{-1}(\mathcal{R}(M))}$. Then, the solution u obtained in Theorem 3.10 is continuous at $t = 0$ in the norm of X, i.e. $u \in C([0,T]; X)$ with $u(0) = \lim_{t \to 0} u(t) = u_0$.*

PROOF. Note that in this case $\mathcal{D}(A) = \mathcal{D}(M^{-1}L) = L^{-1}(\mathcal{R}(M))$.

A more general problem of the form

(D-E.2′)
$$\begin{cases} M\dfrac{du}{dt} = Lu + f(t), & 0 < t \leq T, \\ u(0) = u_0, \end{cases}$$

can be likely handled in the same manner, provided that $f(t)$ is a C^1 function with values in Y. In fact, a change of the unknown function to $\tilde{u}(t) = u(t) - (\gamma M - L)^{-1} f(t)$, $\gamma \in \rho_M(L)$, yields that

(3.16)
$$\begin{cases} M\dfrac{d\tilde{u}}{dt} = L\tilde{u} - M(\gamma M - L)^{-1}\{f'(t) - \gamma f(t)\}, & 0 < t \leq T, \\ \tilde{u}(0) = u_0 - (\gamma M - L)^{-1} f(0). \end{cases}$$

Then, as an immediate consequence of Theorem 3.10, the following result is verified.

THEOREM 3.12. *Let (3.13) be satisfied with $2\alpha + \beta > 2$. For any $f \in C^{1+\sigma}([0; T]; Y)$, $\frac{2-\alpha-\beta}{\alpha} < \sigma(\leq 1)$, and any $u_0 \in X$, (D-E.2′) possesses a unique strict solution u such that $u \in C^1((0,T]; X)$, $Lu \in C((0,T]; Y)$ and that $u(t)$ is continuous at $t = 0$ in the seminorm sense that*

$$\|(\gamma M - L)^{-1} M\{u(t) - u_0\}\|_X \to 0 \quad as \ t \to 0.$$

THEOREM 3.13. *Let $u_0 \in \mathcal{D}(L)$ with $Lu_0 + f(0) \in \mathcal{R}(M)$. Then the solution u obtained in Theorem 3.12 is continuous at $t = 0$ in the norm of X, i.e. $u \in C([0,T]; X)$ with $u(0) = u_0$.*

PROOF. Since

$$L(u_0 - (\gamma M - L)^{-1} f(0)) = Lu_0 + f(0) - \gamma M(\gamma M - L)^{-1} f(0),$$

the assumption yields that $\tilde{u}_0 = u_0 - (\gamma M - L)^{-1} f(0) \in L^{-1}(\mathcal{R}(M))$. Then, by Theorem 3.11, the solution $\tilde{u}$ to (3.16) is continuous at $t = 0$ with $\tilde{u}(0) = \tilde{u}_0$.

REMARK. In view of Theorem 3.3, we can generalize both Theorems 3.9 and 3.11 and obtain similar results for more general initial values u_0. Indeed, if $u_0 \in \mathcal{D}([-A]^\theta) \cup X_A^\theta$ with $\theta > 1 - \beta$, then the solution $Mv(t)$

(resp. $u(t)$) to (D-E.1) (resp. (D-E.2)) is continuous at $t = 0$ in the norm of X.

REMARK. When $A = LM^{-1} - \gamma$, the interpolation spaces X_A^θ, $0 < \theta < 1$, are redefined as

$$(3.17) \qquad X_A^\theta = \{u \in X; \sup_{\xi > 0} \xi^\theta \|(L - \gamma M)((\xi + \gamma)M - L)^{-1}u\|_X < \infty\}$$

in terms of M and L. Similarly, when $A = M^{-1}L - \gamma$, we obtain that

$$(3.18) \qquad X_A^\theta = \{u \in X; \sup_{\xi > 0} \xi^\theta \|\{1 - \xi((\xi + \gamma)M - L)^{-1}M\}u\|_X < \infty\}.$$

3.4 MAXIMAL REGULARITY OF SOLUTIONS IN THE OPTIMAL CASE

In this section we assume that a m. l. operator A in X satisfies Condition (P) with optimal exponents $\alpha = \beta = 1$. Our aim is to establish the maximal regularity of the solutions to (E) which is quite analogous to the one for the non degenerate case.

Therefore, e^{tA}, $t \geq 0$, is bounded as $t \to 0$ and is strongly continuous at $t = 0$ in the norm of X on the subspace $\overline{\mathcal{D}(A)}$. Let X_A^θ, $0 < \theta < 1$, be the interpolation spaces in Section 1.4. Since $u \in X_A^\theta$ implies $\|(J_n - 1)u\|_X \leq \|u\|_{X_A^\theta} n^{-\theta}$, $J_n u \to u$ in X; in particular, $X_A^\theta \subset \overline{\mathcal{D}(A)}$. In virtue of Propositions 3.2 and 3.4 and Theorem 3.5, we know that, for $0 < \theta < 1$, the following estimates are true:

$$(3.19) \qquad \|e^{tA}u\|_{X_A^\theta} \leq C_\theta t^{-\theta}\|u\|, \quad u \in X,$$

$$(3.20) \qquad \|A^\circ e^{tA}u\|_X \leq C_\theta t^{\theta-1}\|u\|_{X_A^\theta}, \quad u \in X_A^\theta,$$

$$(3.21) \qquad \|(e^{tA} - 1)u\|_X \leq C_\theta t^\theta\|u\|_{X_A^\theta}, \quad u \in X_A^\theta.$$

PROPOSITION 3.14. *For each* $0 < \theta < 1$,

$$\left\|\int_0^t A^\circ e^{(t-\tau)A}\{f(t) - f(\tau)\}d\tau\right\|_{X_A^\theta} \leq C_\theta\|f\|_{\mathcal{C}^\theta([0,t];X)}, \quad f \in \mathcal{C}^\theta([0,t]; X),$$

$$\left\|\int_0^t A^\circ e^{(t-\tau)A}f(\tau)d\tau\right\|_{X_A^\theta} \leq C_\theta\|f\|_{\mathcal{B}([0,t];X_A^\theta)},$$

$$f \in \mathcal{C}([0,t]; X) \cap \mathcal{B}([0,t]; X_A^\theta).$$

PROOF. Since

$$A^\circ(\xi - A)^{-1}\int_0^t A^\circ e^{(t-\tau)A}\{f(t) - f(\tau)\}d\tau$$

$$= -\frac{1}{2\pi i}\int_0^t \int_\Gamma \frac{\lambda^2}{\xi - \lambda}e^{(t-\tau)\lambda}(\lambda - A)^{-1}\{f(t) - f(\tau)\}d\lambda d\tau,$$

its X-norm is estimated by

$$C \int_0^t \int_\Gamma \left| \frac{\lambda}{\xi - \lambda} \right| e^{(t-\tau)\Re\lambda} (t - \tau)^\theta |d\lambda| d\tau \|f\|_{C^\theta([0,t];X)}$$

$$\leq C \int_0^t \int_0^\infty \frac{\eta \tau^\theta}{\xi + \eta} e^{-\tau\eta} d\tau d\eta \|f\|_{C^\theta([0,t];X)}$$

$$\leq C\xi^{-\theta} \int_0^\infty \int_0^\infty \frac{\eta \tau^\theta}{1 + \eta} e^{-\tau\eta} d\tau d\eta \|f\|_{C^\theta([0,t];X)} \leq C_\theta \xi^{-\theta} \|f\|_{C^\theta([0,t];X)},$$

as one easily verifies by a suitable change of variable.

Since

$$A^\circ(\xi - A)^{-1} \int_0^t A^\circ e^{(t-\tau)A} f(\tau) d\tau$$

$$= -\frac{1}{2\pi i} \int_0^t \int_\Gamma \frac{\lambda}{\xi - \lambda} e^{(t-\tau)\lambda} A^\circ(\lambda - A)^{-1} f(\tau) d\lambda d\tau,$$

the second estimate is also proven in a similar way.

Using these results, the solution u to (E) is shown to enjoy the following time and space regularities.

THEOREM 3.15. *If $f \in C^\theta([0,T]; X)$ with $0 < \theta < 1$ and if $u_0 \in \mathcal{D}(A)$ with $\{f(0) + Au_0\} \cap X_A^\theta \neq \emptyset$, then the derivative u' of the solution u to (E) has the regularity*

$$u' \in C^\theta([0,T]; X) \cap \mathcal{B}([0,T]; X_A^\theta).$$

PROOF. As verified in the proof of Theorem 3.7, $u'(t) = f(t) - g(t)$, $0 < t \leq T$, where $g(t)$ is the limit of the functions $A_n u_n(t)$. Therefore, according to the Remark to Theorem 3.7,

$$u'(t) = e^{tA}\{f(0) + g_0\} + e^{tA}\{f(t) - f(0)\} + \int_0^t A^\circ e^{(t-\tau)A}\{f(\tau) - f(t)\} d\tau.$$

Here, $g_0 \in Au_0$ is such that $f(0) + g_0 \in X_A^\theta$; notice that $A^\circ e^{tA} u_0 = e^{tA} g$ with any $g \in Au_0$. Then it is easily deduced from (3.19) and (3.21) that

$$e^{tA}\{f(0) + g_0\}, \ e^{tA}\{f(t) - f(0)\} \in C^\theta([0,T]; X) \cap \mathcal{B}([0,T]; X_A^\theta).$$

Since

$$\int_0^t A^\circ e^{(t-\tau)A}\{f(t) - f(\tau)\} d\tau - \int_0^s A^\circ e^{(s-\tau)A}\{f(s) - f(\tau)\} d\tau$$

$$= \int_s^t A^\circ e^{(t-\tau)A}\{f(t) - f(\tau)\} d\tau + (e^{tA} - e^{(t-s)A})\{f(t) - f(s)\}$$

$$+ (e^{(t-s)A} - 1) \int_0^s A^\circ e^{-(s-\tau)A}\{f(s) - f(\tau)\} d\tau,$$

we obtain by Proposition 3.14 that

$$\int_0^t A^\circ e^{(t-\tau)A}\{f(t) - f(\tau)\}d\tau \in \mathcal{C}^\theta([0,T];X).$$

Finally,

$$\int_0^t A^\circ e^{(t-\tau)A}\{f(t) - f(\tau)\}d\tau \in \mathcal{B}([0,T];X_A^\theta)$$

is a direct consequence of Proposition 3.14.

We can now prove another existence theorem of solutions together with their maximal regularity under assumptions of space regularity for the function f.

THEOREM 3.16. *Let* $0 < \theta < 1$. *For any* $f \in \mathcal{C}([0,T];X) \cap \mathcal{B}([0,T];X_A^\theta)$ *and any* $u_0 \in \mathcal{D}(A)$ *such that* $Au_0 \cap X_A^\theta \neq \emptyset$, *(E) possesses a unique strict solution* u *which satisfies*

$$u' - f \in \mathcal{B}([0,T];X_A^\theta) \cap \mathcal{C}^\theta([0,T];X).$$

PROOF. Let A_n, $n = 1,2,3,...$, be the Yosida approximation of A and let u_n be the function given by (3.6). We shall verify that the function

$$A_n u_n(t) = A_n e^{tA_n} u_0 + \int_0^t A_n e^{(t-\tau)A_n} f(\tau)d\tau$$

converges pointwise on $[0,T]$ as $n \to \infty$. Indeed, from

$$A_n(\xi - A_n)^{-1} = \xi(\xi - A_n)^{-1} - 1$$

$$= -\frac{n}{n+\xi} + \xi\left(\frac{n}{n+\xi}\right)^2\left(\frac{n\xi}{n+\xi} - A\right)^{-1}$$

$$= \frac{n}{n+\xi}A^\circ\left(\frac{n\xi}{n+\xi} - A\right)^{-1}, \quad \xi > 0,$$

it follows that, for each $0 < \theta < 1$,

$$\sup_{\xi > 0} \xi^\theta \|A_n(\xi - A_n)^{-1}u\|_X \leq \|u\|_{X_A^\theta}, \quad u \in X_A^\theta.$$

By the same proof as for (3.20), this then yields that

$$\|A_n e^{tA_n} u\|_X \leq C_\theta t^{\theta-1}\|u\|_{X_A^\theta}, \quad u \in X_A^\theta,$$

C_θ being independent of n. Therefore, for each $0 \le t \le T$, $A_n u_n(t)$ converges to

$$h(t) = e^{tA} g_0 + \int_0^t A^\circ e^{(t-\tau)A} f(\tau) d\tau,$$

where $g_0 \in Au_0 \cap X_A^\theta$. From (3.21), $e^{tA} g_0 \in \mathcal{B}([0,T]; X_A^\theta) \cap \mathcal{C}^\theta([0,T]; X)$. Moreover, with the aid of Proposition 3.14, it is shown that

$$\int_0^t A^\circ e^{(t-\tau)A} f(\tau) d\tau \in \mathcal{B}([0,T]; X_A^\theta) \cap \mathcal{C}^\theta([0,T]; X).$$

Then the first assertion of the theorem is proved by the same argument as in the proof of Theorem 3.7. At the same time it is seen that $u'(t) - f(t) = h(t)$.

These theorems automatically apply to the degenerate equations handled in the previous section.

Consider (D-E.1). Let (3.7) be satisfied with $\alpha = \beta = 1$. Since $A = LM^{-1} - \gamma$, $\{f_0 + Au_0\} \cap X_A^\theta \ne \emptyset$ means that there exist g_0 and v_0 such that $f_0 + g_0 \in X_A^\theta$, $g_0 = Lv_0 - \gamma u_0$ and $Mv_0 = u_0$ hold, or, equivalently,

$$(3.22) \qquad f_0 + Lv_0 \in X_A^\theta \text{ and } Mv_0 = u_0.$$

Since the spaces X_A^θ are given by (3.17), the theorems read as follows.

THEOREM 3.17. *If $f \in \mathcal{C}^\theta([0,T]; X)$, $0 < \theta < 1$, and if $u_0 \in M(\mathcal{D}(L))$ satisfies (3.22) with $f_0 = f(0)$, then the solution v to (D-E.1) enjoys the regularity*

$$\frac{dMv}{dt} \in \mathcal{C}^\theta([0,T]; X) \cap \mathcal{B}([0,T]; X_A^\theta).$$

THEOREM 3.18. *Let $0 < \theta < 1$. For any $f \in \mathcal{C}([0,T]; X) \cap \mathcal{B}([0,T]; X_A^\theta)$ and any $u_0 \in M(\mathcal{D}(L))$ satisfying (3.22) with $f_0 = 0$, (D-E.1) possesses a unique strict solution v which satisfies*

$$Lv \in \mathcal{B}([0,T]; X_A^\theta) \cap \mathcal{C}^\theta([0,T]; X).$$

Consider next (D-E.2). Let (3.13) be satisfied with $\alpha = \beta = 1$. Since $A = M^{-1}L - \gamma$, $\{f_0 + Au_0\} \cap X_A^\theta \ne \emptyset$ is equivalent to the condition that $M(f_0 + g_0) = (\gamma M - L)u_0$ with some $g_0 \in X_A^\theta$, or, equivalently,

$$(3.23) \qquad M(f_0 + g_0) = -Lu_0 \text{ with } g_0 \in X_A^\theta.$$

Since the spaces X_A^θ are given by (3.18), we obtain the following consequences.

THEOREM 3.19. *If $f \in C^{\theta}([0,T];X), 0 < \theta < 1$, and if $u_0 \in L^{-1}(\mathcal{R}(M))$ satisfies (3.23) with $f_0 = f(0)$, then the solution u to (D-E.2) enjoys the regularity*

$$\tfrac{du}{dt} \in C^{\theta}([0,T];X) \cap \mathcal{B}([0,T];X_A^{\theta}).$$

THEOREM 3.20. *Let $0 < \theta < 1$. For any $f \in C([0,T];X) \cap \mathcal{B}([0,T];X_A^{\theta})$ and any $u_0 \in L^{-1}(\mathcal{R}(M))$ satisfying (3.23) with $f_0 = 0$, (D-E.2) possesses a unique strict solution u which satisfies*

$$\tfrac{du}{dt} - f \in \mathcal{B}([0,T];X_A^{\theta}) \quad \text{and} \quad Lu \in C^{\theta}([0,T];Y).$$

Consider (D-E.2′). Let (3.13) be satisfied with $\alpha = \beta = 1$. Since the equation in (D-E.2′) is reduced to (3.16), Theorems 3.19 and 3.20 yield the following results respectively.

THEOREM 3.21. *If $f \in C^{1+\theta}([0,T];Y)$, $0 < \theta < 1$, and if $u_0 \in \mathcal{D}(L)$ satisfies*

$$(3.24) \qquad M\{(\gamma M - L)^{-1}f'(0) - g_0\} = Lu_0 + f(0) \quad \text{with} \quad g_0 \in X_A^{\theta},$$

then the solution u to (D-E.2′) enjoys the regularity

$$\tfrac{du}{dt} \in C^{\theta}([0,T];X) \quad \text{and} \quad \tfrac{du}{dt} - (\gamma M - L)^{-1}f' \in \mathcal{B}([0,T];X_A^{\theta}).$$

PROOF. (3.23) is obviously reduced in the present case to (3.24).

THEOREM 3.22. *Let $0 < \theta < 1$. Then, for any $f \in C^1([0,T];Y) \cap \mathcal{B}([0,T];Y_{M,L}^{\theta})$ with $f' \in \mathcal{B}([0,T];Y_{M,L}^{\theta})$ and for any $u_0 \in \mathcal{D}(L)$ satisfying $Lu_0 + f(0) = Mg_0$ with $g_0 \in X_A^{\theta}$, (D-E.2′) possesses a unique strict solution u which satisfies*

$$\tfrac{du}{dt} \in \mathcal{B}([0,T];X_A^{\theta}), \quad Lu \in C^{\theta}([0,T];Y).$$

Here,

$$Y_{M,L}^{\theta} = \{f \in Y; \sup_{\xi > 0} \xi^{\theta} \|((\xi + \gamma)M - L)^{-1}f\|_X < \infty\}, \quad 0 < \theta < 1,$$

with

$$\|f\|_{Y_{M,L}^{\theta}} = \sup_{\xi > 0} \xi^{\theta} \|((\xi + \gamma)M - L)^{-1}f\|_X.$$

PROOF. Recall that X_A^{θ} was given by (3.18). Since

$$\{1 - \xi((\xi + \gamma)M - L)^{-1}M\}(\gamma M - L)^{-1} = ((\xi + \gamma)M - L)^{-1},$$

it is seen that $(\gamma M - L)^{-1} f \in X_A^\theta$ if and only if $f \in Y_{M,L}^\theta$. So the elements $\tilde{f}$ and $\tilde{u}_0$ in (3.16) satisfy all the assumptions of Theorem 3.20. Then the result is a direct consequence of Theorem 3.20.

REMARK. When X is a reflexive Banach space, we can obtain some further characterization of the spaces $\mathcal{D}(A^\theta)$ and X_A^θ. Indeed, let A be a m. l. operator in a reflexive space X satisfying (P) with $\alpha = \beta = 1$. As mentioned in the Remark to Proposition 2.1, X is the direct sum of $\overline{\mathcal{D}(A)}$ and $A0$. It follows that $\overline{\mathcal{D}(A)} \cap Au$ consists of a single element for every $u \in \mathcal{D}(A)$. In other words, the part $\tilde{A}$ of A in the subspace $\overline{\mathcal{D}(A)}$ defined by

$$\begin{cases} \mathcal{D}(\tilde{A}) = \mathcal{D}(A) \\ \tilde{A}u = \overline{\mathcal{D}(A)} \cap Au \end{cases}$$

is a single valued linear operator acting in $\overline{\mathcal{D}(A)}$. Moreover, $\tilde{A}$ is seen to satisfy (P) with the same $\alpha = \beta = 1$, X being replaced by $\overline{\mathcal{D}(A)}$. Since

$$\xi^\theta A^\circ (\xi - A)^{-1} u = \xi^\theta \tilde{A} (\xi - \tilde{A})^{-1} P u - \xi^\theta (1 - P) u, \quad \xi > 0,$$

where $P : X \to \overline{\mathcal{D}(A)}$ is the corresponding projection, it is verified that $X_A^\theta = X_{\tilde{A}}^\theta$ for $0 < \theta < 1$. Similarly, since $(\lambda - A)^{-1} f = (\lambda - \tilde{A})^{-1} P f$, $\lambda \in \Sigma$, it follows that $A^{-\theta} f = \tilde{A}^{-\theta} P f$; therefore, $\mathcal{R}(A^{-\theta}) = \mathcal{R}(\tilde{A}^{-\theta})$ and $\mathcal{D}(A^\theta) = \mathcal{D}(\tilde{A}^\theta)$ for $\theta > 0$.

3.5 MAXIMAL REGULARITY OF SOLUTIONS IN THE GENERAL CASE

In this section we continue to investigate the maximal regularity of solutions to (E) in the general case $2\alpha + \beta > 1$, but we use another approach, precisely, operator equations method. To this end, the theory of P. Grisvard shall be extended to multivalued operators.

Let $\mathcal{X}$ be a Banach space. Consider the operator equation

$(\mathcal{E})$ $\qquad\qquad\qquad\qquad\qquad \mathcal{T}u \in \mathcal{A}u + f$

in $\mathcal{X}$, where $\mathcal{T}$ is a closed linear operator in $\mathcal{X}$, $\mathcal{A}$ is a m. l. operator in $\mathcal{X}$, and where f is an element of $\mathcal{X}$.

We assume:

$(\mathcal{T})$ The resolvent set $\rho(\mathcal{T})$ contains a half plane $\{\lambda \in \mathbb{C}; \Re e\lambda \leq b\}$ and the resolvent $(\lambda - \mathcal{T})^{-1}$ satisfies

$$\|(\lambda - \mathcal{T})^{-1}\|_{\mathcal{L}(\mathcal{X})} \leq \frac{L}{|\Re e\lambda| + 1}, \quad \Re e\lambda \leq b,$$

with constants $b, L > 0$;

($\mathcal{P}$) The resolvent set $\rho(\mathcal{A})$ contains a region

$$\Sigma = \{\lambda \in \mathbb{C}; \Re\mathrm{e}\lambda \geq -c(|\Im\mathrm{m}\lambda| + 1)^\alpha\}$$

and the resolvent $(\lambda - \mathcal{A})^{-1}$ satisfies

$$\|(\lambda - \mathcal{A})^{-1}\|_{\mathcal{L}(\mathcal{X})} \leq \frac{M}{(|\lambda| + 1)^\beta}, \quad \lambda \in \Sigma,$$

with some exponents $0 < \beta \leq \alpha \leq 1$ and constants c, $M > 0$.

Then the following fundamental theorem can be proved.

THEOREM 3.23. *Let ($\mathcal{T}$) and ($\mathcal{P}$) be satisfied and let $\mathcal{T}$ and $\mathcal{A}$ commute in the sense that $\mathcal{T}^{-1}\mathcal{A}^{-1} = \mathcal{A}^{-1}\mathcal{T}^{-1}$. Then, if $2\alpha + \beta > 2$, for any $f \in (\mathcal{X}, \mathcal{D}(\mathcal{T}))_{\theta,\infty}, \frac{2-\alpha-\beta}{\alpha} < \theta < 1$, ($\mathcal{E}$) possesses a unique solution u.*

PROOF. The question is obviously to seek $(\mathcal{T} - \mathcal{A})^{-1}f$. According to Grisvard's device [113], this can be given at least heuristically by

(3.25)

$$\mathcal{S}f = \frac{1}{2\pi i} \int_\Gamma (\lambda + \mathcal{A})^{-1}(\lambda + \mathcal{T})^{-1}f d\lambda$$

$$= -\mathcal{A}^{-1}f + \frac{1}{2\pi i} \int_\Gamma \lambda^{-1}(\lambda + \mathcal{A})^{-1}\mathcal{T}(\lambda + \mathcal{T})^{-1}f d\lambda,$$

where $\Gamma{:}\lambda = c(|\eta|+1)^\alpha + i\eta$, $-\infty < \eta < \infty$, is an integral contour encircling the spectrum of $\mathcal{A}$. Note that this is nothing more than the Cauchy integral formula applied to a function $(\lambda + \mathcal{T})^{-1}f$ which is holomorphic in the half plane $\Re\mathrm{e}\lambda > -b$.

It is an easy matter to verify that the Dunford integral (3.25) certainly exists for every $f \in \mathcal{X}$, since $2\alpha + \beta > 2$ implies $\alpha + \beta > 1$. Since

$$(\lambda + \mathcal{T})^{-1} = (\Re\mathrm{e}\lambda + \mathcal{T})^{-1} - i\Im\mathrm{m}\lambda(\lambda + \mathcal{T})^{-1}(\Re\mathrm{e}\lambda + \mathcal{T})^{-1},$$

$f \in (\mathcal{X}, \mathcal{D}(\mathcal{T}))_{\theta,\infty}$ guarantees by Theorem 1.12 that

(3.26)

$$\|\mathcal{T}(\lambda + \mathcal{T})^{-1}f\|_{\mathcal{X}} \leq \frac{C}{(|\eta| + 1)^{\alpha\theta+\alpha-1}}\|f\|_{(\mathcal{X},\mathcal{D}(\mathcal{T}))_{\theta,\infty}}, \quad \eta = \Im\mathrm{m}\lambda, \lambda \in \Gamma.$$

On the other hand, it is clear from ($\mathcal{P}$) that

$$\|\mathcal{A}^\circ(\lambda + \mathcal{A})^{-1}\|_{\mathcal{L}(\mathcal{X})} \leq C(|\eta| + 1)^{1-\beta}, \quad \eta = \Im\mathrm{m}\lambda, \lambda \in \Gamma.$$

These two estimates then yield that $\mathcal{S}f - \mathcal{A}^{-1}f \in \mathcal{D}(\mathcal{A})$ with

$$-\frac{1}{2\pi i} \int_\Gamma \lambda^{-1}\mathcal{A}^\circ(\lambda + \mathcal{A})^{-1}\mathcal{T}(\lambda + \mathcal{T})^{-1}f d\lambda \in \mathcal{A}(\mathcal{S}f + \mathcal{A}^{-1}f).$$

Moreover, since $T(\lambda + \mathcal{A})^{-1} = (\lambda + \mathcal{A})^{-1}T$ on $\mathcal{D}(T)$, it is seen that

$$\tfrac{1}{2\pi i}\int_{\Gamma}\lambda^{-1}\mathcal{A}^{\circ}(\lambda + \mathcal{A})^{-1}T(\lambda + T)^{-1}fd\lambda = T\mathcal{S}f.$$

Hence, we have shown that $\mathcal{S}f$ is just a solution to $(\mathcal{E})$.

Let us now verify the uniqueness of the solution. Let $(T - \mathcal{A})u \ni 0$. We set $v = T^{-1}u$. Then v is also a solution to $(T - \mathcal{A})v \ni 0$; in other words, $\mathcal{A}v \ni g$ with $u - g = 0$. While,

$$\mathcal{S}g = -\mathcal{A}^{-1}g + \tfrac{1}{2\pi i}\int_{\Gamma}\lambda^{-1}T(\lambda + T)^{-1}(\lambda + \mathcal{A})^{-1}gd\lambda$$

$$= -v + \tfrac{1}{2\pi i}\int_{\Gamma}\lambda^{-1}(\lambda + T)^{-1}\mathcal{A}^{\circ}(\lambda + \mathcal{A})^{-1}Tvd\lambda = -v + \mathcal{S}u.$$

Therefore, $v = \mathcal{S}0 = 0$; hence, $u = 0$.

Furthermore we prove the following result of maximal regularity.

THEOREM 3.24. *Let u be the solution to $(\mathcal{E})$ with $f \in (\mathcal{X}, \mathcal{D}(T))_{\theta,\infty}$, $\frac{2-\alpha-\beta}{\alpha} < \theta < 1$, which was obtained in Theorem 3.23. Then, $Tu \in (\mathcal{X}, \mathcal{D}(T))_{\omega,\infty}$, where $\omega = \alpha\theta + \alpha + \beta - 2$, with an estimate*

$$\|Tu\|_{(\mathcal{X},\mathcal{D}(T))_{\omega,\infty}} \le C\|f\|_{(\mathcal{X},\mathcal{D}(T))_{\theta,\infty}}.$$

PROOF. As shown in the proof of Theorem 3.23,

$$Tu = T\mathcal{S}f = \tfrac{1}{2\pi i}\int_{\Gamma}(\lambda + \mathcal{A})^{-1}T(\lambda + T)^{-1}fd\lambda.$$

So that, for $\xi > 2c$, it follows by Theorem 1.8 that

$$(\xi + T)^{-1}Tu = \tfrac{1}{2\pi i}\int_{\Gamma}\frac{1}{\lambda - \xi}(\lambda + \mathcal{A})^{-1}T\{(\xi + T)^{-1} - (\lambda + T)^{-1}\}fd\lambda$$

$$= -\tfrac{1}{2\pi i}\int_{\Gamma}\frac{1}{\lambda - \xi}(\lambda + \mathcal{A})^{-1}T(\lambda + T)^{-1}fd\lambda$$

$$= \tfrac{1}{2\pi i}\int_{\Gamma}\frac{\lambda}{\lambda - \xi}(\lambda + \mathcal{A})^{-1}(\lambda + T)^{-1}fd\lambda.$$

Therefore, in view of (3.26),

$$\|T(\xi + T)^{-1}Tu\|_{\mathcal{X}} \le C\int_{0}^{\infty}\frac{d\eta}{(\eta + \xi)\eta^{\alpha\theta+\alpha+\beta-2}}\|f\|_{(\mathcal{X},\mathcal{D}(T))_{\theta,\infty}}$$

$$\le C\xi^{-\omega}\|f\|_{(\mathcal{X},\mathcal{D}(T))_{\theta,\infty}}.$$

Hence the proof is complete.

Consider now the multivalued problem

(E)
$$\begin{cases} \dfrac{du}{dt} \in Au + f(t), \quad 0 < t \le T, \\ u(0) = u_0, \end{cases}$$

in a Banach space X. We assume that A is a m. l. operator satisfying Condition (P) of Section 3.1 with $2\alpha + \beta > 2$.

Let us introduce the Banach space

$$\mathcal{X} = \{ f \in \mathcal{C}([0,T]; X); f(0) = 0 \},$$

and define the linear operator

$$\begin{cases} \mathcal{D}(\mathcal{T}) = \{ u \in \mathcal{C}^1([0,T]; X); u(0) = u'(0) = 0 \} \\ \mathcal{T}u = \left(\dfrac{d}{dt} + \varepsilon \right) u \end{cases}$$

together with the m. l. operator

$$\begin{cases} \mathcal{D}(\mathcal{A}) = \{ u \in \mathcal{X}; \text{ there exists } f \in \mathcal{X} \text{ such that } f(t) \in (A + \varepsilon)u(t) \\ \qquad\qquad\qquad \text{for every } 0 \le t \le T \} \\ \mathcal{A}u = \{ f \in \mathcal{X}; f(t) \in (A + \varepsilon)u(t) \text{ for every } 0 \le t \le T \}, \end{cases}$$

where $\varepsilon > 0$ is a suitable positive number. Indeed, if ε is sufficiently small, then $\mathcal{T}$ and $\mathcal{A}$ are seen to satisfy the Conditions $(\mathcal{T})$ and $(\mathcal{P})$ respectively. It is also known (see P. Grisvard [114] and Proposition 0.7) that

(3.27) $(\mathcal{X}, \mathcal{D}(\mathcal{T}))_{\theta,\infty} = \{ u \in \mathcal{C}^\theta([0,T]; X); u(0) = 0 \}, \quad 0 < \theta < 1.$

Then, we are led to the main theorem of this section.

THEOREM 3.25. Let A satisfy (P) with $2\alpha + \beta > 2$. If $f \in \mathcal{C}^\theta([0,T]; X)$, $\frac{2-\alpha-\beta}{\alpha} < \theta < 1$, and $u_0 \in \mathcal{D}(A)$, $f(0)$ satisfy $\{ f(0) + Au_0 \} \cap \{ \mathcal{D}([-A]^\varphi) \cup X_A^\varphi \} \ne \emptyset$ with $\varphi = \alpha^2\theta + (2 - \alpha - \beta)(1 - \alpha)$, then the strict solution u to (E) enjoys the regularity

$$\frac{du}{dt} \in \mathcal{C}^\omega([0,T]; X), \quad \omega = \alpha\theta + \alpha + \beta - 2.$$

PROOF. Noting the formula $(e^{tA} - 1)A^{-1} = \int_0^t e^{(t-\tau)A} d\tau$, u is given by

$$u(t) = e^{tA}u_0 + \int_0^t e^{(t-\tau)A} f(\tau) d\tau$$

$$= -A^{-1}f(0) + e^{tA}\{ u_0 + A^{-1}f(0) \} + \int_0^t e^{(t-\tau)A} f_0(\tau) d\tau,$$

where $f_0(t) = f(t) - f(0)$. Theorem 3.23 jointed with 3.24 then yields that there esists a solution $\bar{u}$ to $(\mathcal{T} - \mathcal{A})\bar{u} = f_0$ such that $\bar{u}' \in C^\omega([0, T]; X)$. Clearly, by the uniqueness of strict solutions to (E) (Theorem 3.7),

$$\bar{u}(t) = \int_0^t e^{(t-\tau)A} f_0(\tau) d\tau, \quad 0 \le t \le T.$$

On the other hand, let $v_0 \in \{f(0) + Au_0\} \cap \{\mathcal{D}([-A]^\varphi) \cup X_A^\varphi\}$. Then

$$\frac{d}{dt} e^{tA}\{u_0 + A^{-1}f(0)\} = -e^{tA}v_0, \quad t > 0.$$

Theorem 3.5 in addition shows that this derivative is in $C^\omega([0, T]; X)$, because of $\alpha + \beta + \varphi - 2 = \alpha\omega$.

We note that in the theorem above some results obtained in Theorem 3.15 have been generalized to the case $2\alpha + \beta > 2$.

Let us proceed to apply it to the degenerate equations (D-E.1), (D-E.2) and (D-E.2$'$).

THEOREM 3.26. *Let L, M satisfy (3.7) with $2\alpha + \beta > 2$ and let $A = LM^{-1} - \gamma$. If $f \in C^\theta([0, T]; X)$, $\frac{2-\alpha-\beta}{\alpha} < \theta < 1$, and if $u_0 \in M(\mathcal{D}(L))$ satisfies the relation*

$$(3.28) \quad Lv_0 + f(0) = g_0, \ u_0 = Mv_0 \ \text{with} \ g_0 \in \mathcal{D}([-A]^\varphi) \cup X_A^\varphi, \ v_0 \in \mathcal{D}(L),$$

where $\varphi = \alpha^2\theta + (2 - \alpha - \beta)(1 - \alpha)$, then the strict solution v to (D-E.1) enjoys the regularity

$$\frac{dMv}{dt}, \ Lv \in C^\omega([0, T]; X), \ \text{where} \ \omega = \alpha\theta + \alpha + \beta - 2.$$

PROOF. The assumption of Theorem 3.25 on the initial value $u_0 \in \mathcal{D}(A)$ is easily read as $g_0 = f(0) + Lv_0 - \gamma u_0$, $u_0 = Mv_0$ with some $g_0 \in \mathcal{D}([-A]^\varphi) \cup X_A^\varphi$ and $v_0 \in \mathcal{D}(L)$, which is obviously equivalent to (3.28).

THEOREM 3.27. *Let L, M satisfy (3.13) with $2\alpha + \beta > 2$ and let $A = LM^{-1} - \gamma$. If $f \in C^\theta([0, T]; X)$, $\frac{2-\alpha-\beta}{\alpha} < \theta < 1$, and if $u_0 \in L^{-1}(\mathcal{R}(M))$ satisfies the relation*

$$(3.29) \qquad Lu_0 + Mf(0) = Mg_0 \ \text{with} \ g_0 \in \mathcal{D}([-A]^\varphi) \cup X_A^\varphi,$$

where $\varphi = \alpha^2\theta + (2 - \alpha - \beta)(1 - \alpha)$, then the strict solution u to (D-E.2) enjoys the regularity

$$\frac{du}{dt} \in C^\omega([0, T]; X), \ Lu \in C^\omega([0, T]; Y), \ \text{where} \ \omega = \alpha\theta + \alpha + \beta - 2.$$

PROOF. The assumption of Theorem 3.25 on $u_0 \in \mathcal{D}(A)$ is read as $M\{f(0) - g_0\} = (\gamma M - L)u_0$ with some $g_0 \in \mathcal{D}([-A]^\varphi) \cup X_A^\varphi$, which is obviously equivalent to (3.29).

THEOREM 3.28. *Let L, M satisfy (3.13) with $2\alpha + \beta > 2$ and let $A = LM^{-1} - \gamma$. If $f \in \mathcal{C}^{1+\theta}([0,T];Y)$, $\frac{2-\alpha-\beta}{\alpha} < \theta < 1$, and if $u_0 \in \mathcal{D}(L)$ satisfies the relation*

(3.30) $Lu_0 + f(0) = M\{(\gamma M - L)^{-1} f'(0) - g_0\}$ *with* $g_0 \in \mathcal{D}([-A]^\varphi) \cup X_A^\varphi$,

where $\varphi = \alpha^2\theta + (2 - \alpha - \beta)(1 - \alpha)$, then the strict solution u to (D-E.2′) enjoys the regularity

$$\frac{du}{dt} \in \mathcal{C}^\omega([0,T];X), \ Lu \in \mathcal{C}^\omega([0,T];Y), \ \text{where} \ \omega = \alpha\theta + \alpha + \beta - 2.$$

PROOF. As noticed, (D-E.2′) is reduced to the problem (3.16) of the form (D-E.2). Therefore, the result follows immeadiately from Theorem 3.27. Note that the condition (3.29) is read as (3.30).

REMARK. As a matter of fact, Theorem 3.23 together with Theorem 3.25 gives an alternative proof of existence and uniqueness of solution to (E) (without using the semigroup e^{tA} explicitly) provided that the initial value $u_0 \in \mathcal{D}(A)$ satisfies the condition $\{f(0) + Au_0\} \cap \mathcal{D}(A) \neq \emptyset$. In fact, let $u_1 \in \{f(0) + Au_0\} \cap \mathcal{D}(A)$ and let $f_1 \in Au_1$. Then $\tilde{f}(t) = f(t) - f(0) + tf_1 \in \mathcal{C}^\theta([0,T];X)$ with $\tilde{f}(0) = 0$; therefore, there exists a unique solution $\tilde{u} \in C^{1+\omega}([0,T];X)$ to the equation $T\tilde{u} - A\tilde{u} \ni \tilde{f}$. It is then easily observed that $u(t) = \tilde{u}(t) + u_0 + tu_1$ is actually the strict solution to (E).

3.6 SOME RESULTS OF TROTTER AND KATO TYPE

Let A_n, $n = 1, 2, 3, ...$, be a sequence of m. l. operators in a Banach space X. We shall study convergence of solutions u_n to the problems

(E$_n$)
$$\begin{cases} \dfrac{du_n}{dt} \in A_n u + f(t), & 0 < t \leq T, \\ u_n(0) = u_0. \end{cases}$$

A_n are assumed to satisfy Condition (P) with uniform exponents $0 < \beta \leq \alpha \leq 1$ and with uniform constants c, $M > 0$.

Let us first verify convergence of the semigroups e^{tA_n} generated by A_n. The following result is a generarization of Kato's result, see T. Kato [117; Chapter IX], to the case of m. l. operators.

THEOREM 3.29. *Let A_n, $n = 1, 2, 3, ...$, satisfy (P) with uniform α, β and uniform c, M. Let Σ be the region defined in (P) which is contained in every resolvent set $\rho(A_n)$. Assume that there exists a point $\lambda_0 \in \Sigma$ such that, as $n \to \infty$,*

(3.31) $(\lambda_0 - A_n)^{-1} f$ *is convergent in X for all $f \in X$.*

Then there exists a unique m. l. operator A in X such that $\Sigma \subset \rho(A)$ and that, as $n \to \infty$, the resolvent $(\lambda - A_n)^{-1}$ converges to $(\lambda - A)^{-1}$ strongly in X for all $\lambda \in \Sigma$. Consequently, A satisfies the Condition (P) with the expoment, too.

PROOF. Let us consider the subset $\Delta = \{\lambda \in \Sigma;\ (\lambda - A_n)^{-1}f$ is convergent in X for all $f \in X\}$. Δ is then seen to be an open set of Σ. Indeed, let $\lambda \in \Delta$. There is a neighborhood V of λ such that $|\mu - \lambda|\|(\lambda - A_n)^{-1}\|_{\mathcal{L}(X)} \leq \frac{1}{2}$ for all n and all $\mu \in V$. Then, for $\mu \in V$,

$$(\mu - A_n)^{-1} = \sum_{n=0}^{\infty} (-1)^n (\mu - \lambda)^n (\lambda - A_n)^{-(n+1)}.$$ From this formula it follows that $(\mu - A_n)^{-1}f$ is convergent for all $f \in X$; hence, $V \subset \Delta$. On the other hand, Δ is seen to be a closed subset of Σ. Indeed, let $\mu \in \overline{\Delta}$. Then there is a point $\lambda \in \Delta$ such that $|\mu - \lambda|\|(\lambda - A_n)^{-1}\|_{\mathcal{L}(X)} \leq \frac{1}{2}$ for all n. Then by the same argument it is observed that $(\mu - A_n)^{-1}f$ is convergent for all $f \in X$; hence, $\mu \in \Delta$. In this way we conclude that $\Delta = \Sigma$ (note that Δ is not empty and that Σ is connected).

Let us then define a family of bounded operators $R(\lambda)$, $\lambda \in \Sigma$, on X by $R(\lambda)f = \lim_{n \to \infty}(\lambda - A_n)^{-1}f$, $f \in X$. Since each $(\lambda - A_n)^{-1}$ satisfies the resolvent equation in Theorem 1.8, it follows that

(3.32) $$R(\lambda) = R(\mu) - (\lambda - \mu)R(\mu)R(\lambda), \quad \lambda, \mu \in \Sigma.$$

For each $\lambda \in \Sigma$, introduce the m. l. operator $A_\lambda = \lambda - R(\lambda)^{-1}$. Then it is easy to verify that A_λ is independent of λ. Indeed, $f \in A_\lambda u$ if and only if $u = R(\lambda)(\lambda u - f)$. In addition, (3.32) yields that

$$R(\mu)u = R(\mu)R(\lambda)(\lambda u - f) = \frac{R(\lambda) - R(\mu)}{\mu - \lambda}(\lambda u - f),$$

so that

$$u = R(\lambda)(\lambda u - f) = R(\mu)(\mu u - f).$$

Hence, $f \in A_\lambda u$ if and only if $f \in A_\mu u$. Let A be the common m. l. operator. Then $(\lambda - A)^{-1} = R(\lambda)$, $\lambda \in \Sigma$, and this A is the m. l. operator to be sought. The proof of the theorem is complete.

Let us study in more details the limit m. l. operator A. Since A satisfies (P), there exists an infinitely differentiable semigroup e^{tA}, $t \geq 0$, generated by A given by the formula

$$e^{tA} = \frac{1}{2\pi i}\int_{\Gamma} e^{t\lambda}(\lambda - A)^{-1}d\lambda, \quad t > 0,$$

where $\Gamma{:}\lambda = -c(1 + |\eta|)^\alpha + i\eta$, $-\infty < \eta < \infty$, is an integral contour lying in Σ. Then, by Theorem 3.29, as $n \to \infty$, e^{tA_n} converges to e^{tA} strongly on X for all $t \geq 0$ and the convergence is uniform in t on any closed interval $[\delta, \infty)$, $\delta > 0$. The same is true for its derivatives $\frac{d^k}{dt^k} e^{tA_n}$, $k = 1, 2, 3, ...$, because of the formulas

$$\frac{d^k}{dt^k} e^{tA} = \frac{1}{2\pi i} \int_\Gamma \lambda^k e^{t\lambda} (\lambda - A)^{-1} d\lambda, \quad t > 0.$$

As a result we obtain convergence of solutions u_n to (E_n) to the solution u to

(E)
$$\begin{cases} \dfrac{du}{dt} \in Au + f(t), & 0 < t \leq T, \\ u(0) = u_0. \end{cases}$$

THEOREM 3.30. *Let A_n satisfy (P) with uniform α, β such that $\alpha + \beta > 1$ and with uniform c, M. Let (3.31) be satisfied and let A be the limit m. l. operator obtained in Theorem 3.29. For any $f \in C([0, T]; X)$ and any $u_0 \in X$, the sequence of functions u_n, $n = 1, 2, 3, ...$, given by*

(3.33) $\qquad u_n(t) = e^{tA_n} u_0 + \displaystyle\int_0^t e^{(t-\tau)A_n} f(\tau) d\tau, \quad 0 < t \leq T,$

converges at each point t to a function u given by

(3.34) $\qquad u(t) = e^{tA} u_0 + \displaystyle\int_0^t e^{(t-\tau)A} f(\tau) d\tau, \quad 0 < t \leq T,$

in the norm of X.

PROOF. The proof is very easy if we notice from (3.3) ($\theta = 0$) that $\|e^{tA_n}\|_{\mathcal{L}(X)} \leq Ct^{\frac{\beta-1}{\alpha}}$ for all n with $\frac{\beta-1}{\alpha} > -1$.

We recall that by Theorem 3.7 any solution to (E_n) (resp. (E)) is of the form (3.33) (resp. (3.34)) and that, when $2\alpha + \beta > 2$, the function given by (3.33) (resp. (3.34)) is in fact the solution to (E_n) (resp. (E)) for any $f \in C^\sigma([0, T]; X)$, $\frac{2-\alpha-\beta}{\alpha} < \sigma$.

Let us next verify that the convergence $u_n \to u$ takes place in a stronger topology. For sake of simplicity we confine our attention to the case $f(0) = u_0 = 0$. Then, (E_n) and (E) are rewritten in the form

$(\mathcal{E}_n)$ $\qquad\qquad\qquad\qquad \mathcal{T} u_n \in \mathcal{A}_n u_n + f$

and

$(\mathcal{E})$ $\qquad\qquad\qquad\qquad \mathcal{T} u \in \mathcal{A} u + f$

respectively as operational equations in the Banach space $\mathcal{X} = \{f \in \mathcal{C}([0,T]; X); f(0) = 0\}$. Here, as in Section 3.5, $\mathcal{T}$ is the differential operator

$$\begin{cases} \mathcal{D}(\mathcal{T}) = \{u \in \mathcal{C}^1([0,T]; X); u(0) = u'(0) = 0\} \\ \mathcal{T}u = (\frac{d}{dt} + \varepsilon)u \end{cases}$$

in $\mathcal{X}$, $\mathcal{A}_n$ and $\mathcal{A}$ are m. l. operators in $\mathcal{X}$ defined by

$$\mathcal{A}_n u = \{f \in \mathcal{X}; f(t) \in (A_n + \varepsilon)u(t) \text{ for every } 0 \le t \le T\},$$
$$\mathcal{A}u = \{f \in \mathcal{X}; f(t) \in (A + \varepsilon)u(t) \text{ for every } 0 \le t \le T\}$$

respectively, where ε is a small constant such that $0 < \varepsilon < \frac{c}{2}$.

If A_n satisfy (P) with uniform α, β and uniform c, M, then also $\mathcal{A}_n$ satisfy Condition $(\mathcal{P})$ stated in Section 3.5 with the same α, β and with some suitable c, M which are uniform in n. Moreover, $\{(\lambda - \mathcal{A}_n)^{-1}f\}(t) = (\lambda - \varepsilon - A_n)^{-1}f(t)$, $0 \le t \le T$. Then, by virtue of Theorem 3.29, if (3.31) is satisfied, then

$$(3.35) \qquad (\lambda - \mathcal{A}_n)^{-1} \to (\lambda - \mathcal{A})^{-1} \text{ strongly on } \mathcal{X} \text{ for all } \lambda \in \Sigma.$$

We are then led to prove the following result.

THEOREM 3.31. *Let A_n satisfy (P) with uniform α, β such that $2\alpha + \beta > 2$ and with uniform c, M. Let A_n be convergent in the sense of (3.31) and let A be the limit m. l. operator obtained in Theorem 2.29. Then, for any $f \in \mathcal{C}^\theta([0,T]; X)$, $\frac{2-\alpha-\beta}{\alpha} < \theta < 1$, the solution u_n to (E_n) with $f(0) = u_0 = 0$ converges to the solution u to (E) in the sense that $\frac{du_n}{dt} \to \frac{du}{dt}$ in $\mathcal{C}^{\omega'}([0,T]; X)$ as $n \to \infty$ with any $\omega' < \omega = \alpha\theta + \alpha + \beta - 2$.*

PROOF. As verified in the proof of Theorem 3.24,

$$\mathcal{T}(u_n - u) = \frac{1}{2\pi i} \int_\Gamma \left\{ (\lambda + \mathcal{A}_n)^{-1} - (\lambda + \mathcal{A})^{-1} \right\} \mathcal{T}(\lambda + \mathcal{T})^{-1} f d\lambda,$$

where $\Gamma : \lambda = c(|\eta| + 1)^\alpha + i\eta$, $-\infty < \eta < \infty$. Here, (3.26) holds and (3.31) implies (3.35). Then, by Lebesgue's convergence theorem, it is concluded that $\mathcal{T}u_n$ converges to $\mathcal{T}u$ in $\mathcal{X}$. Let next $0 < \sigma < \omega$. It is then known by the reiteration theorem (see e.g. [172; p. 62]) that

$$\|\cdot\|_{(\mathcal{X}, \mathcal{D}(\mathcal{T}))_{\sigma, \infty}} \le C \|\cdot\|_{\mathcal{X}}^{1-\frac{\sigma}{\omega}} \|\cdot\|_{(\mathcal{X}, \mathcal{D}(\mathcal{T}))_{\omega, \infty}}^{\frac{\sigma}{\omega}}.$$

On the other hand, from Theorem 3.24, $\|\mathcal{T}u_n\|_{(\mathcal{X}, \mathcal{D}(\mathcal{T}))_{\omega, \infty}}$ are uniformly bounded. Therefore, $\mathcal{T}u_n \to \mathcal{T}u$ in $(\mathcal{X}, \mathcal{D}(\mathcal{T}))_{\sigma, \infty}$. This completes the proof.

On the basis of these theorems we can of course proceed to prove convergence theorems of Trotter and Kato type for the degenerate equations (D-E.1), (D-E.2) and (D-E.2$'$). But, since the arguments are now quite routine, we will leave the results to the reader, cf. the arguments in Sections 3.3, 3.4 and 3.5.

3.7 APPLICATIONS

i) Abstract equations in a Hilbert space.

Let X be a Hilbert space with the inner product $(\cdot, \cdot)$. Let M (resp. L) be a densely defined non negative (resp. positive) selfadjoint operator in X such that $\mathcal{D}(L) \subset \mathcal{D}(M)$. Assume that L and M satisfy:

(3.36) M is L-bounded with L-bound 0 (in the sense of Kato [117; p. 190]);

(3.37) $(Lu, Mu) \leq 0$ for all $u \in \mathcal{D}(L)$.

Then (3.36) yields that the ranges of $\lambda M - L$ are closed for all $\Re\lambda \geq 0$, and (3.37) implies that

$$\|(\lambda M - L)u\|_X^2 \leq |\lambda|^2 \|Mu\|_X^2 + \|Lu\|_X^2, \quad u \in \mathcal{D}(L),$$

for all $\Re\lambda \geq 0$. Therefore, $\lambda M - L$, $\Re\lambda \geq 0$, have bounded inverses on each subspace $\mathcal{R}(\lambda M - L)$. But from $(\lambda M - L)^* = \bar{\lambda}M - L$ it is observed that $\{\mathcal{R}(\lambda M - L)\}^\perp = \{0\}$ and that $\mathcal{R}(\lambda M - L) = X$. In this way Condition (3.7) is seen to be valid with $\alpha = \beta = 1$, $\gamma = 0$. Therefore all our results, including Theorems 3.8, 3.17 and 3.18, can be applied in the present situation.

A typical example of the operators M and L is easily constructed as follows. Let P be a densely defined positive selfadjoint operator in X with $P \geq \delta > 0$. Set $M = P - \varepsilon$ with $\varepsilon \leq \delta$, and set $L = -\sum_{i=0}^{k} a_i P^i$ with $a_i \geq 0$ and $a_k > 0$, k being an integer ≥ 2. Then it is an easy matter to verify that (3.36) and (3.37) hold.

EXAMPLE 3.1. Consider

$$\begin{cases} \dfrac{\partial}{\partial t}\left\{\left(1 - \dfrac{\partial^2}{\partial x^2}\right)v\right\} = -\dfrac{\partial^4 v}{\partial x^4} + f(x,t), & 0 \leq x \leq 1, 0 < t \leq T, \\[2mm] v(0,t) = v(1,t) = \dfrac{\partial^2 v}{\partial x^2}(0,t) = \dfrac{\partial^2 v}{\partial x^2}(1,t) = 0, & 0 < t \leq T, \\[2mm] \left(1 - \dfrac{\partial^2}{\partial x^2}\right)v(x,0) = u_0(x), & 0 \leq x \leq 1. \end{cases}$$

In the space $X = L^2((0,1))$, let $P = -\dfrac{d^2}{dx^2}$ with $\mathcal{D}(P) = H^2((0,1)) \cap H_0^1((0,1))$. P is a positive selfadjoint operator in X. Obviously this problem is regarded as (D-E.1) with $M = P + 1$ and $L = -P^2$. Therefore, as shown above, these M and L satisfy Condition (3.7) with $\alpha = \beta = 1$.

EXAMPLE 3.2. Changing the boundary conditions, consider now

$$\begin{cases} \dfrac{\partial}{\partial t}\left\{\left(1 - \dfrac{\partial^2}{\partial x^2}\right)v\right\} = -\dfrac{\partial^4 v}{\partial x^4} + f(x,t), & 0 \leq x \leq 1, 0 < t \leq T, \\[2mm] v(0,t) = v(1,t) = \dfrac{\partial v}{\partial x}(0,t) = \dfrac{\partial v}{\partial x}(1,t) = 0, & 0 < t \leq T, \\[2mm] \left(1 - \dfrac{\partial^2}{\partial x^2}\right)v(x,0) = u_0(x), & 0 \leq x \leq 1. \end{cases}$$

In $X = L^2((0,1))$ this problem is again regarded as (D-E.1) in which $M = 1 - \frac{d^2}{dx^2}$ with $\mathcal{D}(M) = H^2((0,1)) \cap H_0^1((0,1))$ and $L = -\frac{d^4}{dx^4}$ with $\mathcal{D}(L) = H^4((0,1)) \cap H_0^2((0,1))$ respectively. In the present case (3.37) is not valid, and as a consequence (3.7) is no longer satisfied with $\alpha = \beta = 1$. However, since the energy equality

$$\lambda(\|u\|_X^2 + \|u'\|_X^2) + \|u''\|_X^2 = \int_0^1 f(x)\overline{u(x)}dx$$

holds for $(\lambda M - L)u = f$, we have:

$$(\Re\lambda)(\|u\|_X^2 + \|u'\|_X^2) + \|u''\|_X^2 \le \|f\|_X\|u\|_X,$$
$$|\Im\lambda|(\|u\|_X^2 + \|u'\|_X^2) \le \|f\|_X\|u\|_X.$$

Therefore, $\|u\|_X^2 + \|u'\|_X^2 \le \frac{\|f\|_X^2}{|\Im\lambda|^2}$. In addition, from this it follows that

$$\|u''\|_X^2 \le C\frac{\|f\|_X^2}{|\lambda|} \quad \text{for all } \lambda \text{ such that } \Re\lambda \ge -c|\Im\lambda|, \ c > 0.$$

On the other hand, because we know that $\mathcal{D}([-L]^{\frac{1}{2}}) \subset \mathcal{D}(M)$, (3.36) holds in this case. In this way we verify that (3.7) is valid with $\alpha = 1$, $\beta = \frac{1}{2}$.

ii) Differential equations of elliptic-parabolic type.

EXAMPLE 3.3. Consider the Poisson-heat equation

(3.38)
$$\begin{cases} \dfrac{\partial m(x)v}{\partial t} = \Delta v + f(x,t), & (x,t) \in \Omega \times (0,T], \\ v = 0, & (x,t) \in \partial\Omega \times (0,T], \\ m(x)v(x,0) = u_0(x), & x \in \Omega, \end{cases}$$

in a bounded region $\Omega \subset \mathbb{R}^n$ with a smooth boundary $\partial\Omega$. Here, $m(x) \ge 0$ in Ω is a given function in $L^\infty(\Omega)$.

Obviously, this problem is formulated as a problem of the form (D-E.1) in which M is the multiplication operator by the function $m(x)$ and L is Δ with the Dirichlet boundary conditions. As the underlying space X, we may take both $H^{-1}(\Omega)$ and $L^2(\Omega)$.

Take first $X = H^{-1}(\Omega)$. Then $M:L^2(\Omega) \to L^2(\Omega) \subset X$ and $L:H_0^1(\Omega) \to H^{-1}(\Omega)$. For $\lambda \in \mathbb{C}$, consider the sesquilinear form

$$a_\lambda(u,v) = \lambda \int_\Omega m(x)u\overline{v}dx + \int_\Omega \nabla u \cdot \nabla\overline{v}dx, \quad u, v \in H_0^1(\Omega),$$

defined on $H_0^1(\Omega)$. Obviously this form is continuous on $H_0^1(\Omega)$. In addition, for any $\frac{\pi}{2} < \omega < \pi$ and for some suitable $c > 0$, the following estimate

$$|a_\lambda(u, u)| \geq \delta(\|u\|_{H_0^1}^2 + |\lambda| \|\sqrt{m}u\|_{L^2}^2), \quad u \in H_0^1(\Omega),$$

is observed to hold for each $\lambda \in \Sigma = \{\lambda \in \mathbb{C}; |\arg \lambda| \leq \omega \text{ or } |\lambda| \leq c\}$ with some uniform $\delta > 0$. Then, since

$$a_\lambda(u, v) = \langle(\lambda M - L)u, v\rangle_{H^{-1} \times H_0^1}, \quad u, v \in H_0^1(\Omega),$$

the Lax-Milgram Theorem yields that $\lambda M - L$, $\lambda \in \Sigma$, has a bounded inverse from $H^{-1}(\Omega)$ to $H_0^1(\Omega)$ with an estimate

$$(3.39) \quad \delta(\|u\|_{H_0^1}^2 + |\lambda| \|\sqrt{m}u\|_{L^2}^2) \leq \|\varphi\|_{H^{-1}} \|u\|_{H_0^1} \text{ if } u = (\lambda M - L)^{-1}\varphi.$$

Hence, $\|Lu\|_{H^{-1}} \leq C\|u\|_{H_0^1} \leq C\|\varphi\|_{H^{-1}}$. Moreover, noting the identity $\lambda M(\lambda M - L)^{-1} = 1 + L(\lambda M - L)^{-1}$, we obtain that

$$(3.40) \qquad \|M(\lambda M - L)^{-1}\varphi\|_{H^{-1}} \leq C|\lambda|^{-1}\|\varphi\|_{H^{-1}}, \quad \varphi \in H^{-1}(\Omega).$$

(3.7) is thus valid with $\alpha = \beta = 1$, $\gamma = 0$.

Therefore, for any $f \in \mathcal{C}^\sigma([0, T]; H^{-1}(\Omega))$, $\sigma > 0$, and any $u_0 \in H^{-1}(\Omega)$, (3.38) possesses a unique solution such that $mv \in \mathcal{C}^1((0, T]; H^{-1}(\Omega))$, $v \in \mathcal{C}((0, T]; H_0^1(\Omega))$. From Theorem 3.9, mv is continuous at $t = 0$ if $u_0 = mv_0$ with some $v_0 \in L^2(\Omega)$ (note that $mv_0 \in \overline{M(\mathcal{D}(L))}$).

In addition, Theorems 3.17 and 3.18 yield the following regularity results. If $f \in \mathcal{C}^\theta([0, T]; H^{-1}(\Omega))$, $0 < \theta < 1$, and if u_0 is equal to mv_0 with some $v_0 \in H_0^1(\Omega)$ such that $f(x, 0) + \Delta v_0 \in X_A^\theta$, then

$$\frac{\partial mv}{\partial t} \in \mathcal{C}^\theta([0, T]; H^{-1}(\Omega)) \cap \mathcal{B}([0, T]; X_A^\theta).$$

If $f \in \mathcal{C}([0, T]; H^{-1}(\Omega)) \cap \mathcal{B}([0, T]; X_A^\theta)$, $0 < \theta < 1$, and if u_0 is equal to mv_0 with some $v_0 \in H_0^1(\Omega)$ such that $\Delta v_0 \in X_A^\theta$, then $\Delta v \in \mathcal{C}^\theta([0, T]; H^{-1}(\Omega)) \cap \mathcal{B}([0, T]; X_A^\theta)$. Here, the interpolation space X_A^θ is given by

$$X_A^\theta = \{v \in H^{-1}(\Omega); \sup_{\xi > 0} \xi^\theta \|\Delta(m\xi - \Delta)^{-1}v\|_{H^{-1}} < \infty\}.$$

Take next $X = L^2(\Omega)$. If $(\lambda M - L)u = f \in L^2(\Omega)$, then $u \in H^2(\Omega) \cap H_0^1(\Omega)$. So that, $\lambda M - L$, $\lambda \in \Sigma$, has a bounded inverse from $L^2(\Omega)$ to $H^2(\Omega) \cap H_0^1(\Omega)$. The estimate (3.39) then yields that

$$|\lambda| \|mu\|_{L^2}^2 \leq |\lambda| \|m\|_{L^\infty} \|\sqrt{m}u\|_{L^2}^2 \leq C\|f\|_{H^{-1}}^2 \text{ if } u = (\lambda M - L)^{-1}f.$$

Therefore,
(3.41)
$$\|M(\lambda M - L)^{-1}f\|_{L^2} \le C|\lambda|^{-\frac{1}{2}}\|f\|_{H^{-1}} \le C|\lambda|^{-\frac{1}{2}}\|f\|_{L^2}, \quad f \in L^2(\Omega),$$

which shows that (3.7) is valid with $\alpha = 1$, $\beta = \frac{1}{2}$ ($\gamma = 0$).

Therefore, for any $f \in \mathcal{C}^\sigma([0,T]; L^2(\Omega))$, $\sigma > \frac{1}{2}$, and any $u_0 \in L^2(\Omega)$, (3.38) possesses a unique solution such that $mv \in \mathcal{C}^1((0,T]; L^2(\Omega))$, $v \in \mathcal{C}((0,T]; H^2(\Omega) \cap H_0^1(\Omega))$. From Theorem 3.9, mv is continuous at $t = 0$ if $u_0 = mv_0$ with some $v_0 \in H^2(\Omega) \cap H_0^1(\Omega)$.

In addition, Theorem 3.26 yields that, if $f \in \mathcal{C}^\theta([0,T]; L^2(\Omega))$, $\frac{1}{2} < \theta < 1$, and if u_0 is equal to mv_0 with some $v_0 \in H^2(\Omega) \cap H_0^1(\Omega)$ such that $f(x,0) + \Delta v_0 \in X_A^\theta$, then $\frac{\partial mv}{\partial t} \in \mathcal{C}^{\theta - \frac{1}{2}}([0,T]; L^2(\Omega))$. Here,

$$X_A^\theta = \{v \in L^2(\Omega); \sup_{\xi > 0} \xi^\theta \|\Delta(m\xi - \Delta)^{-1}v\|_{L^2} < \infty\}.$$

If we assume more smoothness and some order of vanishing of the function m on the closed region $\overline{\Omega}$, a better exponent β is obtained. Indeed, let $m \in \mathcal{C}^1(\overline{\Omega})$ with

(3.42) $$|\nabla m(x)| \le Cm(x)^\rho, \quad x \in \overline{\Omega},$$

where $0 < \rho \le 1$. Considering the inner product

$$((\lambda M - L)u, mu) = (f, mu) \text{ for } u = (\lambda M - L)^{-1}f,$$

we observe that

$$\left|\lambda\|mu\|_{L^2}^2 - \|\sqrt{m}\nabla u\|_{L^2}^2\right| \le |(f, mu)_{L^2}| + |(\nabla u, u\nabla m)_{L^2}|$$
$$\le \|f\|_{L^2}\|mu\|_{L^2} + C\|\nabla u\|_{L^2}\|u\|_{L^2}^{1-\rho}\|mu\|_{L^2}^\rho.$$

For λ such that $|\arg \lambda| \le \omega$,

$$\lambda\|mu\|_{L^2}^2 \le C(\|f\|_{L^2}\|u\|_{L^2}^{1-\rho} + \|u\|_{H^1}\|u\|_{L^2}^{1-\rho})\|mu\|_{L^2}^\rho,$$

hence,

$$\|mu\|_{L^2} \le C|\lambda|^{-\frac{1}{2-\rho}}\|f\|_{L^2}.$$

Thus (3.42) yields that $\beta = \frac{1}{2-\rho}$.

EXAMPLE 3.4. Consider the adjoint problem

(3.43)
$$\begin{cases} m(x)\dfrac{\partial u}{\partial t} = \Delta u + m(x)f(x,t), & (x,t) \in \Omega \times (0,T], \\ u = 0, & (x,t) \in \partial\Omega \times (0,T], \\ u(x,0) = u_0(x), & x \in \Omega, \end{cases}$$

of (3.38), with $\Omega \subset \mathbb{R}^n$ and where $m(x)$ is a given function as above. This problem is then regarded as a problem of the form (D-E.2) in which M is the multiplication operator by the function $m(x)$ and L is Δ with the Dirichlet boundary conditions. As underlying space Y, we take $H^{-1}(\Omega)$. Then, $L:H_0^1(\Omega) \to H^{-1}(\Omega)$. As for the space X, we can take either $H_0^1(\Omega)$ or $L^2(\Omega)$.

Take first $X = H_0^1(\Omega)$. Then $M:X \to L^2(\Omega) \subset Y$. As verified above, $(\lambda M - L)^{-1}$, $\lambda \in \Sigma$, exists as a bounded operator from Y to X. Consider the scalar product

$$\langle (\lambda M - L)^{-1} M u, \varphi \rangle_{H_0^1 \times H^{-1}} = \langle u, M(\lambda M - L)^{-1} \varphi \rangle_{H_0^1 \times H^{-1}}.$$

Then it follows from (3.40) that

$$|\langle (\lambda M - L)^{-1} M u, \varphi \rangle_{H_0^1 \times H^{-1}}| \leq C|\lambda|^{-1} \|u\|_{H_0^1} \|\varphi\|_{H^{-1}}.$$

This immediately yields that (3.13) is valid with $\alpha = \beta = 1$, $\gamma = 0$.

Therefore, for any $f \in \mathcal{C}^\sigma([0,T]; H_0^1(\Omega))$, $\sigma > 0$, and any $u_0 \in H_0^1(\Omega)$, (3.43) possesses a unique solution u such that $u \in \mathcal{C}^1((0,T]; H_0^1(\Omega)) \cap \mathcal{C}((0,T]; H^2(\Omega) \cap H_0^1(\Omega))$. From Theorem 3.11, u is continuous at $t = 0$ if $\Delta u_0 = m u_1$ with some $u_1 \in L^2(\Omega)$ (note that $m u_1$ belongs to the closure of $\mathcal{R}(M)$ in Y).

In addition, Theorems 3.19 and 3.20 yield the following results respectively. If $f \in \mathcal{C}^\theta([0,T]; H_0^1(\Omega))$, $0 < \theta < 1$, and if $u_0 \in H_0^1(\Omega)$ satisfies $-\Delta u_0 = m\{f(x,0) - g_0\}$ with some $g_0 \in X_A^\theta$, then $\frac{\partial u}{\partial t} \in \mathcal{C}^\theta([0,T]; H_0^1(\Omega)) \cap \mathcal{B}([0,T]; X_A^\theta)$. If $f \in \mathcal{C}([0,T]; H_0^1(\Omega)) \cap \mathcal{B}([0,T]; X_A^\theta)$, $0 < \theta < 1$, and if $u_0 \in H_0^1(\Omega)$ satisfies $\Delta u_0 = m g_0$ with some $g_0 \in X_A^\theta$, then $\frac{\partial u}{\partial t} - f \in \mathcal{B}([0,T]; X_A^\theta)$. Here,

$$(3.44) \qquad X_A^\theta = \{u \in H_0^1(\Omega); \sup_{\xi > 0} \xi^\theta \|\{1 - \xi(m\xi - \Delta)^{-1} m\} u\|_{H_0^1} < \infty\}.$$

Take next $X = L^2(\Omega)$. Then, by a similar duality argument, it is observed that

$$\|(\lambda M - L)^{-1} M f\|_{L^2} \leq C|\lambda|^{-\frac{1}{2}} \|f\|_{L^2}, \quad f \in L^2(\Omega).$$

That is, (3.13) is fulfilled with $\alpha = 1$, $\beta = \frac{1}{2}$ ($\gamma = 0$).

As a result, for any $f \in \mathcal{C}^\sigma([0,T]; L^2(\Omega))$, $\sigma > \frac{1}{2}$, and any $u_0 \in L^2(\Omega)$, (3.43) possesses a unique solution such that

$$u \in \mathcal{C}^1((0,T]; L^2(\Omega)) \cap \mathcal{C}((0,T]; H^2(\Omega) \cap H_0^1(\Omega)).$$

From Theorem 3.11, if $u_0 \in H^2(\Omega) \cap H_0^1(\Omega)$ satisfies $\Delta u_0 = m u_1$ with some $u_1 \in L^2(\Omega)$, then u is continuous at $t = 0$.

In addition, according to Theorem 3.27, if $f \in \mathcal{C}^\theta([0,T]; L^2(\Omega))$, $\frac{1}{2} < \theta < 1$, and if $u_0 \in H_0^1(\Omega)$ satisfies $-\Delta u_0 = m\{f(x,0) - g_0\}$ with some $g_0 \in X_A^\theta$, then $\frac{\partial u}{\partial t} \in \mathcal{C}^{\theta - \frac{1}{2}}([0,T]; L^2(\Omega))$. Here,

$$(3.45) \qquad X_A^\theta = \{u \in L^2(\Omega); \sup_{\xi > 0} \xi^\theta \|\{1 - \xi(m\xi - \Delta)^{-1}m\}u\|_{L^2} < \infty\}.$$

In Example 3.3 the condition (3.42) on $m(x)$ ensured a better exponent $\beta = \frac{1}{2-\rho}$ than $\frac{1}{2}$ in the space $L^2(\Omega)$. As a direct consequence of that result, the stronger assumption (3.42) of course yields that $\beta = \frac{1}{2-\rho}$ in the present case, too.

EXAMPLE 3.5. Consider a more general problem

$$(3.46) \qquad \begin{cases} m(x)\dfrac{\partial u}{\partial t} = \Delta u + f(x,t), & (x,t) \in \Omega \times (0,T], \\ u = 0, & (x,t) \in \partial\Omega \times (0,T], \\ u(x,0) = u_0(x), & x \in \Omega, \end{cases}$$

than (3.43), where $m(x) \geq 0$ in Ω with $m \in L^\infty(\Omega)$. This problem is regarded as a problem of the form (D-E.2′). M, L are linear operators as in Example 3.4.

Take first $X = H_0^1(\Omega)$, $Y = H^{-1}(\Omega)$. It was seen that (3.13) is valid with $\alpha = \beta = 1, \gamma = 0$. Therefore, by Theorem 3.12, for any $f \in \mathcal{C}^{1+\sigma}([0,T]; H^{-1}(\Omega))$, $\sigma > 0$, and any $u_0 \in H_0^1(\Omega)$, (3.46) possesses a unique solution u such that $u \in \mathcal{C}^1((0,T]; H^1(\Omega))$. From Theorem 3.13, if $u_0 \in H_0^1(\Omega)$ satisfies $\Delta u_0 + f(x,0) = m(x)u_1$ with some $u_1 \in H_0^1(\Omega)$, then u is continuous at $t = 0$ with $u(0) = u_0$ in $H_0^1(\Omega)$. In addition, Theorem 3.21 yields that, if $f \in \mathcal{C}^{1+\theta}([0,T]; H^{-1}(\Omega))$, $0 < \theta < 1$, and if $\Delta u_0 + f(x,0) = m(x)(g_0 - L^{-1}\frac{\partial f}{\partial t}(x,0))$ with some $g_0 \in X_A^\theta$, then $u \in \mathcal{C}^{1+\theta}([0,T]; H_0^1(\Omega))$ and $\frac{du}{dt} + L^{-1}\frac{\partial f}{\partial t} \in \mathcal{B}([0,T]; X_A^\theta)$. For the space X_A^θ, see (3.44).

Take next $X = L^2(\Omega)$, $Y = H^{-1}(\Omega)$. (3.13) was seen to be valid with $\alpha = 1, \beta = \frac{1}{2}, \gamma = 0$. Theorem 3.12 then yields that, for any $f \in \mathcal{C}^{1+\sigma}([0,T]; H^{-1}(\Omega))$, $\sigma > \frac{1}{2}$, and any $u_0 \in L^2(\Omega)$, (3.46) possesses a unique solution u such that $u \in \mathcal{C}([0,T]; H_0^1(\Omega)) \cap \mathcal{C}^1((0,T]; L^2(\Omega))$. From Theorem 3.13, u is continuous at $t = 0$ with $u(0) = u_0$ in $L^2(\Omega)$ if $u_0 \in H_0^1(\Omega)$ satisfies $\Delta u_0 + f(x,0) = m(x)u_1$ with some $u_1 \in L^2(\Omega)$. In addition, Theorem 3.28 yields that, if $f \in \mathcal{C}^{1+\theta}([0,T]; H^{-1}(\Omega))$, $\frac{1}{2} < \theta < 1$, and if $\Delta u_0 + f(x,0) = m(x)(g_0 - L^{-1}\frac{\partial f}{\partial t}(x,0))$ with some $g_0 \in X_A^\theta$, then $\frac{du}{dt} \in \mathcal{C}^{\theta - \frac{1}{2}}([0,T]; L^2(\Omega))$ and $u \in \mathcal{C}^{\theta - \frac{1}{2}}([0,T]; H_0^1(\Omega))$. For the space X_A^θ, see (3.45).

EXAMPLE 3.6. It is also possible to handle the problems (3.38), (3.43) and (3.46) in the space $L^p(\Omega)$, $1 < p < \infty$. Consider, for example,

$$\begin{cases} \dfrac{\partial m(x)v}{\partial t} = \Delta v - cv + f(x,t), & (x,t) \in \Omega \times (0,T], \\ v = 0, & (x,t) \in \partial\Omega \times (0,T], \\ m(x)v(x,0) = u_0(x), & x \in \Omega, \end{cases}$$

in a bounded region $\Omega \subset \mathbb{R}^n$. Here, $m(x) \geq 0$ with $m \in L^\infty(\Omega)$ is the same function as above and $c > 0$ is a constant. Of course the hypothesis on c can be weakened. We are then led to consider (D-E.1) in $X = L^p(\Omega)$ in which M is the multiplication operator by the function $m(x)$ and L is $\Delta - c$ with the Dirichlet boundary conditions. More precisely, $L \colon W_p^2(\Omega) \cap \overset{\circ}{W}{}_p^1(\Omega) \to L^p(\Omega)$.

Let $u \in \mathcal{D}(L)$ and $(\lambda M - L)u = f$. Multiplying this equality by $\bar{u}|u|^{p-2}$ and integrating the product in Ω, we have:

$$\Re\lambda \int_\Omega m|u|^p dx - \Re \int_\Omega \Delta u \bar{u}|u|^{p-2} dx + c \int_\Omega |u|^p dx = \Re \int_\Omega f\bar{u}|u|^{p-2} dx,$$

$$\Im m\lambda \int_\Omega m|u|^p dx - \Im m \int_\Omega \Delta u \bar{u}|u|^{p-2} dx = \Im m \int_\Omega f\bar{u}|u|^{p-2} dx.$$

As it is well known (see e.g. R. Martin [134; pp. 310-311] for the one dimensional case),

$$\left| \Im m \int_\Omega \Delta u \bar{u}|u|^{p-2} dx \right|$$
$$\leq C \left(-\Re \int_\Omega \Delta u \bar{u}|u|^{p-2} dx + c \int_\Omega |u|^p dx \right), \quad u \in \mathcal{D}(L).$$

Hence,

$$|\Im m\lambda| \int_\Omega m|u|^p dx$$
$$\leq \int_\Omega |f||u|^{p-1} dx + C \left(-\Re \int_\Omega \Delta u \bar{u}|u|^{p-2} dx + c \int_\Omega |u|^p dx \right).$$

Therefore we obtain that

$$((C+1)\Re\lambda + |\Im m\lambda|) \int_\Omega m|u|^p dx - \Re \int_\Omega \Delta u \bar{u}|u|^{p-2} dx$$
$$+ c \int_\Omega |u|^p dx \leq (C+2) \int_\Omega |f||u|^{p-1} dx.$$

Here notice that, as observed in A. Pazy [145; p. 215], $-\Re e \int_\Omega \Delta u \bar{u} |u|^{p-2} dx \geq 0$. Hence, for $(C+1)\Re e\lambda + |\Im m\lambda| \geq \varepsilon_0 > 0$,

$$|\lambda| \||m^{\frac{1}{p}} u\|_{L^p}^p + \|u\|_{L^p}^p \leq C\|f\|_{L^p}^p.$$

Since $m \in L^\infty(\Omega)$, it immediately follows that, for $(\lambda M - L)u = f$,

$$\|mu\|_{L^p} \leq C|\lambda|^{-\frac{1}{p}} \|f\|_{L^p}.$$

As for the proof of $\mathcal{R}(\lambda M - L) = X$, we can appeal to the theory of elliptic differential equations (cf. [168; pp. 77-82]).

In this way we prove that (3.7) is fulfilled with $\alpha = 1$, $\beta = \frac{1}{p}$, $\gamma = 0$.

When $2 < p < \infty$, the stronger condition (3.42) on $m(x)$ can yield a better exponent β. In fact, (3.42) yields that $\beta = \frac{2}{p(2-\rho)} \left(> \frac{1}{p} \right)$; for the proof see A. Favini and A. Yagi [96: Example 5].

We can replace Δ with the general second order elliptic differential operator

$$L(x; D)v = \sum_{i,j=1}^n \frac{\partial}{\partial x_i} \left(a_{ij}(x) \frac{\partial u}{\partial x_j} \right) - a_0 v,$$

where $a_{ij} \in \mathcal{C}(\overline{\Omega})$ satisfy

$$\sum_{i,j=1}^n a_{ij}(x)\xi_i\xi_j \geq c_0|\xi|^2, \quad \xi = (\xi_1, ..., \xi_n) \in \mathbb{R}^n, \; x \in \Omega,$$

with some constant $c_0 > 0$ and where $a_0 \geq 0$ is a constant. The operator L is then defined as

$$\begin{cases} \mathcal{D}(L) = W_p^2(\Omega) \cap \mathring{W}_p^1(\Omega) \\ Lu = L(x; D)u \end{cases}$$

Repeating the same argument as above, (3.7) is still shown to be valid with $\alpha = 1$, $\beta = \frac{1}{p}$, $\gamma = 0$.

EXAMPLE 3.7. It is possible to consider (3.38), (3.43) and (3.46) in an unbounded region Ω provided that a damping term is added. Consider for example the simplest case

$$\begin{cases} \dfrac{\partial m(x)v}{\partial t} = \Delta v - cv + f(x,t), & (x,t) \in \mathbb{R}^n \times (0,T], \\ m(x)v(x,0) = u_0(x), & x \in \mathbb{R}^n, \end{cases}$$

in $\mathbb{R}^n$. Here, $m(x) \geq 0$ in $\mathbb{R}^n$ with $m \in L^\infty(\mathbb{R}^n)$. $c > 0$ is a constant.

Take $X = H^{-1}(\mathbb{R}^n)$ and consider (D-E.1) in X with the multiplication operator M by the function $m(x)$ and with $L = \Delta - c : H^1(\mathbb{R}^n) \to H^{-1}(\mathbb{R}^n)$.

As in Example 3.3, the operator $\lambda M - L$ is determined by the sesquilinear form

$$a_\lambda(u, v) = \lambda \int_{\mathbb{R}^n} m(x) u \bar{v} dx + \int_\Omega \nabla u \cdot \nabla \bar{v} dx + c \int_\Omega u \bar{v} dx, \quad u, v \in H^1(\mathbb{R}^n).$$

Then it is easy to see that, for any $\frac{\pi}{2} < \omega < \pi$ and for some suitable $\varepsilon > 0$,

$$|a_\lambda(u, u)| \geq \delta(|\lambda| \|\sqrt{m} u\|_{L^2}^2 + \|\nabla u\|_{L^2}^2 + \|u\|_{L^2}^2), \quad u \in H^1(\mathbb{R}^n),$$

holds for each $\lambda \in \Sigma = \{\lambda \in \mathbb{C}; |\arg \lambda| \leq \omega \text{ or } |\lambda| \leq \varepsilon\}$ with some uniform $\delta > 0$. This then means that we can argue in the same way as in Example 3.3 to verify that (3.7) is valid with $\alpha = \beta = 1$.

Similarly, when we take $X = L^2(\mathbb{R}^n)$, (3.7) is shown to be valid with $\alpha = 1$, $\beta = \frac{1}{2}$.

iii) Degenerate parabolic equations.

EXAMPLE 3.8. Consider the degenerate equation

(3.47)
$$\begin{cases} \dfrac{\partial u}{\partial t} = \Delta\{a(x)u\} + f(x, t), & (x, t) \in \Omega \times (0, T], \\ a(x)u(x, t) = 0, & (x, t) \in \partial\Omega \times (0, T], \\ u(x, 0) = u_0(x), & x \in \Omega, \end{cases}$$

in a bounded region $\Omega \subset \mathbb{R}^n$ with a smooth boundary $\partial\Omega$. Here, the function $a(x) \geq 0$ on $\overline{\Omega}$ and $a(x) > 0$ almost everywhere in Ω is a given function in $L^\infty(\Omega)$.

If we change the unknown functions from $u(x, t)$ to $v(x, t) = a(x)u(x, t)$, then (3.47) is rewritten in the form (3.38) using $m(x) = \frac{1}{a(x)}$. Therefore, (3.47) is regarded as a problem of the form (D-E.1) in which M is the multiplication operator by the function $m(x) = \frac{1}{a(x)}$ and L is Δ with the Dirichlet condition.

Let us first take $X = H^{-1}(\Omega)$ and set $L : H_0^1(\Omega) \to H^{-1}(\Omega)$. $a(x)$ is assumed to satisfy:

(3.48)
$$\begin{cases} a^{-1} \in L^1(\Omega), & \text{when } n = 1, \\ a^{-1} \in L^r(\Omega) \text{ with some } r > 1, & \text{when } n = 2, \\ a^{-1} \in L^{\frac{n}{2}}(\Omega), & \text{when } n \geq 3. \end{cases}$$

For $\lambda \in \mathbb{C}$, consider the sesquilinear form

$$a_\lambda(u, v) = \lambda \int_\Omega m(x) u \bar{v} dx + \int_\Omega \nabla u \cdot \nabla \bar{v} dx, \quad u, v \in H_0^1(\Omega),$$

on $H_0^1(\Omega)$. Since

$$(3.49) \qquad H_0^1(\Omega) \subset \begin{cases} \mathcal{C}(\overline{\Omega}) \text{ when } n = 1, \\ L^p(\Omega) \text{ for any finite } 1 \leq p < \infty \text{ when } n = 2, \\ L^{\frac{2n}{n-2}}(\Omega) \text{ when } n \geq 3, \end{cases}$$

the assumption (3.48) implies in all the cases that

$$\int_\Omega m(x)|u\overline{v}|dx \leq C\|u\|_{H_0^1}\|v\|_{H_0^1},$$

and this shows that $a_\lambda(\cdot, \cdot)$ is bounded on $H_0^1(\Omega)$. On the other hand, for any $\frac{\pi}{2} < \omega < \pi$ and for some suitable $c > 0$ it is seen that

$$|a_\lambda(u, u)| \geq \delta(\|u\|_{H_0^1}^2 + |\lambda|\|\sqrt{m}u\|_{L^2}^2), \quad u \in H_0^1(\Omega),$$

holds for each $\lambda \in \Sigma = \{\lambda \in \mathbb{C}; |\arg \lambda| \leq \omega \text{ or } |\lambda| \leq c\}$ with some uniform $\delta > 0$. Then, since

$$a_\lambda(u, v) = \langle (\lambda M - L)u, v \rangle_{H^{-1} \times H_0^1}, \quad u, v \in H_0^1(\Omega),$$

the Lax-Milgram theorem yields that $\lambda M - L$, $\lambda \in \Sigma$, has a bounded inverse from $H^{-1}(\Omega)$ to $H_0^1(\Omega)$ with an estimate

$$\delta(\|u\|_{H_0^1}^2 + |\lambda|\|\sqrt{m}u\|_{L^2}^2) \leq \|\varphi\|_{H^{-1}}\|u\|_{H_0^1} \text{ if } u = (\lambda M - L)^{-1}\varphi.$$

So that,

$$\|(\lambda M - L)^{-1}\varphi\|_{H_0^1} \leq \delta^{-1}\|\varphi\|_{H^{-1}},$$

$$(3.50) \qquad \|M(\lambda M - L)^{-1}\varphi\|_{H^{-1}} \leq |\lambda|^{-1}\|\varphi\|_{H^{-1}}, \quad \varphi \in H^{-1}(\Omega),$$

which shows that (3.7) is valid with $\alpha = \beta = 1$, $\gamma = 0$.

Under (3.48), let $f \in \mathcal{C}^\sigma((0, T]; H^{-1}(\Omega))$, $\sigma > 0$, and let $u_0 \in H^{-1}(\Omega)$. Then, (3.47) possesses a unique solution such that $u \in \mathcal{C}^1((0, T]; H^{-1}(\Omega))$, $au \in \mathcal{C}((0, T]; H_0^1(\Omega))$. From Theorem 3.9, u is continuous at $t = 0$ if $u_0 = a^{-1}v_0$ with some $v_0 \in H_0^1(\Omega)$.

In addition, Theorems 3.17 and 3.18 yield the following results respectively. If $f \in \mathcal{C}^\theta([0, T]; H^{-1}(\Omega))$, $0 < \theta < 1$, and if u_0 satisfies $au_0 \in H_0^1(\Omega)$ with $f(x, 0) + \Delta(au_0) \in X_A^\theta$, then $\frac{\partial u}{\partial t} \in \mathcal{C}^\theta([0, T]; H^{-1}(\Omega)) \cap \mathcal{B}([0, T]; X_A^\theta)$. If $f \in \mathcal{C}([0, T]; H^{-1}(\Omega)) \cap \mathcal{B}([0, T]; X_A^\theta)$ and if u_0 satisfies $au_0 \in H_0^1(\Omega)$ with $\Delta(au_0) \in X_A^\theta$, then $\Delta(au) \in \mathcal{B}([0, T]; X_A^\theta) \cap \mathcal{C}^\theta([0, T]; H^{-1}(\Omega))$. Here,

$$X_A^\theta = \{u \in H^{-1}(\Omega); \sup_{\xi > 0} \xi^\theta \|\Delta(a^{-1}\xi - \Delta)^{-1}u\|_{H^{-1}} < \infty\}.$$

In order to take $X = L^2(\Omega)$ we need a stronger assumption on $a(x)$:

(3.51) $\qquad a^{-1} \in L^r(\Omega) \begin{cases} \text{with some } r \geq 2 \text{ when } n = 1, \\ \text{with some } r > 2 \text{ when } n = 2, \\ \text{with some } r \geq n \text{ when } n \geq 3. \end{cases}$

Then, from the Hölder inequality $\|mu\|_{L^2} \leq \|m\|_{L^r} \|u\|_{L^{\frac{2r}{r-2}}}$ and from the Sobolev embedding theorem (see (3.49)) $\|u\|_{L^{\frac{2r}{r-2}}} \leq C\|u\|_{H_0^1}^{\frac{n}{r}} \|u\|_{L^2}^{1-\frac{n}{r}}$, it follows that

(3.52) $\qquad \|mu\|_{L^2} \leq C\|u\|_{H_0^1}^{\frac{n}{r}} \|u\|_{L^2}^{1-\frac{n}{r}}, \quad u \in H_0^1(\Omega).$

In particular, this implies that, if $(\lambda M - L)u = f \in L^2(\Omega)$, then $Lu \in L^2(\Omega)$ so that $u \in H^2(\Omega)$; therefore, it turns out that $\lambda M - L, \lambda \in \Sigma$, has a bounded inverse from $L^2(\Omega)$ to $H^2(\Omega) \cap H_0^1(\Omega)$. Since

$$\delta(\|u\|_{H_0^1}^2 + |\lambda| \|\sqrt{m}u\|_{L^2}^2) \leq \|f\|_{L^2} \|u\|_{L^2} \text{ if } u = (\lambda M - L)^{-1}f,$$

we have: $|\lambda| \|\sqrt{m}u\|_{L^2} \leq C\|f\|_{L^2}$. So that,

$$\|(\lambda M - L)^{-1}f\|_{L^2} \leq C|\lambda|^{-1}\|f\|_{L^2}.$$

On the other hand, considering the inner product $((\lambda M - L)u, aLu) = (f, aLu)$ we obtain that

$$\delta(|\lambda| \|\nabla u\|_{L^2}^2 + \|\sqrt{a}Lu\|_{L^2}^2) \leq \|\sqrt{a}f\|_{L^2} \|\sqrt{a}Lu\|_{L^2}.$$

Therefore,

$$\|(\lambda M - L)^{-1}f\|_{H_0^1} \leq C|\lambda|^{-\frac{1}{2}}\|f\|_{L^2}.$$

These two estimates combined with (3.52) enable us to conclude that

(3.53) $\qquad \|M(\lambda M - L)^{-1}f\|_{L^2} \leq C|\lambda|^{-\frac{2r-n}{2r}}\|f\|_{L^2}, \quad f \in L^2(\Omega),$

which shows that (3.7) is valid with $\alpha = 1, \beta = \frac{2r-n}{2r}$.

Under (3.51), let $f \in \mathcal{C}^\sigma((0,T]; L^2(\Omega)), \sigma > \frac{n}{2r}$, and let $u_0 \in L^2(\Omega)$. Then, (3.47) possesses a unique solution such that $u \in \mathcal{C}^1((0,T]; L^2(\Omega))$, $au \in \mathcal{C}((0,T]; H^2(\Omega) \cap H_0^1(\Omega))$. From Theorem 3.9, u is continuous at $t = 0$ if $u_0 = a^{-1}v_0$ with some $v_0 \in H^2(\Omega) \cap H_0^1(\Omega)$.

In addition, Theorem 3.26 yields that, if $f \in \mathcal{C}^\theta([0,T]; L^2(\Omega)), \frac{n}{2r} < \theta < 1$, and if u_0 satisfies $au_0 \in H^2(\Omega) \cap H_0^1(\Omega)$ with $f(x,0) + \Delta(au_0) \in X_A^\theta$, then $\frac{\partial u}{\partial t} \in \mathcal{C}^{\theta - \frac{n}{2r}}([0,T]; L^2(\Omega))$. Here,

$$X_A^\theta = \{u \in L^2(\Omega); \sup_{\xi > 0} \xi^\theta \|\Delta(a^{-1}\xi - \Delta)^{-1}u\|_{L^2} < \infty\}.$$

EXAMPLE 3.9. Consider the adjoint problem

(3.54)
$$\begin{cases} \dfrac{\partial u}{\partial t} = a(x)\Delta u + f(x,t), & (x,t) \in \Omega \times (0,T], \\ u(x,t) = 0, & (x,t) \in \partial\Omega \times (0,T], \\ u(x,0) = u_0(x), & x \in \Omega, \end{cases}$$

of (3.47) in $\Omega \subset \mathbb{R}^n$, where $a(x)$ is a given function as above. This problem is then regarded as an abstract problem (D-E.2) with the multiplication operator M by the function $m(x) = \frac{1}{a(x)}$ and with $L = \Delta : H_0^1(\Omega) \to H^{-1}(\Omega)$. We set $Y = H^{-1}(\Omega)$. As concerns the space X, we can take $H_0^1(\Omega)$ or $L^2(\Omega)$.

Let $X = H_0^1(\Omega)$. We assume that $a(x)$ satisfies (3.48); then, $M : X \to Y$. As verified above, $(\lambda M - L)^{-1}$, $\lambda \in \Sigma$, exists as a bounded operator on Y. Consider the scalar product

$$\langle (\lambda M - L)^{-1} M u, \varphi \rangle_{H_0^1 \times H^{-1}} = \langle u, M(\lambda M - L)^{-1}\varphi \rangle_{H_0^1 \times H^{-1}}.$$

Then it follows from (3.50) that

$$|\langle (\lambda M - L)^{-1} M u, \varphi \rangle_{H_0^1 \times H^{-1}}| \le C|\lambda|^{-1}\|u\|_{H_0^1}\|\varphi\|_{H^{-1}}.$$

This immediately yields that (3.13) is valid with $\alpha = \beta = 1$, $\gamma = 0$.

Under (3.48), let $f \in \mathcal{C}^\sigma((0,T]; H_0^1(\Omega))$, $\sigma > 0$, and let $u_0 \in H_0^1(\Omega)$. Then, (3.54) possesses a unique solution such that $u \in \mathcal{C}^1((0,T]; H_0^1(\Omega))$, $a\Delta u \in \mathcal{C}((0,T]; H_0^1(\Omega))$. From Theorem 3.11, u is continuous at $t = 0$ if u_0 satisfies $\Delta u_0 = a^{-1}u_1$ with some $u_1 \in H_0^1(\Omega)$.

In addition, Theorems 3.19 and 3.20 yield the following results respectively. If $f \in \mathcal{C}^\theta([0,T]; H_0^1(\Omega))$, $0 < \theta < 1$, and if $u_0 \in H_0^1(\Omega)$ satisfies $\Delta u_0 = a^{-1}\{u_1 - f(x,0)\}$ with some $u_1 \in X_A^\theta$, then $\frac{\partial u}{\partial t} \in \mathcal{C}^\theta([0,T]; H_0^1(\Omega)) \cap \mathcal{B}([0,T]; X_A^\theta)$. If $f \in \mathcal{C}([0,T]; H_0^1(\Omega)) \cap \mathcal{B}([0,T]; X_A^\theta)$, $0 < \theta < 1$, and if $u_0 \in H_0^1(\Omega)$ satisfies $\Delta u_0 = a^{-1}u_1$ with $u_1 \in X_A^\theta$, then $\frac{\partial u}{\partial t} - f \in \mathcal{B}([0,T]; X_A^\theta)$. Here,

$$X_A^\theta = \{u \in H_0^1(\Omega); \sup_{\xi>0} \xi^\theta \|\{1 - \xi(a^{-1}\xi - \Delta)^{-1}a^{-1}\}u\|_{H_0^1} < \infty\}.$$

Let next $X = L^2(\Omega)$. Assume the stronger condition (3.51); then, $M : X \to Y$. By using a similar duality argument, it easily follows from (3.53) that

$$\|(\lambda M - L)^{-1}Mf\|_{L^2} \le C|\lambda|^{-\frac{2r-n}{2r}}\|f\|_{L^2}, \quad f \in L^2(\Omega),$$

which shows that (3.13) is valid with $\alpha = 1$, $\beta = \frac{2r-n}{2r}$.

Under (3.51), let $f \in \mathcal{C}^{\sigma}((0,T]; L^2(\Omega))$, $\sigma > \frac{n}{2r}$, and let $u_0 \in L^2(\Omega)$. Then, (3.54) possesses a unique solution such that $u \in \mathcal{C}((0,T]; H_0^1(\Omega)) \cap \mathcal{C}^1((0,T]; L^2(\Omega))$, $a\Delta u \in \mathcal{C}((0,T]; L^2(\Omega))$. From Theorem 3.11, if $u_0 \in H_0^1(\Omega)$ satisfies $\Delta u_0 = a^{-1} u_1$ with some $L^2(\Omega)$, then u is continuous at $t = 0$.

In addition, according to Theorem 3.27, if $f \in \mathcal{C}^\theta([0,T]; L^2(\Omega))$, $\frac{n}{2r} < \theta < 1$, and if $u_0 \in H_0^1(\Omega)$ satisfies $\Delta u_0 = a^{-1}\{u_1 - f(x,0)\}$ with some $u_1 \in X_A^\theta$, then $\frac{\partial u}{\partial t} \in \mathcal{C}^{\theta - \frac{n}{2r}}([0,T]; L^2(\Omega))$. Here,

$$X_A^\theta = \{u \in L^2(\Omega); \sup_{\xi > 0} \xi^\theta \|\{1 - \xi(a^{-1}\xi - \Delta)^{-1}a^{-1}\}u\|_{L^2} < \infty\}.$$

iv) Differential equations of Sobolev type.

Let K be a densely defined closed linear operator in a Banach space X. Assume that -1 is an eigenvalue of K of multiplicity one, that is $\lambda = 0$ is a simple pole for the resolvent $(\lambda - 1 - K)^{-1}$. Then there exists $\delta > 0$ such that

$$\|(\lambda - 1 - K)^{-1}\|_{\mathcal{L}(X)} \leq \frac{C}{|\lambda|}, \quad 0 < |\lambda| \leq \varepsilon.$$

We are concerned with the equation

$$\begin{cases} \dfrac{d}{dt}\{(1 + K)v\} = Kv + f(t), & 0 < t \leq T, \\ (1 + K)v(0) = (1 + K)v_0, \end{cases}$$

in X, where $f(t)$ is a given function, $v_0 \in \mathcal{D}(K)$ is an initial value, and where $v = v(t)$ is the unknown function. By the change of the unknown function to $\tilde{v}(t) = e^{-(\nu+1)t}v(t)$ with some exponent ν to be specified below, our problem is transformed into

$$\begin{cases} \dfrac{d}{dt}\{(1 + K)\tilde{v}\} = -\nu(1 + K)\tilde{v} - \tilde{v} + f_\nu(t), & 0 < t \leq T, \\ (1 + K)\tilde{v}(0) = (1 + K)v_0, \end{cases}$$

where $f_\nu(t) = e^{-(\nu+1)t} f(t)$. So that this equation is viewed as an equation of the form (D-E.1) with $M = 1 + K$ and $L = -\nu(1 + K) - 1$. Since

$$\lambda M - L = (\lambda + \nu)\left(\frac{1}{\lambda + \nu} + M\right),$$

the M resolvent exists if $|\frac{1}{\lambda+\nu}| \leq \varepsilon$ or if $|\lambda + \nu| \geq \frac{1}{\varepsilon}$. In addition, since

$$M(\lambda M - L)^{-1} = \frac{1}{\lambda + \nu}\left\{1 - \frac{1}{\lambda + \nu}\left(\frac{1}{\lambda + \nu} + M\right)^{-1}\right\},$$

we observe that Condition (3.7) is valid with $\alpha = \beta = 1$, $\gamma = 0$, provided that $\nu = \frac{2}{\varepsilon}$.

EXAMPLE 3.10. As a simple example of iv) we give:

$$\begin{cases} \dfrac{\partial}{\partial t}\left(1+\dfrac{\partial^2}{\partial x^2}\right)v = \dfrac{\partial^2 v}{\partial x^2} + f(x,t), & 0 \le x \le \ell\pi,\ 0 < t \le T, \\[2mm] v(0,t) = v(\ell\pi, t) = 0, & 0 < t \le T, \\[2mm] \left(1+\dfrac{\partial^2}{\partial x^2}\right)v(x,0) = \left(1+\dfrac{\partial^2}{\partial x^2}\right)v_0(x), & 0 \le x \le \ell\pi, \end{cases}$$

where ℓ is a positive integer. Although various choices of the underlying spaces are possible, we here confine ourselves to the space of the continuous functions $X = \{f \in \mathcal{C}([0, \ell\pi]; \mathbb{C}); f(0) = f(\ell\pi) = 0\}$. K is then given by

$$\begin{cases} \mathcal{D}(K) = \{v \in \mathcal{C}^2([0, \ell\pi]; \mathbb{C}); v(0) = v(\ell\pi) = v''(0) = v''(\ell\pi) = 0\} \\[2mm] Kv = \dfrac{d^2 v}{dx^2}. \end{cases}$$

v) The Stokes equation.

EXAMPLE 3.11. Consider the Cauchy problem

(3.55)
$$\begin{cases} \dfrac{\partial u}{\partial t} = \nu \Delta u - \nabla p + f(x,t), & (x,t) \in \Omega \times (0,T], \\[2mm] \mathrm{div}\, u = 0, & (x,t) \in \Omega \times (0,T], \\[2mm] u = 0, & (x,t) \in \partial\Omega \times (0,T], \\[2mm] u(x,0) = u_0(x), & x \in \Omega, \end{cases}$$

for the Stokes system in a bounded region $\Omega \subset \mathbb{R}^n$ with a smooth boundary $\partial\Omega$. Here, $f(x,t)$ is a given function, $u_0(x)$ is an initial function. $u = (u_1(x,t), ..., u_n(x,t))$ and $p = p(x,t)$ are the unknown functions. We shall verify that this problem can be formulated by using a multivalued equation.

Take $X = \{L^2(\Omega)\}^n$. It is well known that X is the orthogonal direct sum of the two subspaces:

$$\begin{cases} X_s = \text{the closure of } \{u \in \{\mathcal{C}_0^\infty(\Omega)\}^n; \mathrm{div}\, u = 0 \text{ in } \Omega\} \text{ in } X, \\[2mm] X_g = \{\nabla p; p \in H^1(\Omega)\}. \end{cases}$$

In X we define a m. l. operator A as follows

$$\begin{cases} \mathcal{D}(A) = \{H^2(\Omega)\}^n \cap \{H_0^1(\Omega)\}^n \cap X_s, \\[2mm] Au = \Delta u + X_s. \end{cases}$$

Then (3.55) is written in the form

$$\begin{cases} \dfrac{du}{dt} \in \nu Au + f(t), & 0 < t \le T, \\ u(0) = u_0, \end{cases}$$

with $f:[0,T] \to X$ and $u_0 \in X_s$.

As a matter of fact, this formulation is essentially equivalent to the classical one which is written as

$$\begin{cases} \dfrac{du}{dt} = \nu A_s u + Pf(t), & 0 < t \le T, \\ u(0) = u_0, \end{cases}$$

in the subspace X_s by using a linear section A_s of A such that

$$\begin{cases} \mathcal{D}(A_s) = \mathcal{D}(A), \\ A_s = P\Delta u, \end{cases}$$

P being the orthogonal projection on X_s, $P : X \to X_s$. This A_s is called the Stokes operator.

For $u \in \mathcal{D}(A)$, $f \in (\lambda - \nu A)u$ if and only if $(\lambda - \nu A_s)u = Pf$. Therefore, $(\lambda - \nu A)^{-1} = (\lambda - \nu A_s)^{-1}P$ with $\rho(A) = \rho(A_s)$; since νA_s is the generator of an analytic semigroup on X_s, the same is true for νA on X with $e^{t\nu A} = e^{t\nu A_s}P$, $t > 0$. In particular, it follows that $e^{t\nu A}g = e^{t\nu A_s}g$, $t > 0$, for $g \in X_s = \overline{\mathcal{D}(A)}$ and that $e^{t\nu A}h = 0$, $t > 0$, for $h \in X_g = A0$.

vi) Some stability results.

EXAMPLE 3.12. Consider the problems

(3.56)
$$\begin{cases} \dfrac{\partial m_\varepsilon(x)v_\varepsilon}{\partial t} = \Delta v_\varepsilon + f(x,t), & (x,t) \in \Omega \times (0,T], \\ m_\varepsilon(x)v_\varepsilon(x,t) = 0, & (x,t) \in \partial\Omega \times (0,T], \\ m_\varepsilon(x)v_\varepsilon(x,0) = u_0(x), & x \in \Omega, \end{cases}$$

in a bounded region $\Omega \subset \mathbb{R}^n$ with a smooth boundary $\partial\Omega$. Here, ε is a parameter varying in $0 < \varepsilon \le 1$. $m(x) \ge 0$ in Ω is a given function in $L^\infty(\Omega)$ and $m_\varepsilon(x) = m(x) + \varepsilon$. Hence the problems (3.56) are not degenerate. We are concerned with convergence of solutions v_ε to the solution v of the degenerate problem (3.38) in Example 3.3 as $\varepsilon \to 0$.

Take $X = L^2(\Omega)$. As was done in Example 3.3, (3.56) is written as

(3.57)
$$\begin{cases} \dfrac{du_\varepsilon}{dt} = A_\varepsilon u_\varepsilon + f(t), & 0 < t \le T, \\ u_\varepsilon(0) = u_0 \end{cases}$$

in X, where $A_\varepsilon = LM_\varepsilon^{-1}$ with $L = \Delta : H^2(\Omega) \cap H_0^1(\Omega) \to L^2(\Omega)$ and with the multiplication operator $M_\varepsilon : X \to X$ by the function $m_\varepsilon(x)$, and where $u_\varepsilon = m_\varepsilon v_\varepsilon$. On the other hand, the problem (3.38) was written as a problem of the form (E) in X in which A is the m. l. operator LM^{-1}, $M : X \to X$ being the multiplication operator by the function $m(x)$.

Let us verify the assumptions of Theorem 3.29. The proof given in Example 3.3 for the case when $X = L^2(\Omega)$ shows that L and M_ε satisfy (3.7) uniformly with $\alpha = 1$, $\beta = \frac{1}{2}$, $\gamma = 0$. As a consequence (in view of Theorem 1.14), A_ε satisfy Condition (P) uniformly with the same α, β. In addition, since

$$A_\varepsilon^{-1} = M_\varepsilon L^{-1} = A^{-1} + \varepsilon L^{-1},$$

it is clear that (3.31) holds with $\lambda_0 = 0$. From Theorem 3.29 it therefore follows that $(\lambda - A_\varepsilon)^{-1}$ converges strongly to $(\lambda - A)^{-1}$ in X for all $\lambda \in \Sigma$. Moreover, $\|e^{tA_\varepsilon}\|_{\mathcal{L}(X)} \le Ct^{-\frac{1}{2}}$, $t > 0$, with a uniform constant C; and e^{tA_ε}, $t > 0$, converges to e^{tA} strongly in X, the convergence being uniform in any interval of the form $[\delta, \infty)$, $\delta > 0$.

Let $u_0 \in L^2(\Omega)$ and $f \in \mathcal{C}([0, T]; L^2(\Omega))$. Then, as proved by Theorem 3.7, the strict solution to (3.57) is unique and is necessarily given by

$$u_\varepsilon(t) = e^{tA_\varepsilon} u_0 + \int_0^t e^{(t-\tau)A_\varepsilon} f(\tau) d\tau, \quad 0 \le t \le T.$$

Similarly, the strict solution to (E) with $A = LM^{-1}$ is necessarily given by

$$u(t) = e^{tA} u_0 + \int_0^t e^{(t-\tau)A} f(\tau) d\tau, \quad 0 \le t \le T.$$

Let $0 < \eta < T$. Then, for $0 \le t \le \eta$,

$$\left\| \int_0^t e^{(t-\tau)A_\varepsilon} f(\tau) d\tau - \int_0^t e^{(t-\tau)A} f(\tau) d\tau \right\|_X$$
$$\le C \int_0^t (t-\tau)^{-\frac{1}{2}} \|f\|_C \le C\eta^{\frac{1}{2}} \|f\|_C.$$

On the other hand, for $\eta \le t \le T$,

$$\left\| \int_0^t e^{(t-\tau)A_\varepsilon} f(\tau) d\tau - \int_0^t e^{(t-\tau)A} f(\tau) d\tau \right\|_X$$
$$\le \int_{t-\eta}^t \|e^{(t-\tau)A_\varepsilon} - e^{(t-\tau)A}\|_{\mathcal{L}(X)} \|f(\tau)\|_X d\tau$$
$$+ \int_0^{t-\eta} \|\{e^{(t-\tau)A_\varepsilon} - e^{(t-\tau)A}\} f(\tau)\|_X d\tau$$
$$\le C\eta^{\frac{1}{2}} \|f\|_C + C \sup_{g \in f([0,T]), \eta \le \tau \le T} \|\{e^{\tau A_\varepsilon} - e^{\tau A}\} g\|_X.$$

This then means that $\int_0^t e^{(t-\tau)A_\varepsilon} f(\tau)d\tau$ is convergent as a sequence of functions to $\int_0^t e^{(t-\tau)A} f(\tau)d\tau$ in the space $\mathcal{C}([0,T];X)$. (Fix first $\eta > 0$ sufficiently small. Then let ε tend to 0.) So that, as $\varepsilon \to 0$, u_ε converges to u in X pointwisely on $(0,T]$; the convergence is uniform on any interval of the form $[\delta, T]$, $\delta > 0$.

Of course, if there are no further regularity assumptions on the data, as in Theorem 3.8, we do not know that $u = m(x)v$ and v solves in the strict sense problem (3.38). An answer is given by Theorem 3.31 above. For further results concerning this matter, we indicate V. Barbu and A. Favini [20] and A. Favini and M. Fuhrman [86].

EXAMPLE 3.13. Consider the problems

(3.58)
$$\begin{cases} \dfrac{\partial u_\varepsilon}{\partial t} = \Delta\{a_\varepsilon(x)u_\varepsilon\} + f(x,t), & (x,t) \in \Omega \times (0,T], \\ a_\varepsilon(x)u_\varepsilon(x,t) = 0, & (x,t) \in \partial\Omega \times (0,T], \\ u_\varepsilon(x,0) = u_0(x), & x \in \Omega, \end{cases}$$

in a bounded region $\Omega \subset \mathbb{R}^n$ with a smooth boundary. Here, $0 < \varepsilon \leq 1$ is a parameter, $a(x) > 0$ almost everywhere is a given function in $L^\infty(\Omega)$ which satisfies (3.51) in Example 3.8 and $a_\varepsilon(x) = a(x) + \varepsilon$. We are concerned with convergence of the solutions u_ε of the non degenerate problems (3.58) to the solution u of the degenerate problem (3.47) handled in Example 3.8.

Take $X = L^2(\Omega)$. As done in Example 3.8, (3.58) are regarded as problems

$$\begin{cases} \dfrac{du_\varepsilon}{dt} = A_\varepsilon u_\varepsilon + f(t), & 0 < t \leq T, \\ u_\varepsilon(0) = u_0, \end{cases}$$

in X, where $A_\varepsilon = LB_\varepsilon$ with $L = \Delta : H^2(\Omega) \cap H_0^1(\Omega) \to L^2(\Omega)$ and with the multiplication operators $B_\varepsilon : X \to X$ by the functions $a_\varepsilon(x)$. On the other hand, the problem (3.47) was written as a problem of the form (E) in X in which $A = LB$, $B : X \to X$ being the multiplication operator by the function $a(x)$. By the same argument as in Example 3.8 (the case when $X = L^2(\Omega)$), (3.51) yields that A_ε satisfy Condition (P) uniformly with $\alpha = 1$, $\beta = \frac{2r-n}{2r}$. In addition, since

$$A_\varepsilon^{-1} = B_\varepsilon^{-1}L^{-1} = BB_\varepsilon^{-1}B^{-1}L^{-1} = (1 - \varepsilon B_\varepsilon^{-1})A^{-1},$$

and since $\varepsilon B_\varepsilon^{-1} f$ tends to 0 in X as $\varepsilon \to 0$, for all $f \in X$, A_ε^{-1} converges to A^{-1} strongly in X. By Theorem 3.29, it then follows that $(\lambda - A_\varepsilon)^{-1}$ converges to $(\lambda - A)^{-1}$ strongly in X for all $\lambda \in \Sigma$. As a consequence, $\|e^{tA_\varepsilon}\|_{\mathcal{L}(X)} \leq Ct^{-\frac{n}{2r}}$, $t > 0$, with a uniform constant C, and e^{tA_ε} converges to e^{tA} strongly in X, the convergence being uniform in any interval of the form $[\delta, \infty)$, $\delta > 0$.

Moreover, let $f \in \mathcal{C}([0,T]; L^2(\Omega))$ and $u_0 \in L^2(\Omega)$. Then, repeating the same argument as in the preceding example, we obtain also that the sequence of functions

$$u_\varepsilon(t) = e^{tA_\varepsilon} u_0 + \int_0^t e^{(t-\tau)A_\varepsilon} f(\tau) d\tau, \quad 0 \le t \le T,$$

converges to the function

$$u(t) = e^{tA} u_0 + \int_0^t e^{(t-\tau)A} f(\tau) d\tau, \quad 0 \le t \le T,$$

in X pointwisely on the interval $(0,T]$ and uniformly on any interval of the form $[\delta, T]$, $\delta > 0$. Remember that by Theorem 3.7 the strict solutions to (3.58) and the strict solution to (3.47) are necessarily given by these expressions respectively, but only further regularity of f_n and f guarantees that u_n and u are in fact strict solutions.

CHAPTER IV

DEGENERATE EQUATIONS OF PARABOLIC TYPE, II: THE NON AUTONOMOUS CASE

The methods devised for autonomous degenerate equations will be generalized for studying non autonomous degenerate equations. Precisely, in Section 4.1 we formulate a number of assumptions on $A(t)$, similar to well known ones in semigroup theory, which allow to show that if f has the Hölder property (with exponent θ) and either $f(0) + A(0)u_0 \cap \mathcal{D}([-A(0)]^\theta) \neq \emptyset$ or $f(0) + A(0)u_0 \cap X^\theta_{A(0)} \neq \emptyset$, then the Cauchy problem $\frac{du}{dt} \in A(t)u + f(t)$, $0 < t \leq T$, $u(0) = u_0$ has precisely one strict solution (enjoying some further regularity, as well). Maximal regularity of the solutions is discussed in Section 4.2. Although in a first step we assume $u_0 = 0 = f(0)$, we shall state the general result in Theorem 4.12 after a suitable reduction to this case.

Section 4.3 contains applications of the preceding results to the degenerate Cauchy problems $\frac{d}{dt}M(t)v = L(t)v + f(t)$, $0 < t \leq T$, $M(0)v(0) = u_0$ and $M(t)\frac{du}{dt} = L(t)u + M(t)f(t)$, $0 < t \leq T$, $u(0) = u_0$. In Section 4.4 we shall consider the equations $\frac{d}{dt}t^\ell v = Lv + f(t)$ and $t^\ell \frac{du}{dt} = Lu + f(t)$, where $0 < \ell < \infty$ and L generates an analytic semigroup in the Banach space X.

Last section, as usual, contains some examples of partial differential equations which can be handled according to our theory.

4.1 THE FUNDAMENTAL SOLUTION TO $\frac{du}{dt} \in A(t)u + f(t)$

Consider the Cauchy problem for the non autonomous multivalued equation

(E)
$$\begin{cases} \dfrac{du}{dt} \in A(t)u + f(t), & 0 < t \leq T, \\ u(0) = u_0, \end{cases}$$

of parabolic type in a Banach space X. "Non autonomous" means that $A(t)$ depend explicitly on the time variable t. Here, $A(t)$, $0 \leq t \leq T$, are

m. 1. operators which generate infinitely differentiable semigroups on X, $f:[0,T] \to X$ is a given continuous function, $u_0 \in X$ is an initial value of the problem.

In this section we discuss the construction of a fundamental solution (called often an evolution operator) $U(t,s)$, $0 \le s \le t \le T$, for the problem (E).

On the analogy to the autonomous case studied in Chapter III, we define a strict solution to (E).

DEFINITION. *A strict solution* to (E) *is a function* $u:(0,T] \to X$, $u \in C^1((0,T];X)$, $u(t) \in \mathcal{D}(A(t))$ *for* $0 < t \le T$, *which satisfies the equation in* (E) *at each point* $0 < t \le T$ *and which satisfies the initial condition in* (E) *with respect to the seminorm* $p_{A(0)}(\cdot) = \|A(0)^{-1}\cdot\|_X$ *(that is,* $\lim_{t \to 0} p_{A(0)}(u(t) - u_0) = 0$*).*

In view of Theorem 3.1, $A(t)$, $0 \le t \le T$, are assumed to satisfy:

(P) The resolvent sets $\rho(A(t))$ contain a region $\Sigma = \{\lambda \in \mathbb{C}; \Re e \lambda \ge -c(|\Im m \lambda| + 1)^\alpha\}$ and the resolvents $(\lambda - A(t))^{-1}$ are estimated by

$$\|(\lambda - A(t))^{-1}\|_{\mathcal{L}(X)} \le \frac{M}{(|\lambda| + 1)^\beta}, \quad \lambda \in \Sigma, 0 \le t \le T,$$

with some exponents $0 < \beta \le \alpha \le 1$ and constants c, $M > 0$.

In addition the following regularity condition is assumed:

(A.i) $A(\cdot)^{-1}$ satisfies a Hölder condition

$$\|A(t)^\circ(\lambda - A(t))^{-1}\{A(t)^{-1} - A(s)^{-1}\}\|_{\mathcal{L}(X)}$$
$$\le \frac{K|t-s|^\mu}{(|\lambda| + 1)^\nu}, \quad \lambda \in \Sigma, \ 0 \le t \le T,$$

with some exponents $0 < \mu, \nu \le 1$.

(Ex.i) The exponents satisfy the relation: $2(\alpha + \beta) + \alpha\mu + \nu > 5$.

Let $J_n(t) = (1 - n^{-1}A(t))^{-1}$, $n = 1, 2, 3, ...$, and let

$$A_n(t) = n\{n(n - A(t))^{-1} - 1\} = n(J_n(t) - 1), \quad 0 \le t \le T,$$

be the Yosida approximation of $A(t)$. Since $A_n(t)$ is a Hölder continuous function with values in $\mathcal{L}(X)$, there exists an evolution operator $U_n(t,s)$, $0 \le s \le t \le T$, for $A_n(t)$ and the strict solution to

$$\begin{cases} \dfrac{du_n}{dt} = A_n(t)u_n + f(t), & 0 < t \le T, \\ u_n(0) = u_0, \end{cases}$$

is given uniquely in the form

$$u_n(t) = U_n(t,0)u_0 + \int_0^t U_n(t,\tau)f(\tau)d\tau, \quad 0 \le t \le T.$$

Let us first prove the convergence of $U_n(t,s)$.

THEOREM 4.1. *Let (P), (A.i) and (Ex.i) be satisfied. Then, for each* $0 \leq s < t \leq T$, *both* $U_n(t,s)$ *and* $U_n(t,s)J_n(s)$ *strongly converge to a bounded operator* $U(t,s)$. $\{U(t,s); 0 \leq s < t \leq T\}$ *enjoys the properties: a)* $U(t,s)U(s,r) = U(t,r)$, $0 \leq r < s < t \leq T$; *b)* $U(t,s)$ *is strongly continuous in* (t,s) *with an estimate* $\|U(t,s)\|_{\mathcal{L}(X)} \leq C(t-s)^{\frac{\beta-1}{\alpha}}$; *c)* $U(t,s)$ *is strongly differentiable in* t *for* $0 \leq s < t \leq T$ *and the derivative* $\frac{\partial U(t,s)}{\partial t}$ *is strongly continuous in* (t,s) *with an estimate* $\left\|\frac{\partial U(t,s)}{\partial t}\right\|_{\mathcal{L}(X)} \leq C(t-s)^{\frac{\beta-2}{\alpha}}$; *d) the range* $\mathcal{R}(U(t,s))$ *is contained in* $\mathcal{D}(A(t))$ *and there exists a bounded linear section* $A(t)^\circ U(t,s)$ *of* $A(t)U(t,s)$; *and e)* $\frac{\partial U(t,s)}{\partial t} = A(t)^\circ U(t,s) \subset A(t)U(t,s)$.

PROOF. As verified in Section 3.1, there exists an integer n_0 such that the resolvent sets of $A_n(t)$, $n \geq n_0$, contain a region $\tilde{\Sigma} = \{\lambda \in \mathbb{C}; \Re e\lambda \geq -\tilde{c}(|\Im m| + 1)^\alpha\}$ and a uniform estimate

$$\|(\lambda - A_n(t))^{-1}\|_{\mathcal{L}(X)} \leq \frac{\tilde{M}}{(|\lambda| + 1)^\beta}, \quad \lambda \in \tilde{\Sigma}, n \geq n_0,$$

holds with some constant $\tilde{M}$. On the other hand, since

$$A_n(t)(\lambda - A_n(t))^{-1} = \lambda(\lambda - A_n(t))^{-1} - 1 = \frac{n}{n+\lambda}A(t)^\circ \left(\frac{n\lambda}{n+\lambda} - A(t)\right)^{-1},$$

and since $A_n(t)^{-1} = -\frac{1}{n} + A(t)^{-1}$, (A-i) implies that

$$(4.1) \quad \|A_n(t)(\lambda - A_n(t))^{-1}\{A_n(t)^{-1} - A_n(s)^{-1}\}\|_{\mathcal{L}(X)}$$
$$\leq \frac{\tilde{K}|t - s|^\mu}{(|\lambda| + 1)^\nu}, \quad \lambda \in \tilde{\Sigma}, n \geq n_0,$$

with some uniform constant $\tilde{K}$. In addition, as $n \to \infty$,

$$A_n(t)(\lambda - A_n(t))^{-1}\{A_n(t)^{-1} - A_n(s)^{-1}\}$$
$$\to A(t)^\circ(\lambda - A(t))^{-1}\{A(t)^{-1} - A(s)^{-1}\}$$

in $\mathcal{L}(X)$.

We introduce now the operators

$$V_n(t,s) = U_n(t,s)[-A_n(s)]^{1-\rho}, \quad 0 \leq s \leq t \leq T,$$

where ρ is fixed, but $3 - \alpha - \beta - \alpha\mu < \rho < \nu$. In the same way as in the proof of Proposition 3.1 of [184], the following integral equations are verified:

$$(4.2) \quad U_n(t,s) = e^{(t-s)A_n(s)} + \int_s^t V_n(t,\tau)D_n(\tau,s)A_n(s)e^{(\tau-s)A_n(s)}d\tau,$$

(4.3) $V_n(t, s) = [-A_n(s)]^{1-\rho} e^{(t-s)A_n(s)}$

$$- \int_s^t V_n(t, \tau) D_n(\tau, s) [-A_n(s)]^{2-\rho} e^{(\tau-s)A_n(s)} d\tau,$$

where $D_n(t, s) = [-A_n(t)]^\rho \{A_n(t)^{-1} - A_n(s)^{-1}\}$.

According to Proposition 3.2,

$$\||[-A_n(s)]^{i-\rho} e^{(t-s)A_n(s)}\|_{\mathcal{L}(X)} \leq \tilde{C}(t-s)^{\frac{\beta+\rho-i-1}{\alpha}}, \quad i = 1, 2.$$

On the other hand, since

$$D_n(t, s) = \frac{-1}{2\pi i} \int_\Gamma (-\lambda)^{\rho-1} A_n(t)(\lambda - A_n(t))^{-1} \{A_n(t)^{-1} - A_n(s)^{-1}\} d\lambda,$$

it is estimated by

$$\|D_n(t, s)\|_{\mathcal{L}(X)} \leq \tilde{C}(t-s)^\mu.$$

Therefore, the two integral kernels in (4.3) are seen to have weak integrable singularities at $t = s$ smaller than $(t-s)^{\frac{\alpha+\beta+\rho-2}{\alpha}-1}$ and $(\tau-s)^{\frac{\alpha+\beta+\rho+\alpha\mu-3}{\alpha}-1}$ respectively, which are uniform in $n \geq n_0$. In addition, they are seen to have strong limits as $n \to \infty$. By the theory of integral equations, it is then concluded that the solution $V_n(t, s)$ is also strongly convergent as $n \to \infty$ for each $0 \leq s < t \leq T$. If we note (4.2), this yields furthermore that $U_n(t, s)$ converges strongly to a limit $U(t, s) \in \mathcal{L}(X)$ for each $0 \leq s < t \leq T$. If we multiply from the right hand side both the members of (4.2) by $J_n(s)$, then Proposition 3.6 yields the convergence of $U_n(t, s)J_n(s)$ to the same limit $U(t, s)$.

In order to verify the desired properties of $U(t, s)$, we consider the operators

$$W_n(t, s) = A_n(t)U_n(t, s) - A_n(t)e^{(t-s)A_n(t)}, \quad 0 \leq s \leq t \leq T.$$

One also recognizes that $W_n(t, s)$ satisfies the integral equation

(4.4) $W_n(t, s) = R_n(t, s) + \int_s^t [-A_n(t)]^{2-\rho} e^{(t-\tau)A_n(t)} D_n(t, \tau) W_n(\tau, s) d\tau,$

where

$$R_n(t, s) = \int_s^t [-A_n(t)]^{2-\rho} e^{(t-\tau)A_n(t)} D_n(t, \tau) A_n(\tau) e^{(\tau-s)A_n(\tau)} d\tau.$$

In its turn, $R_n(t, s)$ can be specified in the following form

$$R_n(t, s) = \int_r^t [-A_n(t)]^{2-\rho} e^{(t-\tau)A_n(t)} D_n(t, \tau) A_n(\tau) e^{(\tau-s)A_n(\tau)} d\tau$$

$$- \int_s^r A_n(t)^2 e^{(t-\tau)A_n(t)} \left(e^{(\tau-s)A_n(\tau)} - e^{(\tau-s)A_n(s)} \right) d\tau$$

$$- \int_s^r A_n(t)e^{(t-\tau)A_n(t)} \left(A_n(s)e^{(\tau-s)A_n(s)} - A_n(\tau)e^{(\tau-s)A_n(\tau)} \right) d\tau$$

$$+ A_n(t)e^{(t-r)A_n(t)} \left(e^{(r-s)A_n(s)} - e^{(r-s)A_n(t)} \right),$$

where $r = \frac{s+t}{2}$.

We now need the following lemma.

LEMMA 4.2. *For $\theta \geq 0$,*

$$\|[-A_n(t)]^\theta e^{\tau A_n(t)} - [-A_n(s)]^\theta e^{\tau A_n(s)}\|_{\mathcal{L}(X)}$$
$$\leq \tilde{C}_\theta \tau^{\frac{\beta+\nu-\theta-2}{\alpha}} |t-s|^\mu, \quad \tau > 0, \ 0 \leq s, t \leq T,$$

$\tilde{C}_\theta$ being independent of $n \geq n_0$.

PROOF OF THE LEMMA. Note that

$$[-A_n(t)]^\theta e^{\tau A_n(t)} - [-A_n(s)]^\theta e^{\tau A_n(s)}$$
$$= \frac{1}{2\pi i} \int_\Gamma (-\lambda)^\theta e^{\tau\lambda} \{(\lambda - A_n(t))^{-1} - (\lambda - A_n(s))^{-1}\} d\lambda.$$

Moreover, (4.1) jointed with Theorem 1.9 yields that

$$\|(\lambda - A_n(t))^{-1} - (\lambda - A_n(s))^{-1}\|_{\mathcal{L}(X)} \leq \tilde{C}(|\lambda|+1)^{1-\beta-\nu} |t-s|^\mu, \quad \lambda \in \Gamma.$$

Then the estimate of the lemma follows immediately.

Using this lemma, we establish the uniform estimate of $R_n(t,s)$ by

$$\|R_n(t,s)\|_{\mathcal{L}(X)} \leq \tilde{C}(t-s)^{\frac{2(\alpha+\beta)+\alpha\mu+\nu-5}{\alpha}-1}, \quad 0 \leq s < t \leq T.$$

Notice that (Ex.i) means that the operators $R_n(t,s)$ have a weak integrable singularity at $t = s$ which is uniform in $n \geq n_0$.

Since the strong convergence of $R_n(t,s)$ is evident, we conclude that also the solution $W_n(t,s)$ to (4.4) is strongly convergent in $\mathcal{L}(X)$ as $n \to \infty$ for each $0 \leq s < t \leq T$.

To complete the proof, it now suffices to demonstrate the properties from a) to e) for $U(t,s)$. But these will be immediate consequences of those of $U_n(t,s)$. For example, from $A_n(t)U_n(t,s) = A_n(s)e^{(t-s)A_n(s)} + W_n(t,s)$, it follows that $A_n(t)U_n(t,s), 0 \leq s < t \leq T$, possesses a strong limit in $\mathcal{L}(X)$, say $A(t)^\circ U(t,s)$. Since $U_n(t,s) = A_n(t)^{-1}A_n(t)U_n(t,s)$, $U(t,s) = A(t)^{-1}A(t)^\circ U(t,s)$; this then shows the property d). Furthermore, letting $n \to \infty$ in

$$U_n(t,s) = U_n(s+\varepsilon, s) + \int_{s+\varepsilon}^t A_n(\tau)U_n(\tau,s) d\tau$$

with an arbitrary $\varepsilon > 0$, we verify the property e).

We have thus completed the proof of Theorem 4.1.

Let us next show that $\{U(t,s); 0 \leq s \leq t \leq T\}$, plays the role of the fundamental solution to (E) after defining $U(s,s) = 1$ for $0 \leq s \leq T$.

THEOREM 4.3. *For any* $f \in \mathcal{C}^\sigma([0, T]; X)$, $\sigma > \frac{2-\alpha-\beta}{\alpha}$ *and any* $u_0 \in X$, *the function*

$$(4.5) \qquad u(t) = U(t, 0)u_0 + \int_0^t U(t, \tau)f(\tau)d\tau, \quad 0 \le t \le T,$$

gives a strict solution to (E) satisfying

$$(4.6) \qquad t^{\frac{1-\beta}{\alpha}} u, \quad t^{\frac{2-\beta}{\alpha}} \frac{du}{dt} \in \mathcal{B}((0, T]; X).$$

Conversely, for any $f \in \mathcal{C}([0, T]; X)$ *and any* $u_0 \in X$, *if there exists a strict solution to (E) with an estimate*

$$(4.7) \qquad t^\gamma \frac{du}{dt} \in \mathcal{B}((0, T]; X), \quad \gamma < \frac{\alpha+\nu-1}{\alpha} + \mu,$$

then u *has necessarily the form (4.5).*

PROOF. Consider the sequence of functions

$$u_n(t) = U_n(t, 0)u_0 + \int_0^t U_n(t, \tau)f(\tau)d\tau, \quad 0 \le t \le T,$$

for $n \ge n_0$. Obviously, u_n converges to the function u on $(0, T]$ pointwise. On the other hand, writing

$$A_n(t)u_n(t) = A_n(t)U_n(t, 0)u_0 + \int_0^t A_n(t)U_n(t, \tau)\{f(\tau) - f(t)\}d\tau$$
$$+ \int_0^t W_n(t, \tau)f(t)d\tau + \left(e^{tA_n(t)} - 1\right)f(t),$$

we observe that $A_n u_n$ converges to the function

$$g(t) = A(t)^\circ U(t, 0)u_0 + \int_0^t A(t)^\circ U(t, \tau)\{f(\tau) - f(t)\}d\tau$$
$$+ \int_0^t W(t, \tau)f(t)d\tau + \left(e^{tA(t)} - 1\right)f(t)$$

on $(0, T]$ pointwise. Obviously, g is in $\mathcal{C}((0, T]; X)$. From $u_n(t) = A_n(t)^{-1} \times A_n(t)u_n(t)$, it follows that $u(t) = A(t)^{-1}g(t)$, i.e. $g(t) \in A(t)u(t)$, $0 < t \le T$. On the other hand, from $u_n(t) - u_n(\varepsilon) = \int_\varepsilon^t \{f(\tau) + A_n(\tau)u_n(\tau)\}d\tau$ with an arbitrary $\varepsilon > 0$, $u(t) - u(\varepsilon) = \int_\varepsilon^t \{f(\tau) + g(\tau)\}d\tau$. Thus, u satisfies the equation in (E).

On the other hand, it is easily verified from (4.2) that

(4.8) $\|U(t,0) - e^{tA(0)}\|_{\mathcal{L}(X)} \leq Ct^{\frac{\alpha+2\beta+\rho-4}{\alpha}+\mu} \to 0$ as $t \to 0$.

Hence, in order to show the initial condition in (E) we only need to prove that $e^{-tA(0)}u_0 \to u_0$ as $t \to 0$ in the seminorm $p_{A(0)}(\cdot)$; but this was already seen in Theorem 3.1.

Conversely, consider any strict solution u to (E) with (4.7). We have:

$$u(t) - U_n(t,\varepsilon)u(\varepsilon) = \int_\varepsilon^t U_n(t,\tau)\{-A_n(\tau)u(\tau) + \frac{du}{d\tau}(\tau)\}d\tau$$

$$= \int_\varepsilon^t U_n(t,\tau)\{1 - J_n(\tau)\}g(\tau)d\tau + \int_\varepsilon^t U_n(t,\tau)f(\tau)d\tau, \quad \varepsilon \leq t \leq T,$$

with an arbitrary $\varepsilon > 0$. Note that $J_n(t)g(t) = A_n(t)u(t)$, where $g(t) = \frac{du}{dt} - f(t) \in A(t)u(t)$. By Theorem 4.1 it is then obtained that

$$u(t) - U(t,\varepsilon)u(\varepsilon) = \int_\varepsilon^t U(t,\tau)f(\tau)d\tau, \quad \varepsilon \leq t \leq T.$$

Therefore the proof is reduced to verify that $U(t,\varepsilon)u(\varepsilon) \to U(t,0)u_0$ in X as $\varepsilon \to 0$, or more strongly $u(\varepsilon) - U(\varepsilon,0)u_0 \to 0$ in X. To this end, write

$$u(\varepsilon) - U(\varepsilon,0)u_0 = \{u(\varepsilon) - e^{\varepsilon A(0)}u_0\} + \{e^{\varepsilon A(0)} - U(\varepsilon,0)\}u_0,$$

and observe from (4.8) that the second term tends to 0 as $\varepsilon \to 0$. Therefore, the proof will be completed if we prove the following lemma.

LEMMA 4.4. (4.7) implies

$$\|u(\varepsilon) - e^{\varepsilon A(0)}u_0\|_X \leq C\left(\varepsilon^{\frac{\alpha+\nu-1}{\alpha}+\mu-\gamma} + \varepsilon^{\frac{\alpha+\beta-1}{\alpha}}\right) \to 0 \quad as \quad \varepsilon \to 0.$$

PROOF OF THE LEMMA. For $0 < \delta < \varepsilon$,

$$u(\varepsilon) - e^{(\varepsilon-\delta)A_n(0)}u(\delta) = \int_\delta^\varepsilon e^{(\varepsilon-\tau)A_n(0)}\{-A_n(0)u(\tau) + \frac{du}{d\tau}(\tau)\}d\tau.$$

Letting $n \to \infty$, we obtain that

$$u(\varepsilon) - e^{(\varepsilon-\delta)A(0)}u(\delta) = \int_\delta^\varepsilon A(0)^\circ e^{(\varepsilon-\tau)A(0)}\{A(0)^{-1} - A(\tau)^{-1}\}g(\tau)d\tau$$

$$+ \int_\delta^\varepsilon e^{(\varepsilon-\tau)A(0)}f(\tau)d\tau.$$

Furthermore let $\delta \to 0$. Then, since

$$e^{(\varepsilon-\delta)A(0)}u(\delta) = A(0)^\circ e^{(\varepsilon-\delta)A(0)}A(0)^{-1}u(\delta) \to A(0)^\circ e^{\varepsilon A(0)}A(0)^{-1}u_0,$$

it follows that

$$u(\varepsilon) - e^{\varepsilon A(0)}u_0 = \int_0^\varepsilon A(0)^\circ e^{(\varepsilon-\tau)A(0)}\{A(0)^{-1} - A(\tau)^{-1}\}g(\tau)d\tau$$

$$+ \int_0^\varepsilon e^{(\varepsilon-\tau)A(0)}f(\tau)d\tau.$$

We now observe that (A-i) yields

$$\|A(0)^\circ e^{\tau A(0)}\{A(t)^{-1} - A(0)^{-1}\}\|_{\mathcal{L}(X)} \leq C\tau^{\frac{\nu-1}{\alpha}}t^\mu, \quad \tau > 0, \ 0 \leq t \leq T.$$

In view of this estimate, the lemma is obtained immediately.
Hence we complete the proof of Theorem 4.3.

REMARK. If u_0 satisfies

$$u_0 \in \mathcal{D}([-A(0)]^\theta) \cup X_{A(0)}^\theta, \quad 1 - \beta < \theta < 1,$$

then the solution given by (4.5) is continuous at $t = 0$ in the X-norm. Indeed, by virtue of Theorem 3.3, $e^{tA(0)}u_0 \to u_0$ in X as $t \to 0$; therefore, this combined with (4.8) yields that $U(t,0)u_0 \to u_0$ in X. In other words, the initial condition in (E) is satisfied in the strong sense that $\lim_{t\to 0} u(t) = u_0$ in X.

REMARK. As shown in the proof of Theorem 4.3, for $f \in C^\sigma([0,T]; X)$, $\frac{2-\alpha-\beta}{\alpha} < \sigma$, the derivative of the solution is given by

$$\frac{du}{dt}(t) = e^{tA(t)}f(t) + A(t)^\circ U(t,0)u_0 + \int_0^t A(t)^\circ U(t,\tau)\{f(\tau) - f(t)\}d\tau$$

$$+ \int_0^t W(t,\tau)f(t)d\tau, \quad 0 < t \leq T.$$

Hence, if u_0 satisfies the condition

$$\{f(0) + A(0)u_0\} \cap \{\mathcal{D}([-A(0)]^\theta) \cup X_A^\theta\} \neq \emptyset, \quad 1 - \beta < \theta < 1,$$

then such a function is written in the form

$$\frac{du}{dt}(t) = e^{tA(0)}\{f(0) + f_0\} + \{A(t)^\circ U(t,0)A(0)^{-1} - e^{tA(0)}\}f_0$$

$$+ \{e^{tA(t)}f(t) - e^{tA(0)}f(0)\}$$

$$+ \int_0^t A(t)^\circ U(t,\tau)\{f(\tau) - f(t)\}d\tau + \int_0^t W(t,\tau)f(t)d\tau$$

with a uniquely determined element $f_0 \in A(0)u_0$ satisfying $f(0) + f_0 \in \mathcal{D}([-A(0)]^\theta) \cup X^\theta_{A(0)}$.

Now, by introducing the integral equation

$$A(t)^\circ U(t,s)A(s)^{-1} = A(t)^\circ e^{(t-s)A(t)} A(s)^{-1}$$
$$+ \int_s^t [A(t)^2]^\circ e^{-(t-\tau)A(t)} \{A(t)^{-1} - A(\tau)^{-1}\} A(\tau)^\circ U(\tau,s) A(s)^{-1} d\tau,$$

it is possible to verify that, as $t \to 0$,

$$A(t)^\circ U(t,0)A(0)^{-1} - e^{tA(0)} \to 0 \text{ in } \mathcal{L}(X).$$

Then we conclude that $\lim_{t\to 0} \frac{du}{dt}(t) = f(0) + f_0$. Hence, $u \in \mathcal{C}^1([0,T];X)$ and the equation in (E) holds even at $t = 0$.

We also remark that the condition (A-i) assumed above allows only that the domains $\mathcal{D}(A(t))$ vary temperately with the time variable, since (P) and (A-i) imply that

$$\|A(t)(\lambda - A(t))^{-1}A(s)^{-1}\|_{\mathcal{L}(X)} \le C(|\lambda|^{-\beta} + |\lambda|^{-\nu}).$$

Nevertheless, even in the case where $\mathcal{D}(A(t))$ vary completely we can construct a similar evolution operator. Since the procedure of proof is quite analogous, we only state the theorem. For the detailed proof of it, see Theorem 4.2 of [97].

THEOREM 4.5. *Let $A(t)$, $0 \le t \le T$, satisfy (P) and the following conditions:*

(A.ii) *$A(\cdot)^{-1}$ is strongly continuously differentiable in t and its derivative satisfies the decay estimate*

$$\left\| A(t)^\circ (\lambda - A(t))^{-1} \frac{dA(t)^{-1}}{dt} \right\|_{\mathcal{L}(X)} \le \frac{N}{(|\lambda|+1)^\nu}, \quad \lambda \in \Sigma, \; 0 \le t \le T,$$

with some exponent $0 < \nu \le 1$; and

(Ex.ii) *The exponents satisfy the relation: $2(\alpha + \beta) + \nu > 4$.*

Then, there exists a family of operators $\{U(t,s); 0 \le s < t \le T\}$ which enjoys the properties a) $\sim$ e) announced in Theorem 4.1.

THEOREM 4.6. *Let (P), (A.ii) and (Ex.ii) be satisfied. For any $f \in \mathcal{C}^\sigma([0,T];X)$, $\sigma > \frac{2-\alpha-\beta}{\alpha}$, and any $u_0 \in X$, the function given in the form (4.5) provides a strict solution to (E) with the property (4.6). Conversely, for any $f \in \mathcal{C}([0,T];X)$ and any $u_0 \in X$, if there exists a strict solution to (E) with*

$$t^\gamma u \in \mathcal{B}((0,T];X), \quad \gamma < \frac{\alpha+\beta+\nu-2}{\alpha},$$

then u has necessarily the form (4.5).

4.2 MAXIMAL REGULARITY OF
SOLUTIONS TO $\frac{du}{dt} \in A(t)u + f(t)$

In this section we study the maximal regularity of strict solutions to the Cauchy problem (E). To begin with, we consider the initial value $u_0 = 0$, i.e.

(E) $\qquad \begin{cases} \dfrac{du}{dt} \in A(t)u + f(t), & 0 \le t \le T, \\ u(0) = u_0 = 0, \end{cases}$

where $f \in \mathcal{C}([0,T]; X)$ satisfies $f(0) = 0$. We shall see later on how to remove these restrictions.

In this section we make somewhat different assumptions on the operator coefficients $A(t)$. In fact, in addition to Condition (P), we suppose:

(A.iii.a) $A(\cdot)^{-1}$ is strongly continuously differentiable in t and its derivative satisfies

$$\left\| A(t)^\circ (\lambda - A(t))^{-1} \frac{dA(t)^{-1}}{dt} A(t)^\circ (\lambda - A(t))^{-1} \right\|_{\mathcal{L}(X)}$$
$$\le \frac{N}{(|\lambda| + 1)^\nu}, \quad \lambda \in \Sigma,$$

with some exponent $0 < \nu \le \beta$.

(A.iii.b) $\frac{dA(\cdot)^{-1}}{dt}$ satisfies the Hölder condition

$$\left\| \frac{dA(t)^{-1}}{dt} - \frac{dA(s)^{-1}}{ds} \right\|_{\mathcal{L}(X)} \le K|t - s|^\mu, \quad 0 \le s, t \le T,$$

with some exponent $0 < \mu \le 1$.

(Ex.iii) The exponents satisfy the relation: $2\alpha + \beta > 2$, $\alpha + \nu > 1$ and $\alpha\mu(\alpha + \nu - 1) > (2 - \beta)(2 - \alpha - \beta)$.

By some technical reason (E) is changed in the form

(E$_k$) $\qquad \begin{cases} \dfrac{du}{dt} \in A(t)u - ku + f(t), & 0 < t \le T, \\ u(0) = 0 \end{cases}$

with an arbitrary number $k > 0$. The equivalence of (E) and (E$_k$) is immediately verified by the change of the unknown function to $e^{-kt}u(t)$. Of course, also $f(t)$ becomes $e^{-kt}f(t)$, but $T < \infty$ implies no change in the regularity of the inhomogeneous function.

As it was done in Section 3.5, we treat (E$_k$) as an operator equation

($\mathcal{E}_k$) $\qquad\qquad\qquad ku + \mathcal{T}u - \mathcal{A}u \ni f$

in the Banach space $\mathcal{X} = \{f \in \mathcal{C}([0,T]; X); f(0) = 0\}$ of continuous functions. Here,

$$\begin{cases} \mathcal{D}(\mathcal{T}) = \{u \in \mathcal{C}^1([0,T]; X); u(0) = u'(0) = 0\} \\ \mathcal{T}u = \left(\dfrac{d}{dt} + \varepsilon\right) u. \end{cases}$$

$\mathcal{A}$ is the m. l. operator defined by

$$\begin{cases} \mathcal{D}(\mathcal{A}) = \{u \in \mathcal{X}; \text{there exists } f \in \mathcal{X} \text{ such that } f(t) \in (A(t) + \varepsilon)u(t) \\ \qquad\qquad \text{for every } 0 \le t \le T\} \\ \mathcal{A}u = \{f \in \mathcal{X}; f(t) \in (A(t) + \varepsilon)u(t) \text{ for every } 0 \le t \le T\}, \end{cases}$$

where $\varepsilon > 0$ is a sufficiently small number. As noticed in Section 3.5, $\mathcal{T}$ satisfies Condition $(\mathcal{T})$. On the other hand, (P) implies that $\mathcal{A}$ satisfies Condition $(\mathcal{P})$.

PROPOSITION 4.7. *Let (P),(A.iii.a) and (Ex.iii) be satisfied. Then, if k is sufficiently large, the solution to $(\mathcal{E}_k)$ is unique for any $f \in \mathcal{X}$.*

PROOF. Let us verify that $(k + \mathcal{T} - \mathcal{A})u \ni 0$ implies $u = 0$. Define the linear operator

$$\overline{\mathcal{S}}f = \tfrac{1}{2\pi i} \int_\Gamma (\lambda + k + \mathcal{T})^{-1}(\lambda + \mathcal{A})^{-1} f d\lambda, \quad f \in \mathcal{X},$$

with the integral contour $\Gamma{:}\lambda = c'(|\eta| + 1)^\alpha + i\eta$, $-\infty < \eta < \infty$, c' being a suitable positive constant. Let $g \in \mathcal{A}u$ or $u = \mathcal{A}^{-1}g$ with $(k+\mathcal{T})u - g = 0$. Clearly, $\overline{\mathcal{S}}(k + \mathcal{T})\mathcal{A}^{-1}g - \overline{\mathcal{S}}g = 0$. Moreover,

$$\overline{\mathcal{S}}(k+\mathcal{T})\mathcal{A}^{-1}g = \tfrac{1}{2\pi i} \int_\Gamma \lambda^{-1}(\lambda+k+\mathcal{T})^{-1}(k+\mathcal{T})\{(\lambda+\mathcal{A})^{-1} - \mathcal{A}^{-1}\}g d\lambda$$

$$+ \tfrac{1}{2\pi i} \int_\Gamma (\lambda+k+\mathcal{T})^{-1}[\mathcal{T}, (\lambda+\mathcal{A})^{-1}]\mathcal{A}^{-1}g d\lambda,$$

where $[\mathcal{T}, (\lambda+\mathcal{A})^{-1}] = \mathcal{T}(\lambda+\mathcal{A})^{-1} - (\lambda+\mathcal{A})^{-1}\mathcal{T}$ denotes the commutator. Since

$$\tfrac{1}{2\pi i} \int_\Gamma \lambda^{-1}(k+\mathcal{T})(\lambda+k+\mathcal{T})^{-1}(\lambda+\mathcal{A})^{-1}g d\lambda = -\mathcal{A}^{-1}g + \overline{\mathcal{S}}g$$

and

$$\tfrac{1}{2\pi i} \int_\Gamma \lambda^{-1}(\lambda+k+\mathcal{T})^{-1}(k+\mathcal{T})\mathcal{A}^{-1}g d\lambda = 0,$$

we then conclude that $(1 - \mathcal{V})u = 0$, where $\mathcal{V}$ is the operator given by

$$\mathcal{V}f = \tfrac{1}{2\pi i} \int_\Gamma (\lambda+k+\mathcal{T})^{-1}[\mathcal{T}, (\lambda+\mathcal{A})^{-1}]f d\lambda.$$

We now need the following lemma.

LEMMA 4.8. *For each* $\lambda \in -\Sigma$,

$$[\mathcal{T}, (\lambda + \mathcal{A})^{-1}]f(t) = A(t)^{\circ}(\lambda + A(t))^{-1}\frac{dA(t)^{-1}}{dt}A(t)^{\circ}(\lambda + A(t))^{-1}f(t),$$
$$0 \le t \le T.$$

PROOF OF THE LEMMA. Let $u \in \mathcal{D}(\mathcal{T})$. By definition of $\mathcal{T}$,

$$[\mathcal{T}, (\lambda + \mathcal{A})^{-1}]u(t) = \frac{\partial(\lambda + A(t))^{-1}}{\partial t}u(t).$$

On the other hand, from Theorem 1.9 it is seen that

$$(4.9) \qquad \frac{\partial(\lambda + A(t))^{-1}}{\partial t} = A(t)^{\circ}(\lambda + A(t))^{-1}\frac{dA(t)^{-1}}{dt}A(t)^{\circ}(\lambda + A(t))^{-1}.$$

This lemma combined with (A.iii.a) then yields that

$$(4.10) \qquad \|[\mathcal{T}, (\lambda + \mathcal{A})^{-1}]\|_{\mathcal{L}(\mathcal{X})} \le C(1 + |\lambda|)^{-\nu}, \quad \lambda \in -\Sigma.$$

Therefore,

$$\|\mathcal{V}\|_{\mathcal{L}(\mathcal{X})} \le C\int_0^{\infty}\frac{d\eta}{(\eta + k + 1)^{\alpha}(\eta + 1)^{\nu}} \le Ck^{1-\alpha-\nu}.$$

This shows that $1 - \mathcal{V}$ is invertible if k is sufficiently large; hence, $(1 - \mathcal{V})u = 0$ implies $u = 0$. This completes the proof of the proposition.

PROPOSITION 4.9. *Let* $(P),(A.iii.a),(A.iii.b)$ *and* $(Ex.iii)$ *be satisfied. Then, if* k *is sufficiently large,* $(\mathcal{E}_k)$ *possesses one solution for any* $f \in C^{\sigma}([0,T]; X)$, $\frac{2-\alpha-\beta}{\alpha} < \sigma \le 1$ *with* $f(0) = 0$.

PROOF. On account of (Ex.iii) we fix an exponent θ such that $\frac{2-\alpha-\beta}{\alpha} < \theta < \min\left\{\frac{\mu(\alpha+\nu-1)}{2-\beta}, \sigma\right\}$. For $f \in \{f \in C^{\theta}([0,T]; X); f(0) = 0\} = \mathcal{X}_T^{\theta}$, define

$$\mathcal{S}f = \frac{1}{2\pi i}\int_{\Gamma}(\lambda + \mathcal{A})^{-1}(\lambda + k + \mathcal{T})^{-1}f d\lambda.$$

$\mathcal{S}$ is in fact the same operator that we introduced in the proof of Theorem 3.19. We can then repeat the previous argument and obtain that

$$(4.11) \qquad (k + \mathcal{T} - \mathcal{A})\mathcal{S}f \ni (1 + \mathcal{W})f,$$

where $\mathcal{W}$ is the operator given by

$$\mathcal{W}f = \frac{1}{2\pi i}\int_{\Gamma}[\mathcal{T}, (\lambda + \mathcal{A})^{-1}](\lambda + k + \mathcal{T})^{-1}f d\lambda.$$

We also need another Hölder regularity of the derivative $\frac{\partial}{\partial t}(\lambda + A(t))^{-1}$ established as follows.

LEMMA 4.10.

$$\left\| \frac{\partial(\lambda + A(t))^{-1}}{\partial t} - \frac{\partial(\lambda + A(s))^{-1}}{\partial s} \right\|_{\mathcal{L}(X)}$$
$$\leq C(1 + |\lambda|)^{2-\beta-\nu} |t - s|^{\mu}, \quad 0 \leq s, t \leq T, \ \lambda \in -\Sigma.$$

PROOF OF THE LEMMA. From (4.9) and (A.iii.a),

$$\| (\lambda + A(t))^{-1} - (\lambda + A(s))^{-1} \|_{\mathcal{L}(X)} \leq C(1 + |\lambda|)^{-\nu} |t - s|.$$

(4.9) gives also the expression of $\frac{\partial(\lambda+A(t))^{-1}}{\partial t}$ in terms of $(\lambda + A(t))^{-1}$ and $\frac{dA(t)^{-1}}{dt}$. Then this together with (A.iii.b) immediately yields the desired estimate.

In view of Lemma 4.8, Lemma 4.10 means that $[\mathcal{T}, (\lambda + \mathcal{A})^{-1}]$ is a bounded operator on $\mathcal{X}_T^{\mu}$ with

$$\| [\mathcal{T}, (\lambda + \mathcal{A})^{-1}] \|_{\mathcal{L}(\mathcal{X}_T^{\mu})} \leq C(1 + |\lambda|)^{2-\beta-\nu}.$$

Moreover, interpolating between this and (4.10) $(0 < \theta < \mu)$ and taking into account the reiteration properties for the mean spaces, it follows that

$$(4.12) \qquad \| [\mathcal{T}, (\lambda + \mathcal{A})^{-1}] \|_{\mathcal{L}(\mathcal{X}_T^{\theta})} \leq C(|\lambda| + 1)^{\frac{(2-\beta)\theta}{\mu} - \nu}, \quad \lambda \in -\Sigma.$$

From this it is easy to observe that

$$\| \mathcal{W} \|_{\mathcal{L}(\mathcal{X}_T^{\theta})} \leq Ck^{\frac{(2-\beta)\theta - \mu(\alpha+\nu-1)}{\mu}}.$$

Therefore, if k is sufficiently large, $1 + \mathcal{W}$ is a bijection on $\mathcal{X}_T^{\theta}$. In other words, for any $g \in \mathcal{X}_T^{\theta}$ there exists f such that $g = (1 + \mathcal{W})f$; hence, $(k + \mathcal{T} - \mathcal{A})\mathcal{S}(1 + \mathcal{W})^{-1}g \ni g$ from (4.11). Thus the proposition is proved.

THEOREM 4.11. *Let* (P), (A.iii.a), (A.iii.b) *and* (Ex.iii) *be satisfied. Let* $\frac{2-\alpha-\beta}{\alpha} < \theta < \frac{\mu(\alpha+\nu-1)}{2-\beta}$. *Then, for* $f \in C^{\theta}([0,T]; X)$ *with* $f(0) = 0$, *the solution* u *to* $(\mathcal{E}_k)$ *obtained in Propositio 4.9 actually belongs to* $C^{1+\omega}([0,T]; X)$, *where* $\omega = \alpha\theta + \alpha + \beta - 2$.

PROOF. As it was shown above, u is given by $\mathcal{S}(1 + \mathcal{W})^{-1}f$. Therefore,

$$\mathcal{T}u = \frac{1}{2\pi i} \int_{\Gamma} \mathcal{T}(\lambda + \mathcal{A})^{-1}(\lambda + k + \mathcal{T})^{-1}(1 + \mathcal{W})^{-1}f d\lambda.$$

After some calculation involving commutators, we see that for $\xi > 2c$,

$$T(\xi+T)^{-1}Tu = \tfrac{1}{2\pi i} \int_\Gamma (\lambda+A)^{-1}T^2(\xi+T)^{-1}(\lambda+k+T)^{-1}(1+W)^{-1}f\,d\lambda$$

$$+ \tfrac{1}{2\pi i} \int_\Gamma [T,(\lambda+A)^{-1}]T(\xi+T)^{-1}(\lambda+k+T)^{-1}(1+W)^{-1}f\,d\lambda$$

$$+ \tfrac{1}{2\pi i} \int_\Gamma \xi T(\xi+T)^{-1}[T,(\lambda+A)^{-1}](\xi+T)^{-1}(\lambda+k+T)^{-1}(1+W)^{-1}f\,d\lambda.$$

Here, we used the formula

$$[(\xi+T)^{-1},(\lambda+A)^{-1}] = -(\xi+T)^{-1}[T,(\lambda+A)^{-1}](\xi+T)^{-1}.$$

For the first integral we can repeat the same argument as in the proof of Theorem 3.24. Its $\mathcal{X}$-norm is then estimated by

$$C\xi^{-\omega}\|(1+W)^{-1}f\|_{\mathcal{X}_T^\theta} \le C\xi^{-\omega}\|f\|_{\mathcal{X}_T^\theta}.$$

From (A.iii.a) the $\mathcal{X}$-norm of the second integral is estimated by

$$C\xi^{-\theta} \int_0^\infty \frac{d\eta}{(\eta+k)^\alpha(\eta+1)^\nu}\|(1+W)^{-1}f\|_{\mathcal{X}_T^\theta} \le C\xi^{-\theta}\|f\|_{\mathcal{X}_T^\theta}.$$

Noting (4.12) we estimate also the third integral by

$$C\xi^{-\theta} \int_0^\infty \frac{d\eta}{(\eta+k)^\alpha(\eta+1)^{\nu-\frac{(2-\beta)\theta}{\mu}}}\|(1+W)^{-1}f\|_{\mathcal{X}_T^\theta} \le C\xi^{-\theta}\|f\|_{\mathcal{X}_T^\theta}.$$

Hence the proof is complete.

At last, using Propositions 4.7 and 4.9, we are able to establish existence and uniqueness of the solution u for the general initial condition: $u(0) = u_0$ and an arbitrary $f(0)$. The result reads as follows.

THEOREM 4.12. *Let (P), (A.iii.a), (A.iii.b) and (Ex.iii) be satisfied. Let $f \in C^\theta([0,T];X)$, $\frac{2-\alpha-\beta}{\alpha} < \theta < \frac{\mu(\alpha+\nu-1)}{2-\beta}$, and let $u_0 \in \mathcal{D}(A(0))$ satisfy the condition*

$$\left\{ f(0) + \left(1 - \frac{dA(t)^{-1}}{dt}\bigg|_{t=0}\right) A(0)u_0 \right\} \cap \mathcal{D}(A(0)) \ne \emptyset.$$

Then, for each $f_0 \in A(0)u_0$ such that

$$f(0) + \left(1 - \frac{dA(t)^{-1}}{dt}\bigg|_{t=0}\right) f_0 \in \mathcal{D}(A(0)),$$

(E) possesses a unique strict solution $u \in C^{1+\omega}([0,T];X)$, $\omega = \alpha\theta + \alpha + \beta - 2$, *such that* $u(0) = u_0$ *and* $\frac{du}{dt}(0) + f_0 = f(0)$.

PROOF. With fixed f_0, change the unknown function to $v(t) = u(t) - A(t)^{-1}f_0$. Then (E) is reduced to the equivalent problem

$$\begin{cases} \dfrac{dv}{dt} \in A(t)v + g(t), & 0 < t \le T, \\ v(0) = 0, \end{cases}$$

where $g(t) = f(t) + \left(1 - \frac{dA(t)^{-1}}{dt}\right)f_0$. Since $g(0) \in \mathcal{D}(A(0))$, there exists φ_0 such that $g(0) = A(0)^{-1}\varphi_0$. Then the new change of the unknown function to $w(t) = v(t) - tA(t)^{-1}\varphi_0$ yields

$$\begin{cases} \dfrac{dw}{dt} \in A(t)w + h(t), & 0 < t \le T, \\ w(0) = 0, \end{cases}$$

where $h(t) = g(t) - A(t)^{-1}\varphi_0 + t\left(1 - \frac{dA(t)^{-1}}{dt}\right)\varphi_0$. Because of $h(0) = 0$, Proposition 4.9 and Theorem 4.11 are now applicable.

The solution obtained here obviously satisfies $\frac{du}{dt}(0) = f(0) + f_0$. Then the uniqueness of the solution u with this property follows immediately from Proposition 4.7.

4.3 DEGENERATE EVOLUTION EQUATIONS

Consider the degenerate Cauchy problem

(D-E.1)
$$\begin{cases} \dfrac{dM(t)v}{dt} = L(t)v + f(t), & 0 < t \le T, \\ M(0)v(0) = u_0, \end{cases}$$

in a Banach space X. Here $M(t)$, $0 \le t \le T$, are single valued closed linear operators in X, $L(t)$, $0 \le t \le T$, are single valued closed linear operators in X with inclusion $\mathcal{D}(L(t)) \subset \mathcal{D}(M(t))$, $f:[0,T] \to X$ is a given continuous function, $u_0 \in X$ is an initial value, and $v:[0,T] \to X$ is the unknown function.

By the change of the unknown function to $u(t) = M(t)v(t)$, we rewrite (D-E.1) into the form

(4.13)
$$\begin{cases} \dfrac{du}{dt} \in L(t)M(t)^{-1}u + f(t), & 0 < t \le T, \\ u(0) = u_0. \end{cases}$$

In addition, a change of the unknown function to $u_\gamma(t) = e^{-\gamma t} u(t)$ yields that (4.13) is treated as a multivalued equation of the form (E) in which $A(t)$ are the m. l. operators $L(t)M(t)^{-1} - \gamma$.

Let us assume:

(4.14) The $M(t)$ resolvent sets $\rho_{M(t)}(L(t))$, $0 \leq t \leq T$, contain a region

$$\Sigma_\gamma = \{\lambda \in \mathbb{C}; \Re(\lambda - \gamma) \geq -c(|\Im m\lambda| + 1)^\alpha\}, \quad \gamma \in \mathbb{R},$$

and the $M(t)$ resolvents satisfy

$$\|M(t)(\lambda M(t) - L(t))^{-1}\|_{\mathcal{L}(X)} \leq \frac{C}{(|\lambda - \gamma| + 1)^\beta}, \quad \lambda \in \Sigma_\gamma,$$

with some exponents $0 < \beta \leq \alpha \leq 1$ and constants c, $C > 0$.

By Theorem 1.14, (4.14) is clearly a sufficient condition for the Assumption (P). For simplicity, in what follows we shall assume that (4.14) is satisfied with $\gamma = 0$, and so that we let $A(t) = L(t)M(t)^{-1}$.

Noting that

$$A(t)^\circ(\lambda - A(t))^{-1} = L(t)(\lambda M(t) - L(t))^{-1}$$

and that $A(t)^{-1} = M(t)L(t)^{-1}$, it is not a difficult matter to investigate sufficient conditions which imply (A.i), (A.ii) and (A.iii) respectively. We shall study in some details certain cases where these assumptions translate very easily and appear in concrete applications.

Case that $M(t) \equiv M$ and $\mathcal{D}(L(t)) \equiv \mathcal{D}$ are independent of t. It is seen that

$$A(t)^\circ(\lambda - A(t))^{-1}\{A(t)^{-1} - A(s)^{-1}\}$$
$$= L(t)(\lambda M - L(t))^{-1}M\{L(t)^{-1} - L(s)^{-1}\}$$
$$= M(\lambda M - L(t))^{-1}L(t)\{L(t)^{-1} - L(s)^{-1}\}.$$

Then, the Hölder condition

(4.15) $\|\{L(t) - L(s)\}L(s)^{-1}\|_{\mathcal{L}(X)} \leq C|t - s|^\mu, \quad 0 \leq s, t \leq T,$

in addition to (4.14) implies (A.i) with $\nu = \beta$. Therefore, Theorem 4.3 provides the following result.

PROPOSITION 4.13. *Let $M(t) \equiv M$ and $\mathcal{D}(L(t)) \equiv \mathcal{D}$. Let (4.14) (with $\gamma = 0$) and (4.15) be satisfied with $2\alpha + 3\beta + \alpha\mu > 5$. Then, for any $f \in \mathcal{C}^\sigma([0, T]; X)$, $\sigma > \frac{2 - \alpha - \beta}{\alpha}$ and any $u_0 \in X$, (D-E.1) possesses a unique solution v such that $\frac{dMv}{dt}$, $Lv \in \mathcal{C}((0, T]; X)$, the initial condition being satisfied in some seminorm sense. Moreover, if $u_0 \in \mathcal{D}(A(0)) = M(\mathcal{D}(L(0)))$,* then $Mv \in C([0,T];X)$ *and the initial condition is satisfied in the norm of* X.

Case that $L(t) \equiv L$ are independent of t and that $M(t)$ are bounded operators on X. It is seen that

$$(4.16) \qquad A(t)^{\circ}(\lambda - A(t))^{-1}\frac{dA(t)^{-1}}{dt} = L(\lambda M(t) - L)^{-1}\frac{dM(t)}{dt}L^{-1}.$$

This may then signify that X equiped with the new norm $\|L^{-1}\cdot\|_X$ is another possible underlying space. Let $X_0 \supset X$ be the completion of X with this norm. Since $\|A(t)^{-1}f\|_{X_0} \leq C\|f\|_{X_0}$, $f \in X$, $A(t)^{-1}$ has a unique extension as a bounded operator on X_0. Consequently $A(t)$ is also extended as m. l. operator in X_0 and $A(t)$ shall denote this extended operator, too. From

$$\|(\lambda - A(t))^{-1}f\|_{X_0} = \|L^{-1}M(t)(\lambda M(t) - L)^{-1}f\|_X$$
$$= \|(\lambda M(t) - L)^{-1}M(t)L^{-1}f\|_X \leq C\|(\lambda M(t) - L)^{-1}M(t)\|_{\mathcal{L}(X)}\|f\|_{X_0},$$

$(\lambda - A(t))^{-1}$, $\lambda \in \Sigma$, have also bounded extensions in X_0 and these coincide with the resolvent $(\lambda - A(t))^{-1}$ of $A(t)$ in X_0. Therefore, the condition

$$(4.17) \qquad \|(\lambda M(t) - L)^{-1}M(t)\|_{\mathcal{L}(X)} \leq \frac{C}{(|\lambda| + 1)^{\beta}}, \qquad \lambda \in \Sigma,$$

implies the Assumption (P). Similarly, from (4.16), $M(\cdot) \in C^1([0,T];\mathcal{L}(X))$ with

$$(4.18) \qquad \left\|(\lambda M(t) - L)^{-1}\frac{dM(t)}{dt}\right\|_{\mathcal{L}(X)} \leq \frac{C}{(|\lambda| + 1)^{\nu}}, \qquad \lambda \in \Sigma,$$

implies (A.ii). Hence, Theorem 4.6 provides the following result.

PROPOSITION 4.14. *Let* $L(t) \equiv L$ *and* $M(\cdot) \in C^1([0,T];\mathcal{L}(X))$. *Let* *(4.14) (with* $\gamma = 0$*), (4.17) and (4.18) be satisfied with* $2(\alpha + \beta) + \nu > 4$. *Then, for any* $f \in C^{\sigma}([0,T];X)$, $\sigma > \frac{2-\alpha-\beta}{\alpha}$, *(4.13) possesses a unique solution* u *such that* $\frac{du}{dt} \in C^1((0,T];X_0)$. *If* $u_0 \in M(0)(\mathcal{D}(L))$, *then* $u \in C([0,T];X_0)$ *and the initial condition is satisfied in the norm of* X_0.

Case that $\mathcal{D}(L(t)) \equiv \mathcal{D}$ are independent of t. It is seen that

$$A(t)^{\circ}(\lambda - A(t))^{-1}\frac{dA(t)^{-1}}{dt}A(t)^{\circ}(\lambda - A(t))^{-1}$$

$$= L(t)(\lambda M(t) - L(t))^{-1}\left\{\frac{dM(t)}{dt}L(t)^{-1}+\right.$$

$$\left. M(t)\frac{dL(t)^{-1}}{dt}\right\}L(t)(\lambda M(t) - L(t))^{-1}$$

$$= L(t)(\lambda M(t) - L(t))^{-1}\frac{dM(t)}{dt}(\lambda M(t) - L(t))^{-1}$$

$$+ M(t)(\lambda M(t) - L(t))^{-1}L(t)\frac{dL(t)^{-1}}{dt}L(t)(\lambda M(t) - L(t))^{-1}.$$

Therefore,

(4.19) $$\left\| \frac{dM(t)}{dt} (\lambda M(t) - L(t))^{-1} \right\|_{\mathcal{L}(X)} \leq \frac{C}{(|\lambda| + 1)^{\tilde{\nu}}}, \quad \lambda \in \Sigma,$$

with some $0 < \tilde{\nu} \leq \beta$ implies (A.iii.a) with $\nu = \tilde{\nu} + \beta - 1$. On the other hand,

(4.20) $$M(\cdot), \ L(\cdot) \in C^{1+\mu}([0,T]; \mathcal{L}(\mathcal{D}, X))$$

obviously implies (A.iii.b). Therefore, Theorem 4.12 provides the following result.

PROPOSITION 4.15. *Let* $\mathcal{D}(L(t)) \equiv \mathcal{D}$. *Let* (4.14) *(with* $\gamma = 0$*), (4.19) and (4.20) be satisfied with* $\alpha\mu(\alpha + \beta + \tilde{\nu} - 2) > (2 - \beta)(2 - \alpha - \beta)$. *Then, for any* $f \in C^\theta([0,T]; X)$, $\frac{2-\alpha-\beta}{\alpha} < \theta < \frac{\mu(\alpha+\beta+\tilde{\nu}-2)}{2-\beta}$, *and any* $v_0 \in \mathcal{D}$ *such that*

$$f(0) + \left(1 - \left. \frac{dA(t)^{-1}}{dt} \right|_{t=0} \right) L(0)v_0 \in M(0)(\mathcal{D}),$$

(D-E.1) with $u_0 = M(0)v_0$ *possesses a unique strict solution* v *such that* $\frac{dM(t)v}{dt}, \ L(t)v \in C^{\alpha\theta+\alpha+\beta-2}([0,T]; X)$ *and that* $v(0) = v_0$.

PROOF. It suffices to apply Theorem 4.12 with $f_0 = L(0)v_0$.

Consider next the dual problem

(D-E.2) $$\begin{cases} M(t)\dfrac{du}{dt} = L(t)u + M(t)f(t), & 0 < t \leq T, \\ u(0) = u_0, \end{cases}$$

of (D-E.1) in a Banach space Y. Here, $M(t)$, $0 \leq t \leq T$, are bounded linear operators from X to Y, X being another Banach space such that $X \subset Y$ continuously. $L(t)$, $0 \leq t \leq T$, are single valued closed linear operators in Y with $\mathcal{D}(L(t)) \subset X$. $f:[0,T] \to X$ is a given function. $u_0 \in X$ is an initial value.

Since the equation in (D-E.2) is rewritten as a multivalued equation, (D-E.2) is equivalent to the problem

(4.21) $$\begin{cases} \dfrac{du}{dt} \in M(t)^{-1}L(t)u + f(t), & 0 < t \leq T, \\ u(0) = u_0, \end{cases}$$

in X. Therefore, (D-E.2) is treated as a problem of the form (E) in which the coefficient operators $A(t)$ are m. l. operators $M(t)^{-1}L(t) - \gamma$ in X, γ being determined below.

We assume:

(4.22) The $M(t)$ resolvent sets $\rho_{M(t)}(L(t))$, $0 \le t \le T$, contain a region

$$\Sigma_\gamma = \{\lambda \in \mathbb{C}; \Re e(\lambda - \gamma) \ge -c(|\Im m\lambda| + 1)^\alpha\}, \quad \gamma \in \mathbb{R},$$

and the $M(t)$ resolvents satisfy

$$\|(\lambda M(t) - L(t))^{-1} M(t)\|_{\mathcal{L}(X)} \le \frac{C}{(|\lambda - \gamma| + 1)^\beta}, \quad \lambda \in \Sigma,$$

with some exponents $0 < \beta \le \alpha \le 1$ and constants c, $C > 0$.

Then, since $(\lambda - A(t))^{-1} = ((\lambda + \gamma)M(t) - L(t))^{-1} M(t)$, this condition is a sufficient condition for the Assumption (P). For sake of simplicity, again (4.22) is assumed to hold with $\gamma = 0$, so that $A(t) = M(t)^{-1} L(t)$.

We now indicate some sufficient conditions to imply (A.i), (A.ii) and (A.iii), respectively. They are motivated by various concrete applications.

Case that $M(t) \equiv M$ and $\mathcal{D}(L(t)) \equiv \mathcal{D}$ are independent of t and that the kernel of M is the null space, $\mathcal{K}(M) = \{0\}$. In this case it is seen that

$$A(t)^\circ(\lambda - A(t))^{-1}\{A(t)^{-1} - A(s)^{-1}\}$$
$$= -(\lambda M - L(t))^{-1}\{L(t) - L(s)\}L(s)^{-1}M.$$

This then suggests that X equipped with the new norm $\|M\cdot\|_Y$ is another possible underlying space in considering (4.21). Let $X_1 \supset X$ be the completion of X with respect to the norm $\|M\cdot\|_Y$. Let us assume that

$$(4.23) \qquad \|M(\lambda M - L(t))^{-1}\|_{\mathcal{L}(Y)} \le \frac{C}{(|\lambda| + 1)^\beta}, \quad \lambda \in \Sigma, \ 0 \le t \le T.$$

Then this yields in particular that $A(t)^{-1}$ is bounded on X in the norm $\|\cdot\|_{X_1}$; therefore, $A(t)^{-1}$ has a unique bounded extension on X_1 and its inverse is an extension of $A(t)$ as a m. l. operator acting in X_1. If $A(t)$ denotes also the extended m. l. operator, too, it is also observed that the resolvent $(\lambda M - L(t))^{-1}$ in X has a bounded extension to X_1 and this extension coincides precisely with the resolvent in X_1. In this way (4.23) implies that Assumption (P) is verified in the space X_1.

Similarly, the Hölder condition

$$(4.24) \qquad \|\{L(t) - L(s)\}L(s)^{-1}\|_{\mathcal{L}(Y)} \le C|t - s|^\mu, \quad 0 \le s, t \le T,$$

implies (A.i) with $\nu = \beta$.

Therefore, Theorem 4.3 provides the following result.

PROPOSITION 4.16. *Let $M(t) \equiv M$, $\mathcal{D}(L(t)) \equiv \mathcal{D}$ and $\mathcal{K}(M) = \{0\}$. Let (4.22) (with $\gamma = 0$) (4.23) and (4.24) be satisfied with $2\alpha + 3\beta + \alpha\mu > 5$. Then, for any $f \in C^\sigma([0, T]; X)$, $\sigma > \frac{2-\alpha-\beta}{\alpha}$ and any $u_0 \in X$, (4.21) possesses a unique solution $u \in C^1((0, T]; X_1)$. If $u_0 \in \mathcal{D}(A(0)) = L(0)^{-1}(\mathcal{R}(M))$, then $u \in C^1((0, T]; X_1)$ and the initial condition is satisfied in the norm of X_1.*

Case that $L(t) \equiv L$ are independent of t. It is then seen that

$$A(t)^\circ (\lambda - A(t))^{-1} \frac{dA(t)^{-1}}{dt} = (\lambda M(t) - L(t))^{-1} \frac{dM(t)}{dt}.$$

Therefore,

$$(4.25) \quad \left\| (\lambda M(t) - L)^{-1} \frac{dM(t)}{dt} \right\|_{\mathcal{L}(X)} \leq \frac{C}{(|\lambda| + 1)^\nu}, \quad \lambda \in \Sigma, \ 0 \leq t \leq T,$$

implies (A.ii). Therefore, Theorem 4.6 provides the following result.

PROPOSITION 4.17. *Let $L(t) \equiv L$ and $M(\cdot) \in C^1([0, T]; \mathcal{L}(X, Y))$. Let (4.22) (with $\gamma = 0$) and (4.25) be satisfied with $2(\alpha + \beta) + \nu > 4$. Then, for any $f \in C^\sigma([0, T]; X)$, $\sigma > \frac{2-\alpha-\beta}{\alpha}$ and any $u_0 \in X$, (D-E.2) possesses a unique solution $u \in C^1((0, T]; X)$. If $u_0 \in \mathcal{D}(A(0)) = L^{-1}(\mathcal{R}(M(0)))$, then $u \in C([0, T]; X)$ and the initial condition is satisfied in the norm of X.*

Case that $L(t)$ satisfies the condition

$$(4.26) \quad \left\| \frac{dL(t)^{-1}}{dt} L(t) u \right\|_X \leq C \|u\|_X, \quad u \in \mathcal{D}(L(t)), \ 0 \leq t \leq T.$$

with some constant C. We remark that, if $\mathcal{D}(L(t))$ are dense in Y, this condition implies that the dual operators $L(t)'$ in Y' have domains which are independent of t. In the present case, it is seen that

$$A(t)^\circ (\lambda - A(t))^{-1} = \lambda(\lambda M(t) - L(t))^{-1} M(t) - 1$$
$$\supset (\lambda M(t) - L(t))^{-1} L(t),$$
$$\frac{dA(t)^{-1}}{dt} = \frac{dL(t)^{-1}}{dt} M(t) + L(t)^{-1} \frac{dM(t)}{dt}.$$

Hence,

$$A(t)^\circ (\lambda - A(t))^{-1} \frac{dL(t)^{-1}}{dt} M(t) A(t)^\circ (\lambda - A(t))^{-1}$$
$$= \{\lambda(\lambda M(t) - L(t))^{-1} M(t) - 1\} \frac{dL(t)^{-1}}{dt} L(t) (\lambda M(t) - L(t))^{-1} M(t)$$

and

$$A(t)^\circ(\lambda - A(t))^{-1}L(t)^{-1}\frac{dM(t)}{dt}A(t)^\circ(\lambda - A(t))^{-1}$$

$$= (\lambda M(t) - L(t))^{-1}\frac{dM(t)}{dt}\{\lambda(\lambda M(t) - L(t))^{-1}M(t) - 1\}.$$

Therefore, (4.26) together with the condition

(4.27) $$\left\|(\lambda M(t) - L(t))^{-1}\frac{dM(t)}{dt}\right\|_{\mathcal{L}(X)} \le \frac{C}{(|\lambda| + 1)^{\tilde{\nu}}}, \quad \lambda \in \Sigma,$$

where $0 < \tilde{\nu} \le \beta$, yields (A.iii.a) with $\nu = \beta + \tilde{\nu} - 1$. In addition, $M(\cdot) \in \mathcal{C}^{1+\mu}([0,T]; \mathcal{L}(X,Y))$ and $L(\cdot)^{-1} \in \mathcal{C}^{1+\mu}([0,T]; \mathcal{L}(Y,X))$ clearly yields (A.iii.b). Therefore, Theorem 4.12 provides the following result.

PROPOSITION 4.18. *Let*

$$M(\cdot) \in \mathcal{C}^{1+\mu}([0,T]; \mathcal{L}(X,Y)), \quad L(\cdot)^{-1} \in \mathcal{C}^{1+\mu}([0,T]; \mathcal{L}(Y,X)).$$

Let (4.22) (with $\gamma = 0$), (4.26) and (4.27) be satisfied with $\alpha\mu(\alpha + \beta + \tilde{\nu} - 2) > (2 - \beta)(2 - \alpha - \beta)$. Then, for any $f \in \mathcal{C}^\theta([0,T]; X)$, $\frac{2-\alpha-\beta}{\alpha} < \theta < \frac{\mu(\alpha+\beta+\tilde{\nu}-2)}{2-\beta}$, and any $f_0 \in A(0)u_0 = M(0)^{-1}L(0)u_0$ such that

$$f(0) + \left(1 - \frac{dA(t)^{-1}}{dt}\bigg|_{t=0}\right)f_0 \in L(0)^{-1}(\mathcal{R}(M(0))),$$

(D-E.2) possesses a unique solution

$$u \in \mathcal{C}^{1+\omega}([0,T]; X), \quad L(\cdot)u \in \mathcal{C}^\omega([0,T]; Y),$$

where $\omega = \alpha\theta + \alpha + \beta - 2$, such that $u(0) = u_0$ and $\frac{du}{dt}(0) = f(0) + f_0$.

4.4 DEGENERATE EQUATIONS OF THE FORM $\frac{d}{dt}t^\ell v = Lv + f(t)$ AND $t^\ell\frac{du}{dt} = Lu + f(t)$

Consider the initial value problem

(4.28) $$\begin{cases} \dfrac{d}{dt}t^\ell v = Lv + f(t), & 0 \le t \le T, \\ \lim\limits_{t \to 0} t^\ell v(t) = 0, \end{cases}$$

in a Banach space X, where L is a single valued, closed linear operator in X, $\mathcal{D}(L)$ being dense in X, $f:[0,T] \to X$ is a given function, and $0 < \ell < \infty$ is a positive constant.

More precisely, L is assumed to satisfy (3.7) with $M = I$, $\alpha = \beta = 1$, $\gamma = 0$, and f belongs to $C^{\sigma}([0, T]; X)$ with $0 < \sigma \leq 1$.

The techniques will vary according to the cases $\ell > 1$, $\ell = 1$ and $0 < \ell < 1$. Let first $\ell > 1$. Setting $M(t) = t^{\ell}I$ and $L(t) = L$, we apply Proposition 4.15. Indeed, these $M(t)$ and L satisfy (4.14) with $\alpha = \beta = 1$, $\gamma = 0$. Moreover, from

$$\|t^{\ell-1}(\lambda t^{\ell} - L)^{-1}\|_{\mathcal{L}(X)} \leq \frac{Ct^{\ell-1}}{|t^{\ell}\lambda| + 1} \leq \frac{C}{|\lambda|^{1-\frac{1}{\ell}}}, \quad \Re\lambda \geq -c|\Im m\lambda|,$$

where $c > 0$ is some suitable constant, (4.19) is verified with $\tilde{\nu} = 1 - \frac{1}{\ell}$. (4.20) is obvious with $\mu = \min\{\ell - 1, 1\}$. Therefore, Proposition 4.15 yields that (4.28) possesses a unique solution v such that $t^{\ell}v \in C^1([0, T]; X)$, $v \in C([0, T]; \mathcal{D}(L))$. Notice that, since $M(0) = 0I$, the only possibility of taking v_0 is that $v_0 = L^{-1}f(0)$. At $t = 0$, $\lim_{t\to 0} \frac{d}{dt}t^{\ell}v(t) = 0$ and $Lv(0) + f(0) = 0$.

Consider the case $\ell = 1$. We use Theorem 3.19 by setting

$$\begin{cases} \mathcal{D}(\mathcal{T}) = \{v \in C([0, T]; X); \ tv \in C^1([0, T]; X)\}, \\ \mathcal{T}v = \frac{d}{dt}tv, \end{cases}$$

and

$$\begin{cases} \mathcal{D}(\mathcal{A}) = C([0, T]; \mathcal{D}(L)), \\ \mathcal{A}u = Lu, \end{cases}$$

in the underlying space $\mathcal{X} = C([0, T]; X)$. It is easily observed that $-\mathcal{T}$ is the generator of a C_0 semigroup given by

$$\left(e^{-\tau\mathcal{T}}f\right)(t) = e^{-\tau}f\left(e^{-\tau}t\right), \quad \tau \geq 0, \ f \in \mathcal{X},$$

and that its resolvent is given by

$$\left((\lambda - \mathcal{T})^{-1}f\right)(t) = -\int_0^1 \theta^{-\lambda}f(t\theta)d\theta, \quad \Re\lambda < 1,$$

with an estimate

$$\|(\lambda - \mathcal{T})^{-1}f\|_{\mathcal{X}} \leq \frac{1}{1 - \Re\lambda}\|f\|_{\mathcal{X}}, \quad \Re\lambda < 1.$$

So that, $\mathcal{T}$ satisfies $(\mathcal{T})$, see Section 3.5. Since one has:

$$(\mathcal{X}, \mathcal{D}(\mathcal{T}))_{\theta,\infty} = \{f \in \mathcal{X}; \xi^{-\theta}(e^{-\xi\mathcal{T}} - 1)f \in L^{\infty}((0, \infty); \mathcal{X})\}, \quad 0 < \theta < 1,$$

(cf. [172; pp. 76-78], see Proposition 0.7 too), it is observed directly that

$$C^{\theta}([0, T]; X) \subset (\mathcal{X}, \mathcal{D}(\mathcal{T}))_{\theta,\infty}.$$

On the other hand, it is obvious that $\mathcal{A}$ satisfies $(\mathcal{P})$ with $\alpha = \beta = 1$. Therefore, by Theorem 3.19, there exists a unique solution v to $\mathcal{T}v = \mathcal{A}v + f$ for any $f \in C^\sigma([0,T];X)$, $\sigma > 0$. Since v is continuous at $t = 0$, $\lim_{t\to 0} tv(t) = 0$. Hence, (4.28) possesses a unique solution v such that $tv \in C^1([0,T];X)$, $v \in C([0,T];\mathcal{D}(L))$.

In the present case, neither $\lim_{t\to 0} \frac{d}{dt}tv(t) = f(0)$ nor $Lv(0) + f(0) = 0$ in general. Indeed, let $L = -aI$ with a positive constant $a > 0$. Then the solution v is written as

$$v(t) = t^{-a-1}\int_0^t \tau^a f(\tau)d\tau, \quad 0 < t \le T.$$

So that, $\lim_{t\to 0}\frac{d}{dt}tv(t) = \frac{1}{a+1}f(0)$ and $Lv(0) + \frac{a}{a+1}f(0) = 0$.

Let $0 < \ell < 1$. We change the variable to $\tilde{t} = t^{1-\ell}$, $0 \le \tilde{t} \le T^{1-\ell}$ and change the unknown function to $\tilde{u}(\tilde{t}) = t^\ell v(t)$. Then (4.28) is reduced to

$$\begin{cases} \dfrac{d\tilde{u}}{d\tilde{t}} = \frac{1}{1-\ell}L\tilde{u} + \frac{1}{1-\ell}\tilde{t}^{\frac{\ell}{1-\ell}}f(\tilde{t}^{\frac{1}{1-\ell}}), & 0 < \tilde{t} \le T^{1-\ell}, \\ \tilde{u}(0) = 0, \end{cases}$$

which is an usual evolution equation. Since L is the generator of an analytic semigroup, this problem possesses a unique solution of the form

$$\tilde{u}(\tilde{t}) = \frac{1}{1-\ell}\int_0^{\tilde{t}} e^{\frac{\tilde{t}-\tau}{1-\ell}L}\tau^{\frac{\ell}{1-\ell}}f(\tau^{\frac{1}{1-\ell}})d\tau, \quad 0 \le \tilde{t} \le T^{1-\ell},$$

for any $f \in C^\sigma([0,T];X)$, $\sigma > 0$. Moreover, it is verified that $\tilde{t}^{-\frac{\ell}{1-\ell}}L\tilde{u} \in C([0,T];X)$ with

(4.29) $$\lim_{t\to 0} \tilde{t}^{-\frac{\ell}{1-\ell}}L\tilde{u}(\tilde{t}) = 0.$$

Indeed, the essential thing in the proof is that, as $\tilde{t} \to 0$,

$$\tilde{t}^{-\frac{\ell}{1-\ell}}\int_0^{\tilde{t}} Le^{\frac{\tilde{t}-\tau}{1-\ell}L}(\tilde{t}^{\frac{\ell}{1-\ell}} - \tau^{\frac{\ell}{1-\ell}})f d\tau \to 0 \text{ in } X$$

for every $f \in X$ (use the density of $\mathcal{D}(L)$ in X). Note that $\tilde{t}^{-\frac{\ell}{1-\ell}}\frac{d\tilde{u}}{d\tilde{t}} = \frac{1}{1-\ell}\frac{d}{dt}t^\ell v$ and $\tilde{t}^{-\frac{\ell}{1-\ell}}L\tilde{u} = Lv$. In this way, (4.28) possesses a unique solution v such that $t^\ell v \in C^1([0,T];X)$, $v \in C([0,T];\mathcal{D}(L))$. In addition, from (4.29), $\lim_{t\to 0}\frac{d}{dt}t^\ell v(t) = f(0)$ and $Lv(0) = 0$.

The dual problem of (4.28) is written in the form

(4.30) $$\begin{cases} t^\ell\dfrac{du}{dt} = Lu + t^\ell f(t), & 0 < t \le T, \\ u(0) = 0. \end{cases}$$

The change of unknown function to $v(t) = t^{-\ell}u(t)$, however, yields the same problem as (4.28). Therefore, by the above results, (4.30) possesses a unique solution u such that $u \in \mathcal{C}^1([0,T];X)$, $t^{-\ell}u \in \mathcal{C}([0,T];\mathcal{D}(L))$ for any $f \in \mathcal{C}^\sigma([0,T];X)$, $\sigma > 0$.

Let us now consider the more general problem

(4.31)
$$\begin{cases} t^\ell \dfrac{du}{dt} = Lu + f(t), & 0 < t \le T, \\ u(0) = u_0, \end{cases}$$

in X. The arguments again vary depending on $\ell > 1$, $\ell = 1$ and $0 < \ell < 1$.

Let $\ell > 1$. We assume that $f \in \mathcal{C}^{1+\sigma}([0,T];X)$, $\sigma > 0$. If we change the unknown function to $w(t) = u(t) - (t^\ell - L)^{-1}f(t)$, then the problem (4.31) is reduced to

$$\begin{cases} t^\ell \dfrac{dw}{dt} = Lw + t^\ell g(t), & 0 < t \le T, \\ w(0) = u_0 + L^{-1}f(0), \end{cases}$$

where $g(t) = \left(1 - \frac{d}{dt}\right)(t^\ell - L)^{-1}f(t)$. If $Lu_0 + f(0) = 0$, then this is nothing more than (4.30). Hence we obtain the following result. Let $f \in \mathcal{C}^{1+\sigma}([0,T];X)$, $\sigma > 0$, and let $u_0 \in \mathcal{D}(L)$ with $Lu_0 + f(0) = 0$. Then (4.31) possesses a unique solution u such that $u \in \mathcal{C}^1([0,T];X)$, $t^{-\ell}(Lu + f) \in \mathcal{C}([0,T];X)$. Note here that

$$t^{-\ell}Lw(t) = t^{-\ell}\{Lu(t) + f(t)\} - (t^\ell - L)^{-1}f(t).$$

Let $\ell = 1$. In this case we repeat the same kind of argument as for (4.28). Let $\mathcal{X} = \mathcal{C}([0,T];X)$ and set

$$\begin{cases} \mathcal{D}(\mathcal{T}) = \{u \in \mathcal{X};\ u \in \mathcal{C}^1((0,T];X),\ t\frac{du}{dt} \in \mathcal{X}\} \\ \mathcal{T}u = t\frac{du}{dt} + \varepsilon u, \end{cases}$$

and let us introduce the operator $\mathcal{A}$ by means of

$$\begin{cases} \mathcal{D}(\mathcal{A}) = \mathcal{C}([0,T];\mathcal{D}(L)) \\ \mathcal{A}u = Lu + \varepsilon u, \end{cases}$$

where ε is a sufficiently small positive constant. It is seen that $-\mathcal{T}$ is the generator of the $\mathcal{C}_0$ semigroup

$$\left(e^{-\tau\mathcal{T}}f\right)(t) = e^{-\varepsilon\tau}f(e^{-\tau}t), \quad 0 \le t \le T, \tau \ge 0,$$

on $\mathcal{X}$. Since $\|e^{-\tau\mathcal{T}}\|_{\mathcal{L}(\mathcal{X})} \le e^{-\varepsilon\tau}$, $\tau \ge 0$, $\mathcal{T}$ satisfies $(\mathcal{T})$. Moreover, by a direct calculation,

$$\mathcal{C}^\theta([0,T];X) \subset (\mathcal{X},\mathcal{D}(\mathcal{T}))_{\theta,\infty}, \quad 0 < \theta < 1.$$

On the other hand, $\mathcal{A}$ satisfies $(\mathcal{P})$ with $\alpha = \beta = 1$ provided that ε is sufficiently small. Therefore we can apply Theorem 3.19 and conclude that (4.31) possesses a unique solution u such that $u \in \mathcal{C}([0,T];X) \cap \mathcal{C}^1((0,T];X)$, $t\frac{du}{dt} \in \mathcal{C}([0,T];X)$ for any $f \in \mathcal{C}^\sigma([0,T];X)$, $\sigma > 0$. The initial condition in (4.31) is not imposed.

Let $0 < \ell < 1$. Change the variable to $\tilde{t} = t^{1-\ell}$ and change the unknown function to $\tilde{u}(\tilde{t}) = u(\tilde{t}^{\frac{1}{1-\ell}})$. Then (4.31) reduces to

$$
\begin{cases}
\dfrac{d\tilde{u}}{d\tilde{t}} = \frac{1}{1-\ell} L\tilde{u} + \frac{1}{1-\ell} f(\tilde{t}^{\frac{1}{1-\ell}}), & 0 < \tilde{t} \le T^{1-\ell}, \\
\tilde{u}(0) = u_0,
\end{cases}
$$

which is a usual evolution equation. This problem then possesses a unique solution $\tilde{u} \in \mathcal{C}^1([0,T];X) \cap \mathcal{C}([0,T];\mathcal{D}(L))$ for any $u_0 \in \mathcal{D}(L)$ and any $f \in \mathcal{C}^\sigma([0,T];X)$, $\sigma > 0$. Since $\frac{d\tilde{u}}{d\tilde{t}}(\tilde{t}) = \frac{1}{1-\ell} t^\ell \frac{du}{dt}(t)$, this $\tilde{u}$ gives the unique solution u to (4.31) such that $u \in \mathcal{C}([0,T];\mathcal{D}(L)) \cap \mathcal{C}^1((0,T];X)$, $t^\ell \frac{du}{dt} \in \mathcal{C}([0,T];X)$ for any such intial value u_0 and any such given function $f(t)$.

4.5 APPLICATIONS

EXAMPLE 4.1. Consider the initial value problem in a Banach space X

(4.32)
$$
\begin{cases}
\dfrac{dm(t)v}{dt} = Lv + f(t), & 0 < t \le T, \\
\lim_{t \to 0} m(t)v(t) = 0,
\end{cases}
$$

that is a more general version of (4.28); here $m(t) \ge 0$ is a function in $\mathcal{C}^{1+\mu}([0,T];\mathbb{R}_+)$, $0 < \mu \le 1$, such that $m(0) = 0$. It is assumed that the derivative satisfies

(4.33)
$$
\left| \frac{dm}{dt}(t) \right| \le Cm(t)^\nu, \quad 0 \le t \le T,
$$

with some $0 < \nu \le 1$. As before, L is a single valued closed linear operator satisfying (3.7) with $M = I$, $\alpha = \beta = 1$, $\gamma = 0$; $f \in \mathcal{C}^\sigma([0,T];X)$, $\sigma > 0$, is a given function.

It is seen that

$$
\left\| \frac{dm}{dt}(t)(\lambda m(t) - L)^{-1} \right\|_{\mathcal{L}(X)} \le \frac{Cm(t)^\nu}{|m(t)\lambda| + 1} \le \frac{C}{|\lambda|^\nu}, \quad \Re\lambda \ge -c|\Im\lambda|,
$$

where $c > 0$ is some suitable constant. Then, in the same way as for the case $\ell > 1$ in (4.28), it is concluded that (4.32) possesses a unique solution v such that $m(t)v \in \mathcal{C}^1([0,T];X)$, $v \in \mathcal{C}([0,T];\mathcal{D}(L))$. At $t = 0$, $\lim_{t\to 0} \frac{d}{dt} m(t)v(t) = 0$ and $Lv(0) + f(0) = 0$.

EXAMPLE 4.2. Consider similarly the more general problem

(4.34)
$$\begin{cases} m(t)\dfrac{du}{dt} = Lu + f(t), & 0 < t \le T, \\ u(0) = u_0, \end{cases}$$

than (4.31) in a Banach space X, where $m(t) \ge 0$ is the same function as in Example 4.1 satisfying (4.33). The single valued closed linear operator L satisfies (3.7) with $M = I$, $\alpha = \beta = 1$, $\gamma = 0$; f is assumed to be in $\mathcal{C}^{1+\sigma}([0,T];X)$, $\sigma > 0$; u_0 is in $\mathcal{D}(L)$ with the relation $Lu_0 + f(0) = 0$.

As for (4.31) let us change the unknown function to $w(t) = u(t) - (m(t) - L)^{-1} f(t)$. Then (4.34) is reduced to

$$\begin{cases} m(t)\dfrac{dw}{dt} = Lw + m(t)g(t), & 0 < t \le T, \\ w(0) = 0, \end{cases}$$

where $g(t) = (1 - \frac{d}{dt})(m(t) - L)^{-1}f(t)$. Then Proposition 4.18 is now applicable with taking $X = Y$, $M(t) = m(t)I$, $L(t) \equiv L$, $f(t) = g(t)$ and $u_0 = 0$. Indeed (4.22) and (4.27) are easily verified. Since $\frac{dA(t)^{-1}}{dt}|_{t=0} = 0$, the unique choice for f_0 is just $g(0)$. Therefore, (4.34) possesses a unique solution u such that $u \in \mathcal{C}^1([0,T];X) \cap \mathcal{C}([0,T];\mathcal{D}(L))$. At $t = 0$, $\frac{du}{dt}(0) + L^{-1}\frac{df}{dt}(0) = 0$.

EXAMPLE 4.3. Consider the problem

(4.35)
$$\begin{cases} \dfrac{\partial m(x)v}{\partial x} = a(x,t)\Delta v + f(x,t), & (x,t) \in \Omega \times (0,T], \\ v = 0, & (x,t) \in \partial\Omega \times (0,T], \\ m(x)v(x,0) = u_0(x), & x \in \Omega, \end{cases}$$

in a bounded region $\Omega \subset \mathbb{R}^n$ with a smooth boundary $\partial\Omega$. Here, $m(x) \ge 0$ is a given function in $L^\infty(\Omega)$. $a(x,t) \ge \delta$ belongs to $\mathcal{C}^\mu([0,T];\mathcal{C}^1(\overline{\Omega}))$, $0 < \mu \le 1$, δ being a positive constant.

Let us handle the problem in $X = H^{-1}(\Omega)$. Obviously (4.35) is regarded as a problem of the form (D-E.1) in which $M(t) \equiv M$ is the multiplication operator by the function $m(x)$ and $L(t) = a(x,t)\Delta:H_0^1(\Omega) \to H^{-1}(\Omega)$.

Since $(\lambda M - L(t))u = f$ is equivalent to $(\frac{\lambda m(x)}{a(x,t)} - \Delta)u = -\frac{f}{a(x,t)}$, and since $\|af\|_{H^{-1}} \le C\|a\|_{\mathcal{C}^1}\|f\|_{H^{-1}}$, the result obtained in Example 3.3 immediately yields that M and $L(t)$ satisfy (4.14) with $\alpha = \beta = 1$, $\gamma = 0$. On the other hand, since

$$\{L(t) - L(s)\}L(s)^{-1} = \{a(x,t) - a(x,s)\}a(x,s)^{-1}I,$$

the assumption on $a(x,t)$ implies (4.15) with the same exponent μ. We can then apply Proposition 4.13 to obtain that, for any $f \in \mathcal{C}^\sigma([0,T];H^{-1}(\Omega))$, $\sigma > 0$, and any $u_0 = m(x)v_0$ with $v_0 \in H_0^1(\Omega)$, (4.35) possesses a unique solution v such that $Mv \in \mathcal{C}([0,T];H^{-1}(\Omega)) \cap \mathcal{C}^1((0,T];H^{-1}(\Omega))$, $L(t)v \in \mathcal{C}((0,T];H^{-1}(\Omega))$.

EXAMPLE 4.4. Consider the dual problem

$$(4.36) \quad \begin{cases} m(x)\dfrac{\partial u}{\partial t} = a(x,t)\Delta u + m(x)f(x,t), & (x,t) \in \Omega \times (0,T], \\ u = 0, & (x,t) \in \partial\Omega \times (0,T], \\ u(x,0) = u_0(x), & x \in \Omega, \end{cases}$$

of (4.35). In the present problem we assume that $m(x) > 0$ almost everywhere in Ω with $m \in L^\infty(\Omega)$. $a(x,t) \geq \delta > 0$ is assumed to belong to $C^\mu([0,T]; C^1(\overline{\Omega}))$, $0 < \mu \leq 1$, as above.

Taking $X = H_0^1(\Omega)$, $Y = H^{-1}(\Omega)$, let us consider (4.36) as a problem of the form (D-E.2); M and $L(t)$ are the operators as in Example 4.3. Then, by the same argument as in Example 3.4, it is verified that M and $L(t)$ satisfy (4.22) and (4.23) with $\alpha = \beta = 1$, $\gamma = 0$. On the other hand, (4.24) is already verified. Therefore, Proposition 4.16 is applicable and provides that, for any $f \in C^\sigma([0,T]; H_0^1(\Omega))$, $\sigma > 0$, and any $u_0 \in H_0^1(\Omega)$ for which $a(x,0)\Delta u_0 = m(x)v_0$ with $v_0 \in H_0^1(\Omega)$, (4.36) possesses a unique solution u such that $u \in C([0,T]; X_1) \cap C^1((0,T]; X_1)$. Here, X_1 is the space obtained by the completion of $H_0^1(\Omega)$ with respect to the norm $\|m(x)\cdot\|_{H^{-1}}$.

EXAMPLE 4.5. Consider the problem

$$(4.37) \quad \begin{cases} \dfrac{\partial}{\partial t}m(x,t)v = \Delta v + f(x,t), & (x,t) \in \Omega \times (0,T], \\ v = 0, & (x,t) \in \partial\Omega \times (0,T], \\ m(x,0)v(x,0) = u_0(x), & x \in \Omega, \end{cases}$$

in a bounded region $\Omega \subset \mathbb{R}^n$ with a smooth boundary $\partial\Omega$. $m(x,t) \geq 0$ is assumed to have the regularity

$$(4.38) \qquad m \in C([0,T]; L^\infty(\Omega)) \cap C^1([0,T]; L^1(\Omega))$$

and to satisfy

$$(4.39) \qquad \left|\frac{\partial m}{\partial t}(x,t)\right| \leq Cm(x,t)^\nu, \quad 0 \leq t \leq T, \text{ a.e. } x \in \Omega$$

with some $0 < \nu \leq 1$.

Taking $X = H^{-1}(\Omega)$ let us regard (4.37) as a problem of the form (D-E.1) in which $M(t)$ are the multiplication operators by the functions $m(x,t)$ and $L(t) \equiv L$ is Δ: $H_0^1(\Omega) \to H^{-1}(\Omega)$. We shall argue along the line of Proposition 4.14. Let $A(t) = LM(t)^{-1}$, $0 \leq t \leq T$, be m. l. operators. By the result obtained in Example 3.3, L and $M(t)$ satisfy (4.14) with $\alpha = \beta = 1$, $\gamma = 0$; so that, $A(t)$ satisfy Assumption (P) with $\alpha = \beta = 1$. On the other hand, since $H_0^1(\Omega) \subset L^{p_n}(\Omega)$ (by Sobolev embedding theorem), where $p_n = \infty$ if $n = 1$, $p_n < \infty$ is any finite number if $n = 2$ and $p_n = \frac{2n}{n-2}$ if $n \geq 3$, we observe that

$$|\langle A(t)^{-1}\varphi, u\rangle| \leq \|m(t)\|_{L^{q_n}}\|L^{-1}\varphi\|_{L^{p_n}}\|u\|_{L^{p_n}}$$
$$\leq C\|m(t)\|_{L^{q_n}}\|\varphi\|_{H^{-1}}\|u\|_{H^1_0}, \quad \varphi \in H^{-1}(\Omega),\ u \in H^1_0(\Omega),$$

where $q_n = 1$ if $n = 1$, $q_n > 1$ is any number bigger than 1 and $q_n = \frac{n}{2}$ if $n \geq 3$. It then follows that $\|A(t)^{-1}\|_{\mathcal{L}(H^{-1})} \leq C\|m(t)\|_{L^{q_n}}$. While, from

$$\frac{\partial m}{\partial t} \in \mathcal{C}([0,T]; L^1(\Omega)) \cap \mathcal{B}([0,T]; L^\infty(\Omega)) \subset \mathcal{C}([0,T]; L^{q_n}(\Omega)),$$

it is verified that $m \in \mathcal{C}^1([0,T]; L^{q_n}(\Omega))$. Therefore,

$$A(t)^{-1} \in \mathcal{C}^1([0,T]; \mathcal{L}(H^{-1}(\Omega))$$

with $\frac{dA(t)^{-1}}{dt} = \frac{\partial m}{\partial t}L^{-1}$. In addition, Assumption (A.ii) in Theorem 4.5 is verified. Indeed, in view of (4.10),

$$\left|\langle L(\lambda M(t) - L)^{-1}\frac{dM(t)}{dt}L^{-1}\varphi, u\rangle\right| = |(L^{-1}\varphi, \frac{\partial m}{\partial t}(\bar{\lambda}M(t) - L)^{-1}Lu)|$$
$$\leq C\|L^{-1}\varphi\|_{L^2}\|\frac{\partial m}{\partial t}(\bar{\lambda}M(t) - L)^{-1}\|_{\mathcal{L}(H^{-1}, L^2)}\|Lu\|_{H^{-1}}$$
$$\leq C\|\varphi\|_{H^{-1}}\|m(t)^\nu(\bar{\lambda}M(t) - L)^{-1}\|_{\mathcal{L}(H^{-1}, L^2)}\|u\|_{H^1_0},$$
$$\varphi \in H^{-1}(\Omega),\ u \in H^1_0(\Omega).$$

Moreover, recalling (3.41), we observe that

$$\|m(t)^\nu(\bar{\lambda}M(t) - L)^{-1}\|_{\mathcal{L}(H^{-1}, L^2)} \leq \|m(t)(\bar{\lambda}M(t) - L)^{-1}\|^\nu_{\mathcal{L}(H^{-1}, L^2)}$$
$$\times \|(\bar{\lambda}M(t) - L)^{-1}\|^{1-\nu}_{\mathcal{L}(H^{-1}, L^2)} \leq C|\lambda|^{-\frac{\nu}{2}}.$$

Thus,

$$\left\|A(t)^\circ(\lambda - A(t))^{-1}\frac{dA(t)^{-1}}{dt}\right\|_{\mathcal{L}(H^{-1})} \leq \frac{C}{(|\lambda| + 1)^{\frac{\nu}{2}}}, \quad \lambda \in \Sigma.$$

We can now apply Theorem 4.6 and conclude that, for any

$$f \in \mathcal{C}^\sigma([0,T]; H^{-1}(\Omega)),\ \sigma > 0,$$

and any $u_0 = m(x,0)v_0$ with some $v_0 \in H^1_0(\Omega)$, (4.37) possesses a unique solution v such that

$$M(t)v \in \mathcal{C}([0,T]; H^{-1}(\Omega)) \cap \mathcal{C}^1((0,T]; H^{-1}(\Omega)),\ v \in \mathcal{C}((0,T]; H^1_0(\Omega)).$$

EXAMPLE 4.6. Consider the dual problem

(4.40)
$$\begin{cases} m(x,t)\dfrac{\partial u}{\partial t} = \Delta u + m(x,t)f(x,t), & (x,t) \in \Omega \times (0,T], \\ u = 0, & (x,t) \in \partial\Omega \times (0,T], \\ u(x,0) = u_0(x), & x \in \Omega, \end{cases}$$

of (4.37). Here, $m(x,t)$ is the same function satisfying (4.38) and (4.39) as above.

Let $X = H_0^1(\Omega)$, $Y = H^{-1}(\Omega)$. (4.40) is regarded as a problem of the form (D-E.2), $M(t)$ being the multiplication operators by the functions $m(x,t)$ and $L(t) \equiv L$ being $\Delta : H_0^1(\Omega) \to H^{-1}(\Omega)$. As obtained in Example 3.4, $M(t)$ and L satisfy (4.22) with $\alpha = \beta = 1, \gamma = 0$. On the other hand, we have verified in Example 4.5 that $M(t) \in \mathcal{C}^1([0,T]; \mathcal{L}(H_0^1(\Omega), H^{-1}(\Omega))$ with $\frac{dM(t)}{dt} = \frac{\partial m}{\partial t}I$. Moreover, since

$$\left| \langle (\lambda M(t) - L)^{-1}\frac{dM(t)}{dt}u, \varphi \rangle \right| = \left| (u, \frac{dM(t)}{dt}(\lambda M(t) - L)^{-1}\varphi) \right|$$
$$\leq C\|u\|_{L^2}\|\tfrac{\partial m}{\partial t}(\lambda M(t) - L)^{-1}\varphi\|_{L^2}$$
$$\leq C\|u\|_{H_0^1}\|m(t)^\nu(\lambda M(t) - L)^{-1}\|_{\mathcal{L}(H^{-1},L^2)}\|\varphi\|_{H^{-1}},$$
$$u \in H_0^1(\Omega), \varphi \in H^{-1}(\Omega),$$

(4.25) is seen to hold. Therefore, by Proposition 4.17, for any

$$f \in \mathcal{C}^\sigma([0,T]; H_0^1(\Omega)), \sigma > 0,$$

and any $u_0 \in H^2(\Omega) \cap H_0^1(\Omega)$ such that $\Delta u_0 = m(x,0)u_1$ with some $u_1 \in L^2(\Omega)$, (4.40) possesses a unique solution u such that

$$u \in \mathcal{C}([0,T]; H_0^1(\Omega)) \cap \mathcal{C}^1((0,T]; H_0^1(\Omega)) \cap \mathcal{C}((0,T]; H^2(\Omega)).$$

EXAMPLE 4.7. Consider problem (4.37) in an interval $0 < x < 1$. We now want to take $L^2(0,1)$ as the underlying space X.

To this end we assume that $m(x,t) \geq 0$ is a $\mathcal{C}^1$ function on $[0,1] \times [0,T]$ such that

(4.41) $$\left| \frac{\partial}{\partial x}m(x,t) \right| \leq Cm(x,t)^\rho, \quad 0 < \rho \leq 1,$$

(4.42) $$\left| \frac{\partial}{\partial t}m(x,t) \right| \leq Cm(x,t)^\eta, \quad 0 < \eta \leq 1$$

(4.43) $$\left| \frac{\partial}{\partial t}m(x,t) - \frac{\partial}{\partial s}m(x,s) \right| \leq C|t - s|^\varphi, \quad 0 < \varphi \leq 1.$$

Here, L is the operator defined on $X = L^2(0,1)$ by $\mathcal{D}(L) = H_0^1(0,1) \cap H^2(0,1)$, $Lu = u''$. If $M(t)$ denotes the multiplication operator by $m(\cdot,t)$, from $\lambda M(t)u - Lu = f \in X$, $\lambda \in \mathbb{C}$, we deduce, by taking the inner product of f by $M(t)u$,

$$\lambda \int_0^1 m(x,t)^2 |u(x)|^2 dx + \int_0^1 \frac{\partial m}{\partial x}(x,t)u'(x)\overline{u(x)}dx$$
$$+ \int_0^1 m(x,t)|u'(x)|^2 dx = \int_0^1 m(x,t)f(x)\overline{u(x)}dx$$

and hence, by (4.41),

$$\Re\lambda \|M(t)u\|_X^2 + \int_0^1 m(x,t)|u'(x)|^2 dx$$
$$\leq \|f\|_X \|M(t)u\|_X + C'\|M(t)u\|_X^\rho \|u\|_X^{1-\rho} \|u'\|_X,$$
$$|\Im\lambda| \|M(t)u\|_X^2 \leq \|f\|_X \|M(t)u\|_X + C''\|M(t)u\|_X^\rho \|u\|^{1-\rho} \|u'\|_X.$$

However, this time taking the inner product of f and u in X, we see that $\|u'\|_X \leq C\|f\|_X$. Thus there is a positive constant $c_0 > 0$ such that

$$(\Re\lambda + |\Im\lambda| + c_0)\|M(t)u\|_X^2 \leq \|f\|_X \|M(t)u\|_X + C\|M(t)u\|_X^\rho \|f\|_X^{2-\rho}.$$

Take $\Re\lambda + |\Im\lambda| + c_0 \geq \varepsilon_0 > 0$. Then

$$|\Re\lambda| \|M(t)u\|_X^2 \leq \|f\|_X \|M(t)u\|_X + C\|M(t)u\|_X^\rho \|u\|_X^{1-\rho} \|u'\|_X$$

and also

$$|\lambda| \|M(t)u\|_X^{2-\rho} \leq C\|f\|_X^{2-\rho}$$

for any λ in the sector Σ that we have introduced in Example 4.5. We conclude that for each $\lambda \in \Sigma$ one has

$$\|M(t)(\lambda M(t) - L)^{-1}\| \leq C(1 + |\lambda|)^{-1/(2-\rho)}, \quad \lambda \in \Sigma,$$

where $\beta = (2-\rho)^{-1} > \frac{1}{2}$. Further, using moment inequality, we obtain the estimates

$$\left\|\frac{\partial}{\partial t}(\lambda - A(t))^{-1}\right\|_{\mathcal{L}(X)} \leq C(1 + |\lambda|)^{2-2\beta-\eta},$$
$$\left\|\frac{dA(t)^{-1}}{dt} - \frac{dA(s)^{-1}}{ds}\right\|_{\mathcal{L}(X)} \leq C|t-s|^\varphi,$$

where $A(t) = LM(t)^{-1}$. Then Theorem 4.12 permits to conclude that (notice that $\alpha = 1$, $\beta = (2-\rho)^{-1}$, $\nu = 2\beta + \eta - 1$, $\mu = \varphi$) if $\frac{1-\rho}{2-\rho} <$

$\frac{2(\eta+\rho-1)-\rho\eta}{3-2\rho}\varphi$, then for all $f \in C^\theta([0,T]; X)$, $\theta \in (\frac{1-\rho}{2-\rho}, \frac{2(\eta+\rho-1)-\rho\eta}{3-2\rho}\varphi)$, $v_0 \in$ $H^2(0,1) \cap H_0^1(0,1)$ such that

$$f(x,0) + v_0''(x) - \frac{\partial}{\partial t}m(x,0)v_0(x) = m(x,0)v_1(x)$$

with a certain $v_1 \in H^2(0,1) \cap H_0^1(0,1)$, the problem (4.37) in Example 4.5 with the initial function $u_0 = m(x,0)v_0$ has a unique solution v with the property

$$\frac{\partial}{\partial t}m(\cdot,t)u(\cdot,t) \in C^\omega([0,T]; X), \quad \omega = \theta - \frac{1-\rho}{2-\rho}.$$

4.6 THE OPERATOR EQUATION
$\mathcal{T}\mathcal{M}v - \mathcal{L}v = f$: SOME EXTENSIONS

Let $\mathcal{X}$ be a Banach space. Let $\mathcal{T}$, $\mathcal{M}$ and $\mathcal{L}$ be three linear operators in $\mathcal{X}$. $\mathcal{T}$ is assumed to satisfy Condition $(\mathcal{T})$ announced in Section 3.5, $\mathcal{M}$ and $\mathcal{L}$ are similarly assumed to satisfy Condition $(\mathcal{P})$ in Section 3.5. The operators $\mathcal{T}$, $\mathcal{M}$ and $\mathcal{L}$ do not commute in general.

We consider the operational equation

(4.44) $$(\mathcal{T} + k)\mathcal{M}v - \mathcal{L}v = f,$$

where k is a sufficiently large real parameter, but in the applications we have in mind, a change of variable argument leads (4.44) to the case where $k = 0$ and then we obtain the desired properties.

We begin with a uniqueness result. Let Γ be the contour parametrized by $\lambda = a - c(1 + |y|)^\alpha + iy$, $-\infty < y < \infty$, where $a > c > 0$, $a - c < b$, see Assumption $(\mathcal{T})$, p. 64. Denote by $[\mathcal{T}; (\lambda\mathcal{A}^{-1} - I)^{-1}]$ the commutator

$$[\mathcal{T}; (\lambda\mathcal{A}^{-1} - I)^{-1}] = \mathcal{T}(\lambda\mathcal{A}^{-1} - I)^{-1} - (\lambda\mathcal{A}^{-1} - I)^{-1}\mathcal{T}$$

defined on $\mathcal{D}(\mathcal{T})$, where $\mathcal{A} = \mathcal{L}\mathcal{M}^{-1}$. We shall suppose that

(4.45) $\mathcal{D}(\mathcal{T})$ is everywhere dense in $\mathcal{X}$,

(4.46) $[\mathcal{T}; (\lambda\mathcal{A}^{-1} - I)^{-1}]$ has a bounded extension from $\mathcal{X}$ into itself and there are a positive constant C and $0 \leq \sigma < \alpha$ such that

$$\|[\mathcal{T}; (\lambda\mathcal{A}^{-1} - I)^{-1}]\|_{\mathcal{L}(\mathcal{X})} \leq C(1 + |\lambda|)^\sigma, \quad \lambda \in \Gamma.$$

Then, by a quite similar argument as in the proof of Proposition 4.7, we obtain the following theorem. Indeed, the estimate (4.10) played an essential role in the proof of Proposition 4.7, and in this case (4.46) corresponds to (4.10).

THEOREM 4.19. *Under (T) and (P) in Section 3.5, assume (4.45) and (4.46). Then problem (4.44) has at most one solution provided that k is sufficiently large.*

In order to obtain an existence result for (4.44) some more than (4.46) is needed. For sake of simplicity, we shall use $\mathcal{X}_T^\theta = (\mathcal{X}, \mathcal{D}(T))_{\theta,\infty}$, $0 < \theta < 1$. The additional assumption reads

(4.47) $[T; (\lambda \mathcal{A}^{-1} - I)^{-1}]$ has a bounded extension in $\mathcal{L}(\mathcal{X}_T^\theta)$, $0 < \theta < 1$, and (the same notation being used for this extension)

$$\|[T; (\lambda \mathcal{A}^{-1} - I)^{-1}]\|_{\mathcal{L}(\mathcal{X}_T^\theta)} \leq C(1 + |\lambda|)^\varphi, \quad \lambda \in \Gamma,$$

with some $0 < \varphi \leq 1$.

Then, by a quite similar argument as in the proof of Proposition 4.9, the following result is proved. Indeed, the estimate (4.12) was the essential condition in the proof of Proposition 4.9; while, in this case (4.47) is supposed.

THEOREM 4.20. *Under (T) and (P), let us assume (4.45), (4.46) and (4.47), and suppose that $0 \leq \varphi < \alpha$, $0 < \beta \leq \alpha \leq 1$, $2\alpha + \beta > 2$. Let $\frac{2-\alpha-\beta}{\alpha} < \theta < 1$. Then, for sufficiently large k and any $f \in \mathcal{X}_T^\theta$, the equation (4.44) has exactly one solution.*

This theorem generalizes [96,Theorem 6] and, in particular, the pioneering result by G. Da Prato and P. Grisvard [64,Théorème 6.7] relative to $\alpha = \beta = 1$, $\mathcal{M} = I$.

Assumption (4.47) may seem difficult to check. For this reason we establish the following proposition which can be proved by a direct calculation.

PROPOSITION 4.21. *If, in addition to (4.46), $[T; [T; (\lambda \mathcal{A}^{-1} - I)^{-1}]]$ has a bounded extension to $\mathcal{X}$ for all $\lambda \in \Gamma$ with*

$$\|[T; [T; (\lambda \mathcal{A}^{-1} - I)^{-1}]]\|_{\mathcal{L}(\mathcal{X})} \leq C(1 + |\lambda|)^\tau, \quad \lambda \in \Gamma,$$

with some $\tau \geq 1$, then for all $\theta \in (0, 1)$

$$\|[T; (\lambda \mathcal{A}^{-1} - I)^{-1}]\|_{\mathcal{L}(\mathcal{X}_T^\theta)} \leq C(1 + |\lambda|)^{\sigma + \theta(\tau - \sigma)}, \quad \lambda \in \Gamma.$$

PROOF. It is an immediate consequence of the assumptions and the interpolation properties of the mean spaces.

In view of Proposition 4.21 we could take $\sigma - \theta(\tau - \sigma)$ as the constant φ in (4.47). Since $\varphi < \alpha$ must hold, this implies

$$\frac{2-\alpha-\beta}{\alpha} < \frac{\alpha - \sigma}{\tau - \sigma}, \quad 2\alpha + \beta > 2.$$

If these are verified, then Theorem 4.20 gives an existence result for any $f \in \mathcal{X}_T^\theta$, where $\theta \in (\frac{2-\alpha-\beta}{\alpha}, \frac{\alpha-\sigma}{\tau-\sigma})$.

For example, if $\alpha = 1$, the exponent τ is to be chosen so that $(1-\beta)\tau < 1 - \sigma\beta$; then θ varies in the interval $(1 - \beta, (1 - \sigma)(\tau - \sigma)^{-1})$.

As a simple application, we can consider the equation

$$(4.48) \qquad t\frac{d}{dt}M(t)v = L(t)v + f(t), \quad 0 < t \leq T,$$

in a Banach space X.

Here, $M(t)$ and $L(t)$ are two families of closed linear operators acting in the Banach space X such that $\mathcal{D}(L(t)) \subset \mathcal{D}(M(t))$, $M(t)$ and $L(t)$ satisfies (4.14) with $\alpha = 1$, $\mathcal{D}(L(t)) = \mathcal{D}$ is independent of t, and that $L(t)v$ is twice strongly continuously differentiable on $[0, T]$ for all $v \in \mathcal{D}$. $f(t)$ is a function in $\mathcal{C}([0, T]; X)$.

In applying the above results with $\mathcal{X} = \mathcal{C}([0, T]; X)$, the key role is of course played by the operator $\mathcal{T}$ given by

$$\begin{cases} \mathcal{D}(\mathcal{T}) = \{u \in \mathcal{C}([0,T];X) \cap \mathcal{C}^1((0,T];X); \, tu' \in \mathcal{C}([0,T];X)\} \\ \mathcal{T}u = tu'. \end{cases}$$

The operators $\mathcal{M}$ and $\mathcal{L}$ are also defined in a natural way in the space $\mathcal{X}$ by using $M(t)$ and $L(t)$ respectively. As a result, (4.48) can be viewed as an equation of the form (4.44).

In order to apply Theorem 4.20, we need only to verify (4.46) and (4.47). To this end, we observe that the commutator $[\mathcal{T}, (\lambda \mathcal{A}^{-1} - I)^{-1}]$ coincides with $\mathcal{K}$, where

$$(\mathcal{K}u)(t) = t\left\{\frac{\partial}{\partial t}(\lambda A(t)^{-1} - I)^{-1}\right\}u(t) = \left\{\frac{\partial}{\partial t}(\lambda A(t)^{-1} - I)^{-1}\right\}(tu(t))$$

and hence to have (4.46) it is enough to assume that $M(t)f$ is once strongly continuously differentiable on $[0, T]$ for all $f \in \mathcal{D}$ and

$$(4.49) \qquad \left\|\frac{d}{dt}M(t)f\right\|_X \leq Ct^{-1}\|M(t)\|_X, \quad f \in \mathcal{D}.$$

In fact, then

$$\left\{\frac{\partial}{\partial t}(\lambda A(t)^{-1} - I)^{-1}\right\}(tu(t))$$

$$= -\lambda(\lambda A(t)^{-1} - I)^{-1}\frac{dA(t)^{-1}}{dt}(\lambda A(t)^{-1} - I)^{-1}(tu(t))$$

$$= -\lambda(\lambda A(t)^{-1} - I)^{-1}(M'(t)L(t)^{-1}$$

$$\qquad - A(t)^{-1}L'(t)L(t)^{-1})(\lambda A(t)^{-1} - I)^{-1}(tu(t))$$

yields

$$\sup_{0 \le t \le T} \left\| \left\{ \frac{\partial}{\partial t} (\lambda A(t)^{-1} - I)^{-1} \right\} (tu(t)) \right\|_X \le C(1 + |\lambda|)^{2(1-\beta)} \|u\|_X.$$

Thus (4.46) is verified provided that $\sigma = 2(1 - \beta) < 1$, i. e. $\beta > \frac{1}{2}$.

Concerning (4.47), we shall use the Proposition 4.21. In the present case, the commutator $[\mathcal{T}; [\mathcal{T}; (\lambda \mathcal{A}^{-1} - I)^{-1}]]$ coincides with the operator $\mathcal{R}$ given by

$$(\mathcal{R}u)(t) = t \left\{ \frac{\partial^2}{\partial t^2} (\lambda A(t)^{-1} - I)^{-1} \right\} (tu(t)) + \left\{ \frac{\partial}{\partial t} (\lambda A(t)^{-1} - I)^{-1} \right\} (tu(t)).$$

Hence if we add the condition that $M(t)f$ is twice strongly continuously differentiable on $[0, T]$ for all $f \in \mathcal{D}$ and

(4.50) $\qquad t^2 \|M''(t)f\|_X \le C \|M(t)f\|_X, \quad 0 < t \le T, f \in \mathcal{D},$

we obtain

(4.51) $\quad \displaystyle\sup_{0 \le t \le T} t^2 \left\| \left\{ \frac{\partial^2}{\partial t^2} (\lambda A(t)^{-1} - I)^{-1} \right\} f \right\|_X$

$$\le C(1 + |\lambda|)^{4-3\beta} \|f\|_X, \quad f \in X.$$

Indeed, $\frac{\partial^2}{\partial t^2} (\lambda A(t)^{-1} - I)^{-1} = (1) + (2)$, where

$$(1) = -2\lambda \frac{\partial}{\partial t} (\lambda A(t)^{-1} - I)^{-1} \frac{d}{dt} A(t)^{-1} (\lambda A(t)^{-1} - I)^{-1},$$

$$(2) = -\lambda (\lambda A(t)^{-1} - I)^{-1} \frac{d^2}{dt^2} A(t)^{-1} (\lambda A(t)^{-1} - I)^{-1}.$$

Now,

$$(1) = 2\lambda^2 (\lambda A(t)^{-1} - I)^{-1} \left\{ \frac{d}{dt} A(t)^{-1} (\lambda A(t)^{-1} - I)^{-1} \right\}^2$$

and thus assumption (4.49) furnishes

$$\|(1)\|_{\mathcal{L}(X)} \le C t^{-2} (1 + |\lambda|)^{4-3\beta}, \quad \lambda \in \Sigma,$$

because

$$\frac{d}{dt} A(t)^{-1} (\lambda A(t)^{-1} - I)^{-1}$$
$$= M'(t)L(t)^{-1}(\lambda A(t)^{-1} - I)^{-1} - A(t)^{-1}L'(t)L(t)^{-1}(\lambda A(t)^{-1} - I)^{-1}$$

gives

$$\lambda^2(\lambda A(t)^{-1} - I)^{-1}\left\{\frac{d}{dt}A(t)^{-1}(\lambda A(t)^{-1} - I)^{-1}\right\}^2$$

$$= \lambda^2(\lambda A(t)^{-1} - I)^{-1}\{M'(t)L(t)^{-1}(\lambda A(t)^{-1} - I)^{-1}\}^2$$
$$+ \lambda^2(\lambda A(t)^{-1} - I)^{-1}\{A(t)^{-1}L'(t)L(t)^{-1}(\lambda A(t)^{-1} - I)^{-1}\}^2$$
$$- \lambda^2(\lambda A(t)^{-1} - I)^{-1}M'(t)L(t)^{-1}(\lambda A(t)^{-1} - I)^{-1}A(t)^{-1}$$
$$\times L'(t)L(t)^{-1}(\lambda A(t)^{-1} - I)^{-1}$$
$$- \lambda^2(\lambda A(t)^{-1} - I)^{-1}A(t)^{-1}L'(t)L(t)^{-1}(\lambda A(t)^{-1} - I)^{-1}$$
$$\times M(t)'L(t)^{-1}(\lambda A(t)^{-1} - I)^{-1}$$
$$= (i) + (ii) + (iii) + (iv).$$

But for any $f \in X$

$$|\lambda|^2\|\{M'(t)L(t)^{-1}(\lambda A(t)^{-1} - I)^{-1}\}^2 f\|_X$$
$$\leq C|\lambda|^2\|t^{-1}A(t)^{-1}(\lambda A(t)^{-1} - I)^{-1}\|^2_{\mathcal{L}(X)}\|f\|_X$$
$$\leq Ct^{-2}(1 + |\lambda|)^{2(1-\beta)}\|f\|_X$$

implies

$$\|(i)\|_{\mathcal{L}(X)} \leq C_1 t^{-2}(1 + |\lambda|)^{3(1-\beta)}, \quad \lambda \in \Sigma.$$

Since

$$(ii) = \lambda^2(\lambda A(t)^{-1} - I)^{-1}A(t)^{-1}L'(t)L(t)^{-1}$$
$$\times (\lambda A(t)^{-1} - I)^{-1}A(t)^{-1}L'(t)L(t)^{-1}(\lambda A(t)^{-1} - I)^{-1}$$
$$= \{(\lambda A(t)^{-1} - I)^{-1}(\lambda A(t)^{-1} - I + I)L'(t)L(t)^{-1}\}^2(\lambda A(t)^{-1} - I)^{-1},$$

it follows that

$$\|(ii)\|_{\mathcal{L}(X)} \leq C_2(1 + |\lambda|)^{3(1-\beta)}, \quad \lambda \in \Sigma.$$

Moreover, since no commutativity assumption on the operators is made,

$$\|(iii)\|_{\mathcal{L}(X)} \leq C_3 t^{-1}(1 + |\lambda|)^{4-3\beta}, \quad \lambda \in \Sigma,$$
$$\|(iv)\|_{\mathcal{L}(X)} \leq C_4 t^{-1}(1 + |\lambda|)^{3(1-\beta)}, \quad \lambda \in \Sigma.$$

Thus the estimate for (1) is proven.

Let us consider (2). The expression

$$M''(t)L(t)^{-1} - 2M'(t)L(t)^{-1}L'(t)L(t)^{-1}$$
$$+ 2A(t)^{-1}\{L'(t)L(t)^{-1}\}^2 - A(t)^{-1}L''(t)L(t)^{-1}$$

for the second derivative $\frac{d^2}{dt^2}A(t)^{-1}$ of $A(t)^{-1}$ permits to deduce, in view of (4.49) and (4.50), that

$$\|(2)\|_{\mathcal{L}(X)} \leq C\{t^{-2}(1+|\lambda|)^{2(1-\beta)} + t^{-1}(1+|\lambda|)^{3-2\beta} + (1+|\lambda|)^{2(1-\beta)}\},$$
$$\lambda \in \Sigma.$$

Therefore (4.51) has been achieved.

It now follows that Proposition 4.21 applies with $\tau = 4 - 3\beta$, $\sigma = 2(1 - \beta)$, $\beta > \frac{1}{2}$, since then $\mathcal{R} = [\mathcal{T}; [\mathcal{T}; (\lambda\mathcal{A}^{-1} - I)^{-1}]]$ acts continuously in $\mathcal{X}$. Hence we are compelled to take β in the interval $(\frac{5-\sqrt{13}}{2}, 1]$. Since $(\alpha - \sigma)(\tau - \sigma)^{-1} = (2\beta - 1)(2 - \beta)^{-1}$, in view of Theorem 4.20 we have one $\mathcal{C}^1$ solution v to (4.48) provided that $f \in \mathcal{C}^\theta([0, T]; X)$, $1 - \beta < \theta < \frac{2\beta-1}{2-\beta}$.

We must observe that, if $L(t) \equiv L$, then the estimate (4.51) is improved to

$$\sup_{0 \leq t \leq T} t^2 \left\| \left\{ \frac{\partial^2}{\partial t^2} (\lambda A(t)^{-1} - I)^{-1} \right\} f \right\|_X \leq C(1+|\lambda|)^{3(1-\beta)} \|f\|_X, \quad f \in X,$$

and thus $\tau = \max\{1, 3(1 - \beta)\}$.

CHAPTER V

DEGENERATE EQUATIONS: THE GENERAL CASE

Grisvard's operational approach to handle operator equations is suitably modified to solve the equation $T M v - L v = f$ under growth conditions for the resolvent $(\lambda M - L)^{-1}$ that allow application to degenerate differential equations of hyperbolic type and of singular type as well.

Section 5.2 focuses on both the equations $\frac{d}{dt} M v - L v = f(t)$ and $M \frac{du}{dt} - L u = f(t)$, $0 \leq t \leq T$, while in Section 5.3 various partial differential equations are shown to satisfy the hypotheses above.

Section 5.4 is devoted to $T M v - L v = f$ under a Gevrey type assumption for the modified resolvent, a property "intermediate" between parabolicity and hyperbolicity.

Section 5.5 describes a method to solve $\frac{d}{dt} M v - L v = f(t)$, based upon the properties of the bounded linear operator $B = M L^{-1}$ instead of the multivalued linear operator $A = B^{-1}$. The basic roles are played by the direct sum representation $X = \mathcal{N}(B) \oplus \overline{\mathcal{R}(B)}$ and the restriction of B to $\overline{\mathcal{R}(B)}$.

Next Section 5.6 extends this methods to the case where $X = \mathcal{N}(B^k) \oplus \overline{\mathcal{R}(B^k)}$ with k an integer ≥ 1, a. e. when $\lambda = 0$ is a polar singularity for $(\lambda - B)^{-1} = L(\lambda L - M)^{-1}$.

In Section 5.7 there is discussed the differential equation of order $n > 1$

$$A_n \frac{d^n u}{dt^n} + \cdots + A_0 u = f(t),$$

via its reduction to a first order system in a suitable product space.

Section 5.8 is devoted to the integrated solutions to $\frac{d}{dt} M v - L v = f(t)$, while last section contains some results on convergence of solutions v_n to approximate problems $T M_n v - L_n v = f_n$ as $n \to \infty$.

5.1 THE OPERATIONAL EQUATION
$TMv - Lv = f$: THE GENERAL CASE

We shall consider an operational equation of the form

$$(\mathcal{E}) \qquad\qquad TMv - Lv = f$$

in a Banach space $\mathcal{X}$. Here, T is a closed linear operator in $\mathcal{X}$, M and L are two single valued closed linear operators in $\mathcal{X}$ such that $\mathcal{D}(L) \subset \mathcal{D}(M)$, L^{-1} being single valued and bounded on $\mathcal{X}$, f is an element in $\mathcal{X}$. In this section we in fact extend the approach given in Section 3.5 to embrace the situations less favourable arising, a.e., in problems of hyperbolic type or in highly degenerate ordinary differential equations. For sake of simplicity, we avoid to treat the multivalued version $Tu - \mathcal{A}u \ni f$ of $(\mathcal{E})$ and we take $\mathcal{A} = LM^{-1}$ immediately.

We assume:

(T) The resolvent set $\rho(T)$ contains the logarithmic region

$$\Pi = \{\lambda \in \mathbb{C}; \Re\mathrm{e}\lambda \le a + b\log(1 + |\lambda|)\},$$

where $a > 0$, $b \ge 0$, and the resolvent $(\lambda - T)^{-1}$ satisfies

$$(5.1) \qquad\qquad \|(\lambda - T)^{-1}\|_{\mathcal{L}(\mathcal{X})} \le C(1 + |\lambda|)^p, \qquad \lambda \in \Pi,$$

with some exponent $p \ge 0$ and constant $C > 0$.

(M-L) The resolvent set $\rho_M(L)$ contains the region

$$\Lambda = \{\lambda \in \mathbb{C}; \Re\mathrm{e}\lambda \ge \tilde{a} + b\log(1 + |\lambda|)\},$$

where $0 < \tilde{a} < a$, and the M resolvent satisfies

$$(5.2) \qquad\qquad \|L(\lambda M - L)^{-1}\|_{\mathcal{L}(\mathcal{X})} \le C(1 + |\lambda|)^m, \qquad \lambda \in \Lambda,$$

with some integer $m \ge 0$ and a constant $C > 0$.

In addition, $B = -ML^{-1} \in \mathcal{L}(\mathcal{X})$ is assumed to commute with T according to

$$(5.3) \qquad\qquad T^{-1}B = BT^{-1}.$$

Then we can prove the following existence and uniqueness results for solution to $(\mathcal{E})$.

THEOREM 5.1. *Let (T) and (M-L) be satisfied and let* $T^{-1}B = BT^{-1}$. *If k is an integer greater than $m+p+1$, then $(\mathcal{E})$ has at least one solution for any $f \in \mathcal{D}(T^k)$.*

PROOF. From (M-L) it is seen that $\rho_M(L) \supset \Lambda$. Take the integral contour Γ as the boundary of Λ oriented from downstairs to upstairs and define the operator J_k by

$$J_k g = -\tfrac{1}{2\pi i} \int_\Gamma \lambda^{-k}(\lambda M - L)^{-1}(\lambda - T)^{-1}g \, d\lambda, \quad g \in \mathcal{X}.$$

From (5.1) and (5.2) it is easy to see that J_k is a bounded operator on $\mathcal{X}$. In addition, define operators I_j for $j \geq k$ by means of

$$I_j g = -\tfrac{1}{2\pi i} \int_\Gamma \lambda^{-j}(\lambda B + 1)^{-1}(\lambda - T)^{-1}g \, d\lambda, \quad g \in \mathcal{X}.$$

Clearly, $I_j \in \mathcal{L}(\mathcal{X})$. Take $g = T^k f$, $f \in \mathcal{D}(T^k)$ and set $v = J_k T^k f$. Then, repeating the argument that was used to prove Theorem 3.23, it is verified that $v \in \mathcal{D}(M)$ with

$$Mv = -\tfrac{1}{2\pi i} \int_\Gamma \lambda^{-(k+1)}(\lambda B + 1 - 1)(\lambda B + 1)^{-1}(\lambda - T)^{-1}T^k f \, d\lambda$$

$$= BI_k T^k f = T^{-1}f - I_{k+1}T^k f.$$

On the other hand, since $\int_\Gamma \lambda^{-(k+1)}(\lambda B + 1)^{-1}d\lambda = 0$, it follows that $Mv = T^{-1}f - I_k T^{k-1}f$. Since (5.3) yields $BT = TB$ on $\mathcal{D}(T)$, it thus obtained that

$$TMv = f - I_k T^k f = f + LJ_k T^k f = f + Lu.$$

COROLLARY 5.2. *If $f \in \bigcap_{j=1}^{\infty} \mathcal{D}(T^j)$, then the solution $v = J_k T^k f$ obtained in Theorem 5.1 satisfies $Mv \in \bigcap_{j=1}^{\infty} \mathcal{D}(T^j)$.*

COROLLARY 5.3. *Assume (T), (M-L) and $BT^{-1} = T^{-1}B$. In addition, suppose that*

$$(5.4) \qquad (\lambda M - L)^{-1}T^{-1} = T^{-1}(\lambda M - L)^{-1}, \quad \lambda \in \Gamma.$$

Then, for any $f \in \mathcal{D}(T^h)$, $h > m+p+2$, the element given by $u = J_k T^k f$, where k is the same integer as in Theorem 5.1, is a solution to the equation

$$(5.5) \qquad MTu - Lu = f.$$

Moreover, if $f \in \bigcap\limits_{j=1}^{\infty} \mathcal{D}(T^j)$, then $u \in \bigcap\limits_{j=1}^{\infty} \mathcal{D}(T^j)$.

We may observe that, under assumption (5.3), (5.4) reads equivalently $L^{-1}T^{-1} = T^{-1}L^{-1}$.

REMARK. If the operator T is densely defined in $\mathcal{X}$, and hence $\mathcal{D}(T^j)$ is dense in $\mathcal{X}$ for all $j \geq 1$, then for all $f \in \mathcal{X}$ there is a sequence $f_n \in \mathcal{D}(T^k)$ with $k > m+p+1$ such that $f_n \to f$ in $\mathcal{X}$ as $n \to \infty$. Therefore there exist the limits of $T^{-k}Lv_n$ and $T^{-(k-1)}Mv_n$ as $n \to \infty$, where $v_n = J_k T^k f_n$.

In particular, if (5.4) is the case, then for all $f \in \mathcal{X}$ there is a sequence $v_n \in \mathcal{D}(L)$ such that $Mv_n \in \mathcal{D}(T)$, v_n satisfies $TMv_n - Lv_n \to f$ in $\mathcal{X}$ as $n \to \infty$ and $T^{-k}v_n$ has a limit u in $\mathcal{X}$ as $n \to \infty$.

On the other hand, under all the assumptions (T), $(M\text{-}L)$, (5.3) and (5.4), the equation

$$TMw_k - Lw_k = T^{-k}f$$

has a solution w_k and it is easily verified that, if $TMw_k \in \mathcal{D}(T^k)$ (or equivalently $Lw_k \in \mathcal{D}(T^k)$), then $T^k w_k$ in fact solves $(\mathcal{E})$. We could say that w_k is a k-integrated (with respect to T) solution to $(\mathcal{E})$. Theorem 5.1 then yields that for all $f \in \mathcal{X}$ the equation $(\mathcal{E})$ possesses a k-integrated solution provided that $k > m + p + 1$. This generalizes the well known notion of k-integrated solutions to Cauchy problem, see H. Kellermann and M. Hieber [119], and we shall devote a special paragraph (Section 5.8) just to treat integrated solutions to degenerate initial value problems. An analogous argument applies to (5.5) also. But in this case we need to assume $k > m + p + 2$.

Next we will examine what happens when one changes the role of the operators T and B. Assuming for a moment that $B = -ML^{-1}$ has a single valued inverse, we set $u = Mv$. Then, $(\mathcal{E})$ reads $Tu + B^{-1}u = f$. Therefore we can apply the existence result just obtained. If we apply the existence result described above, we get accordingly

$$u = -\frac{1}{2\pi i}\int_{\Gamma_1} \lambda^{-k}(\lambda - T)^{-1}(\lambda + B^{-1})^{-1}(-B^{-1})^k(-f)d\lambda$$

$$= (-1)^k \frac{1}{2\pi i}\int_{\Gamma_1} \lambda^{-k}B(T - \lambda)^{-1}(\lambda B + 1)^{-1}B^{-k}f d\lambda,$$

where k is a certain positive integer and Γ_1 is a suitable unbounded contour in the complex plane to be specified below. What we have to do is to give some sense to $B^{-k}f$. If $f \in \mathcal{R}(B^k)$, we can hope that the solution v is given by $v = S_k g$, where $f = B^k g$ and

$$S_k g = (-1)^k \frac{1}{2\pi i}\int_{\Gamma_1} \lambda^{-k}L^{-1}(T - \lambda)^{-1}(\lambda B + 1)^{-1}g d\lambda.$$

In this way we are led to prove an alternative existence result which may be viewed as the *"space"* counterpart to Theorem 5.1.

THEOREM 5.4. *Let (T) and (M-L) be satisfied and let $T^{-1}B = BT^{-1}$. For any $f \in \mathcal{R}(B^k)$, where k is an integer greater than $m + p + 1$, there exists a solution v to $(\mathcal{E})$.*

PROOF. Let us first define the contour Γ_1 appearing in the definition of S_k. Γ_1 is indeed determined by composing $\Re e \lambda = \tilde{a} + b \log(1+|\lambda|)$, $|\lambda| \geq a_0$ and $\lambda = a_0 e^{i\phi}$, $|\lambda| \leq a_0$, where $\phi_0 \leq \phi \leq 2\pi - \phi_0$ with

$$\phi_0 = \arccos\left(\frac{\tilde{a} + b\log(1 + a_0)}{a_0}\right)$$

and a_0 is a suitable constant so that both $(\lambda - T)^{-1}$ and $(\lambda B + 1)^{-1}$ exist in $\mathcal{L}(\mathcal{X})$ for every $\lambda \in \Gamma_1$ (remember that L^{-1} is bounded). At last, Γ_1 is oriented for $\Im m \lambda$ increasing along Γ_1.

Let $v = S_k g$ with $f = B^k g$. Then, v satisfies $(\mathcal{E})$. Indeed,

$$Mv = (-1)^{k+1} \frac{1}{2\pi i} \int_{\Gamma_1} \lambda^{-(k+1)} (T - \lambda)^{-1} (\lambda B + 1)^{-1} g d\lambda.$$

So that, $Mv \in \mathcal{D}(T)$ and

$$TMv = (-1)^{k+1} \frac{1}{2\pi i} \int_{\Gamma_1} \lambda^{-(k+1)}(\lambda B + 1)^{-1} g d\lambda + Lv = B^k g + Lv = f + Lv,$$

because

$$\frac{1}{2\pi i} \int_{\Gamma_\varepsilon} \lambda^{-(k+1)}(\lambda B + 1)^{-1} g d\lambda$$

$$= -\frac{1}{2\pi i} \int_{\Gamma_1} \lambda^{-(k+1)}(\lambda B + 1)^{-1} g d\lambda = (-1)^{k+1} B^k g,$$

where $\Gamma_\varepsilon : |\lambda| = \varepsilon$ is counterclockwise oriented.

REMARK. If $f \in \overline{\mathcal{R}(B^k)}$, then f is a limit of $f_n = B^k g_n$ as $n \to \infty$. Let v_n be the solution to $TMv_n - Lv_n = f_n$ which is given by $v_n = S_k g_n$. This means that there exists $v_n \in \mathcal{D}(L)$ such that $TMv_n - Lv_n \to f$ in $\mathcal{X}$. Moreover, from

$$B^k Lv_n = (-1)^k \frac{1}{2\pi i} \int_{\Gamma_1} \lambda^{-k}(T - \lambda)^{-1}(\lambda B + 1)^{-1} f_n d\lambda,$$

$B^k Lv_n = -B^{k-1} Mv_n$ has a limit in $\mathcal{X}$ as $n \to \infty$. Theorem 5.4 implies also that, if $f \in \bigcap_{j=1}^{\infty} \mathcal{R}(B^j)$, then $Lv \in \bigcap_{j=1}^{\infty} \mathcal{R}(B^j)$, for $f = B^j f_j$ yields $f = B^k(B^{j-k} f_j)$ and hence $Lv = B^{j-k} LS_k f_j \in \mathcal{R}(B^{j-k})$. Notice that LS_k and B commute in view of (5.3).

Let us now prove uniqueness of the solution to $(\mathcal{E})$.

THEOREM 5.5. *Let (T) and $(M\text{-}L)$ be satisfied and let $T^{-1}B = BT^{-1}$.*
Then $(\mathcal{E})$ has at most one solution.

PROOF. Let $w = -Lv$. All reduces to verify that $TBw + w = 0$ implies
$w = 0$. Let Γ_1 be the same integral contour used in the proof of Theorem
5.4 and choose a sufficiently large integer k. We have:

$$0 = \tfrac{1}{2\pi i} \int_{\Gamma_1} \lambda^{-k}(\lambda B + 1)^{-1}(T - \lambda)^{-1}(TBw + w)d\lambda$$

$$= \tfrac{1}{2\pi i} \int_{\Gamma_1} \lambda^{-k}T(T - \lambda)^{-1}(\lambda B + 1)^{-1}Bwd\lambda$$

$$\qquad\qquad + \tfrac{1}{2\pi i} \int_{\Gamma_1} \lambda^{-k}(\lambda B + 1)^{-1}(T - \lambda)^{-1}wd\lambda$$

$$= (i) + (ii).$$

Since $Bw = \lambda^{-1}(\lambda Bw + w - w)$, the first integral reduces to

$$(i) = -\tfrac{1}{2\pi i} \int_{\Gamma_1} \lambda^{-(k+1)}T(T - \lambda)^{-1}(\lambda B + 1)^{-1}wd\lambda$$

$$= -\tfrac{1}{2\pi i} \int_{\Gamma_1} \lambda^{-(k+1)}(\lambda B + 1)^{-1}wd\lambda - (ii) = (-1)^k B^k w - (ii).$$

Therefore, $0 = (i) + (ii) = (-1)^k B^k w$ and thus $B^k w = 0$. Since $TBw + w = 0$, it follows that $B^{k-1}w = 0$. Repeating this step by step, we conclude
that $w = 0$.

COROLLARY 5.6. *Assume (T), $(M\text{-}L)$, (5.3) and (5.4). Then the solu-*
tion u to the equation (5.5) is unique.

PROOF. Let $MTu - Lu = 0$. Then $w = -Lu$ satisfies $BLTL^{-1}w + w = 0$. On the other hand, by (5.4), $LTL^{-1}f = g$ is equivalent to $Tf = g$. So
that, $BTw + w = 0$; applying T to this equation, $TB(Tw) + (Tw) = 0$.
Theorem 5.5 then yields that $Tw = 0$, hence $w = 0$.

5.2 THE EQUATIONS $\frac{dMv}{dt} - Lv = f(t)$ AND $M\frac{du}{dt} - Lu = f(t)$: THE GENERAL CASE

Let us apply the results obtained in the preceding section to the Cauchy
problem for the degenerate equation

(D-E.1)
$$\begin{cases} \dfrac{dMv}{dt} - Lv = f(t), & 0 \le t \le T, \\ Mv(0) = Mv_0, \end{cases}$$

in a Banach space X. Here, M and L are closed linear operators in X
with $\mathcal{D}(L) \subset \mathcal{D}(M)$, L^{-1} being single valued and bounded on X. $f \in \mathcal{C}([0, T]; X)$ is a given function, $v_0 \in \mathcal{D}(L)$ is an initial value.

M and L are assumed to satisfy the condition (M-L) in the space X in some logarithmic region Λ ($\tilde{a} > 0$, $b \geq 0$) and with some integer $m \geq 0$.

First, the change of the unknown function to $w = -Lv$ transforms (D-E.1) into the problem

$$\begin{cases} \dfrac{d}{dt}Bw + w = f(t), & 0 \leq t \leq T, \\ Bw(0) = Bw_0, \end{cases}$$

where $B = -ML^{-1} \in \mathcal{L}(X)$ and $w_0 = -Lv_0$. Define furthermore

$$z(t) = w(t) - \sum_{j=0}^{r} \frac{t^j}{j!}w_j,$$

where the r elements $w_1, ..., w_r \in X$ shall be fixed later on. Then, the last problem reads

(5.6)
$$\begin{cases} \dfrac{d}{dt}Bz(t) + z(t) = h(t), & 0 \leq t \leq T, \\ Bz(0) = 0, \end{cases}$$

where $h(t) = f(t) - \displaystyle\sum_{j=0}^{r-1} \frac{t^j}{j!}(Bw_{j+1} + w_j) - \frac{t^r}{r!}w_r$.

This shows that we can use Theorems 5.1, 5.4 and 5.5 by taking the underlying space $\mathcal{X} = \mathcal{C}([0,T]; X)$ endowed with the usual supremum norm. Let $\mathcal{M}$ and $\mathcal{L}$ be the linear operators in $\mathcal{X}$ defined by $(\mathcal{M}v)(t) = Mv(t)$, $0 \leq t \leq T$, with $\mathcal{D}(\mathcal{M}) = \mathcal{C}([0,T]; \mathcal{D}(M))$ and by $(\mathcal{L}v)(t) = Lv(t)$, $0 \leq t \leq T$, with $\mathcal{D}(\mathcal{L}) = \mathcal{C}([0,T]; \mathcal{D}(L))$ respectively. For sake of brevity we shall however identify $\mathcal{M}$ to M and $\mathcal{L}$ to L, and we shall use M and L instead of $\mathcal{M}$ and $\mathcal{L}$, respectively.

THEOREM 5.7. *Let (M-L) be satisfied and let k be an integer such that $k > m + bT + 1$. If $f \in C^k([0,T]; X)$ and if v_0 satisfies the relation*

$$v_0 = \sum_{j=0}^{k-1}(-1)^{j+1}L^{-1}B^j f^{(j)}(0) + v_1$$

(5.7)
$$= -\sum_{j=0}^{k-1}L^{-1}(ML^{-1})^j f^{(j)}(0) + v_1,$$

with $v_1 \in \mathcal{R}(L^{-1}B^k)$, then (D-E.1) possesses a unique solution v such that $Mv \in C^1([0,T]; X)$, $Lv \in C([0,T]; X)$.

PROOF. In order to apply Theorem 5.1 to (5.6) we need to specify the operator $\mathcal{T}$ and to verify Condition $(\mathcal{T})$. We choose

$$
\begin{cases}
\mathcal{D}(\mathcal{T}) = \{u \in C^1([0,T];X); u(0) = 0\} \\
\mathcal{T}u = \dfrac{du}{dt}.
\end{cases}
$$

Then it is easily seen that $(\mathcal{T})$ is satisfied with $a > \tilde{a}$, $b \geq 0$ and (5.1) holds with $p = bT$. Take $r = k$ in the definition of $z(t)$. Then, Theorems 5.1 and 5.5 conclude that (5.6), and hence (D-E.1), possesses a unique solution provided that $f^{(j)}(0) - (Bw_{j+1} + w_j) = 0$ for $j = 0, 1, \ldots, k-1$. But these conditions are written as

$$
w_{k-j} = \sum_{i=0}^{j-1}(-1)^i B^i f^{(k-j+1)}(0) + (-1)^j B^j \tilde{w}, \quad j = 1, \ldots, k;
$$

here w_k is denoted by $\tilde{w}$. In particular,

$$
w_0 = \sum_{j=0}^{k-1}(-1)^j B^j f^{(j)}(0) + (-1)^k B^k \tilde{w},
$$

or

$$
Mv_0 = \sum_{j=0}^{k-1}(-1)^j B^{j+1} f^{(j)}(0) + w, \quad w \in \mathcal{R}(B^{k+1}),
$$

which is nothing more than (5.7).

Obviously, the best situation takes place when $b = 0$, that is, when the m. l. operator $A = LM^{-1}$ generates an integrated semigroup, as we could say in analogy with the case of single valued linear operators.

COROLLARY 5.8. *Let (M-L) be satisfied with $b = 0$ and let M be one-to-one. Then, if $f \in C^{m+2}([0,T];X)$ and if v_0 satisfies the relation*

$$
f^{(j)}(0) + w_j \in \mathcal{D}(LM^{-1}), \quad j = 0, 1, ..., m+1,
$$

with $w_0 = Lv_0$ and $w_j = LM^{-1}(f^{(j-1)}(0) + w_{j-1})$, $j = 1, 2, ..., m+2$, then (D-E.1) possesses a unique solution v.

Theorems 5.4 and 5.5 then yield the following result.

THEOREM 5.9. *Let (M-L) be satisfied and let k be an integer such that $k > m + bT + 1$. If $f = B^k g$ with some $g \in C([0,T];X)$ and if v_0 satisfies*

$Mv_0 = B^{k+1}f_0$ with some $f_0 \in X$, then (D-E.1) possesses a unique solution v such that $Mv \in C^1([0,T];X)$, $Lv \in C([0,T];X)$.

PROOF. The change of the unknown function to $w(t) = Lv_0 - Lv(t)$ yields that

$$\begin{cases} \dfrac{d}{dt}Bw + w = h(t), & 0 \le t \le T, \\ Bw(0) = 0, \end{cases}$$

where $h(t) = f(t) + Lv_0$. Then the results are obtained from Theorems 5.4 and 5.5 in the same manner as in the proof of Theorem 5.7.

Let us now consider the dual problem

(D-E.2′)
$$\begin{cases} M\dfrac{du}{dt} - Lu = f(t), & 0 \le t \le T, \\ u(0) = u_0, \end{cases}$$

of (D-E.1), where M and L are the same linear operators as in (D-E.1), $f \in C([0,T];X)$ is a given function, and $u_0 \in \mathcal{D}(L)$ is an initial value. (M-L), (5.3) and (5.4) are assumed to hold. Changing the unknown function to

$$z(t) = u(t) - \sum_{j=0}^{r} \frac{t^j}{j!}u_j,$$

with some suitable r elements $u_1, ..., u_r \in X$ yields that

(5.8)
$$\begin{cases} M\dfrac{dz}{dt} - Lz = h(t), & 0 < t \le T, \\ z(0) = 0, \end{cases}$$

where $h(t) = f(t) + \sum_{j=0}^{r-1} \dfrac{t^j}{j!}(Lu_j - Mu_{j+1}) + \dfrac{t^r}{r!}Lu_r$.

We then apply Corollary 5.3 and Corollary 5.6 to (5.8) by taking $\mathcal{X} = C([0,T];X)$ and $\mathcal{T} = \frac{d}{dt}$ in the same way as for (D-E.1). It is to be observed that assumptions (5.3) and (5.4) are obviously satisfied even if $\mathcal{D}(\mathcal{T})$ is not dense in this case. Therefore we have the following results for (D-E.2′).

THEOREM 5.10. Assume (M-L), (5.3) and (5.4) to hold in the space X. If $f \in C^h([0,T];X)$, where h is an integer such that $h > m + bT + 2$, and if u_0 satisfies the compatibility relation

$$Lu_0 + \sum_{j=0}^{h-1}(ML^{-1})^j f^{(j)}(0) \in \mathcal{R}((ML^{-1})^h),$$

then (D-E.2′) possesses a unique solution u such that $u \in C^1([0,T];\mathcal{D}(M))$, $Lu \in C([0,T];X)$.

THEOREM 5.11. *Under the same assumptions on L and M as in Theorem 5.10, let h be an integer $> m + bT + 2$. If $f = B^h g$ with some $g \in \mathcal{C}([0,T]; X)$, and if $u_0 \in \mathcal{D}(L)$ satisfies $L u_0 \in \mathcal{R}(B^h)$, then (D-E.$\mathcal{Z}$) possesses a unique solution u such that $u \in \mathcal{C}^1([0,T]; \mathcal{D}(M))$, $Lu \in \mathcal{C}([0,T]; X)$.*

5.3 APPLICATIONS

In this section we list a number of cases where our assumptions on operators L, M are satisfied. Some of them concern again abstract models inspired by concrete partial differential equations. Others take into consideration important classes of differential operators.

EXAMPLE 5.1. Let L be a densely defined closed linear operator acting in a Hilbert space X. Assume that L and its adjoint L^* satisfy

$$\Re e(Lu, u)_X \leq -c\|u\|_X^2, \quad u \in \mathcal{D}(L),$$
$$\Re e(L^* w, w)_X \leq -c\|w\|_X^2, \quad w \in \mathcal{D}(L^*),$$

with some constant $c > 0$. Let M be a non negative bounded self adjoint operator on X. As it is well known, $(\lambda M - L)^* = \bar{\lambda} M - L^*$ for all $\Re e \lambda \geq 0$. Since

$$\Re e((\lambda M - L)u, u)_X \geq c\|u\|_X^2, \quad u \in \mathcal{D}(L), \ \Re e \lambda \geq 0,$$

$\lambda M - L$ has a closed range and is one to one for all $\Re e \lambda \geq 0$. Similarly, $\mathcal{R}(\lambda M - L)^\perp = \mathcal{N}(\bar{\lambda} M - L^*) = \{0\}$ for all $\Re e \lambda \geq 0$. Therefore, $\|(\lambda M - L)^{-1}\|_{\mathcal{L}(X)} \leq c^{-1}$. This means that M and L fulfil (M-L) with $b = 0$ and $m = 1$.

EXAMPLE 5.2. Let $-L$ be a positive self adjoint operator acting in a Hilbert space X. Let M be a densely defined closed linear operator in X with $\mathcal{D}([-L]^{\frac{1}{2}}) \subset \mathcal{D}(M) \cap \mathcal{D}(M^*)$. Assume that M satisfies

$$\Re e(Mu, u)_X \geq 0, \quad u \in \mathcal{D}(M),$$
$$\Re e(M^* w, w)_X \geq 0, \quad w \in \mathcal{D}(M^*).$$

For $\Re e \lambda > 0$, let $\lambda M u - L u = f$. Considering its inner product by $v = \lambda u$, we have:

$$(Mv, v)_X + \frac{\bar{\lambda}}{|\lambda|^2} \|[-L]^{\frac{1}{2}} v\|_X^2 = (f, v)_X.$$

So that,

$$\frac{\Re e \lambda}{|\lambda|^2} \|[-L]^{\frac{1}{2}} v\|_X^2 \leq \|f\|_X \|v\|_X \text{ or } \frac{\Re e \lambda}{|\lambda|} \|[-L]^{\frac{1}{2}} u\|_X^2 \leq \|f\|_X \|u\|_X.$$

Since $-L$ is positive, it follows that

$$\frac{\Re e \lambda}{|\lambda|}\|u\|_X \le C\|f\|_X.$$

On the other hand, it is verified from $\mathcal{D}([-L]^{\frac{1}{2}}) \subset \mathcal{D}(M)$ that

$$\|Mu\|_X \le C(1+|\lambda|)\|(\lambda M - L)u\|_X, \quad u \in \mathcal{D}(L), \ \Re e\lambda \ge \tilde{a},$$

(5.9) $\quad \|Lu\|_X \le C(1+|\lambda|^2)\|(\lambda M - L)u\|_X, \quad u \in \mathcal{D}(L), \ \Re e\lambda \ge \tilde{a},$

with some $\tilde{a} > 0$. In addition, $\mathcal{D}([-L]^{\frac{1}{2}}) \subset \mathcal{D}(M) \cap \mathcal{D}(M^*)$ implies that M and M^* are L-bounded with L-bound 0. Therefore, it is concluded that $\lambda M - L$ is a closed linear operator with $(\lambda M - L)^* = \overline{\lambda}M^* - L$. Repeating the same argument as for (5.9), we observe that $\overline{\lambda}M^* - L$ is injective, so that $\lambda M - L$ has a dense range. In this way (M-L) is fulfilled with $\tilde{a} > 0$, $b = 0$ and $m = 2$.

Notice that this is a typical hyperbolic situation quite different from the one described shortly below. See (5.10).

EXAMPLE 5.3. In Example 5.2, assume that M is a self adjoint operator too, that is let M be a non negative self adjoint operator. In this case one easily recognizes that

$$\|[-L]^{\frac{1}{2}}u\|_X \le C\|(\lambda M - L)u\|_X, \quad u \in \mathcal{D}(L), \ \Re e\lambda > 0,$$
$$\|Mu\|_X \le C\|(\lambda M - L)u\|_X, \quad u \in \mathcal{D}(L), \ \Re e\lambda > 0.$$

This implies that the exponent m can be improved to $m = 1$.

We here remark that, if the (appearently) a bit stronger condition

(5.10) $\qquad\qquad \mathcal{D}([-L]^{\frac{1}{2}}) \subset \mathcal{D}(M^{1+s}), \ s > 0,$

is assumed, then essentially better estimates are obtained implying parabolicity. In fact, it follows from $(\lambda M - L)u = f$ that

$$\lambda\|M^{\frac{1}{2}}u\|_X^2 + \|[-L]^{\frac{1}{2}}u\|_X^2 = (f, u)_X.$$

Hence, for $\Re e\lambda + |\Im m\lambda| \ge \varepsilon_0 > 0$,

$$(\Re e\lambda + |\Im m\lambda|)\|M^{\frac{1}{2}}u\|_X^2 + \|[-L]^{\frac{1}{2}}u\|_X^2$$
$$\le C\|f\|_X\|u\|_X \le \varepsilon\|[-L]^{\frac{1}{2}}u\|_X^2 + C\varepsilon^{-1}\|f\|_X^2,$$

(remember that $-L$ is positive). Therefore,

$$(|\lambda| + 1)\|M^{\frac{1}{2}}u\|_X^2 + \|[-L]^{\frac{1}{2}}u\|_X^2 \le C\|f\|_X^2.$$

This together with (5.10) then yields that

$$\|M^{\frac{1}{2}}(\lambda M - L)^{-1}f\|_X \le C(|\lambda| + 1)^{-\frac{1}{2}}\|f\|_X,$$
$$\|M^{1+s}(\lambda M - L)^{-1}f\|_X \le C\|f\|_X.$$

Using the moment inequality, we thus conclude that

$$\|M(\lambda M - L)^{-1}f\|_X$$
$$\le C(|\lambda| + 1)^{-\frac{s}{1+2s}}\|f\|_X, \quad f \in X, \ \Re\lambda + |\Im\lambda| \ge \varepsilon_0 > 0.$$

This means that M and L satisfy the condition (3.7) in Chapter III.

EXAMPLE 5.4. Let M and L be densely defined closed linear operators in a Hilbert space X with $\mathcal{D}(L) \subset \mathcal{D}(M)$, L^{-1} being single valued and bounded on X. Assume that M (resp. M^*) is L-bounded (L^*-bounded) with L-bound (resp. L^*-bound) 0. In addition, assume that

$$\Re(Lu, Mu)_X \le 0, \quad u \in \mathcal{D}(L),$$
$$\Re(L^*w, M^*w)_X \le 0, \quad w \in \mathcal{D}(L^*).$$

For $\Re\lambda > 0$, consider the inner product of $(\lambda M - L)u = f$ and Lv, where $v = -\lambda u$. Then,

$$-\Re(Mv, Lv)_X + \frac{\Re\lambda}{|\lambda|^2}\|Lv\|_X^2 = -\Re(f, Lv)_X.$$

So that, $\dfrac{\Re\lambda}{|\lambda|^2}\|Lv\|_X \le \|f\|_X$, and hence

$$\|Lu\|_X \le \frac{|\lambda|}{\Re\lambda}\|(\lambda M - L)u\|_X, \quad u \in \mathcal{D}(L), \ \Re\lambda > 0.$$

As before, the hypotheses of relative L-boundedness and L^*-boundedness yield that $\lambda M - L$ is closed with $(\lambda M - L)^* = \bar{\lambda}M^* - L^*$. Moreover, the same estimate holds for $\bar{\lambda}M^* - L^*$ also. Therefore, we conclude that (M-L) is fulfilled with $b = 0$ and $m = 1$.

EXAMPLE 5.5. Let Ω be a bounded domain in $\mathbb{R}^n$ with a smooth boundary $\partial\Omega$. Let $L = \Delta : H^2(\Omega) \cap H_0^1(\Omega) \to L^2(\Omega)$. Let M be the multiplication operator by a positive function $m(x)$. When $n \ge 3$, we assume that $m \in L^n(\Omega)$. Then, since $\|mu\|_{L^2}^2 \le \|m\|_{L^n}^2\|u\|_{L^{\frac{2n}{n-2}}}^2$, $\mathcal{D}(M) = L^{\frac{2n}{n-2}}(\Omega)$. On the other hand, since $H_0^1(\Omega) \subset L^{\frac{2n}{n-2}}(\Omega)$ (by the Sobolev embedding theorem), it follows that $\mathcal{D}([-L]^{\frac{1}{2}}) \subset \mathcal{D}(M)$. Hence, the assumptions of Example 5.3 are valid. When $n = 2$, we assume that $m \in L^q(\Omega)$ with some

$q > 2$. Then, since $H_0^1(\Omega) \subset L^p(\Omega)$ for any finite $1 \leq p < \infty$, it similarly follows that M and L fulfil the assumptions of Example 5.3. In the case where $\Omega = (a, b)$ $(n = 1)$, we assume the different condition that

$$\int_a^b m(x)^2(b - x)(x - a)dx < \infty.$$

Repeating an anologous argument as in Example 3.8, it is again verified that $\mathcal{D}([-L]^{\frac{1}{2}}) \subset \mathcal{D}(M)$ and that Example 5.3 is applicable.

Clearly these facts permit us to handle the Cauchy problem for degenerate equations

$$\frac{\partial v}{\partial t} = \Delta\{a(x)v\} + f(x, t), \quad (x, t) \in \Omega \times (0, T),$$

and

$$\frac{\partial u}{\partial t} = a(x)\Delta u + f(x, t), \quad (x, t) \in \Omega \times (0, T),$$

where $a(x) = \frac{1}{m(x)}$.

EXAMPLE 5.6. Let $G(t, s)$ be a continuous function from $[0, 1]^2$ to $\mathcal{L}(\mathbb{R}^n)$ such that $G(t, s)$ is symmetric: $G(t, s) = G(s, t)^*$, $0 \leq s, t \leq T$, and is non negative in the sense that

$$\int_0^1 \int_0^1 (\varphi(t), G(t, s)\varphi(s))_{\mathbb{R}^n} ds dt \geq 0, \quad \forall \varphi \in L^2([0, 1]; \mathbb{R}^n).$$

Then notice that, since

$$\int_0^1 \int_0^1 (\varphi_2(t), G(t, s)\varphi_1(s))_{\mathbb{R}^n} ds dt$$

$$= \int_0^1 \int_0^1 (\varphi_1(t), G(t, s)\varphi_2(s))_{\mathbb{R}^n} ds dt, \quad \varphi_1, \varphi_2 \in L^2([0, 1]; \mathbb{R}^n),$$

we have:

$$\int_0^1 \int_0^1 (\varphi(t), G(t, s)\varphi(s))_{\mathbb{C}^n} ds dt = \int_0^1 \int_0^1 (\varphi_1(t), G(t, s)\varphi_1(s))_{\mathbb{R}^n} ds dt$$

$$+ \int_0^1 \int_0^1 (\varphi_2(t), G(t, s)\varphi_2(s))_{\mathbb{R}^n} ds dt \geq 0$$

for any $\varphi \in L^2([0, 1]; \mathbb{C}^n)$ such that $\varphi = \varphi_1 + i\varphi_2$, $\varphi_1, \varphi_2 \in L^2([0, 1]; \mathbb{R}^n)$. Let $X = L^2([0, 1]; \mathbb{C}^n)$. $G(t, s)$ then defines a non negative self adjoint operator M from X into itself given by

$$(M\varphi)(t) = \int_0^1 G(t, s)\varphi(s)ds, \quad 0 \leq t \leq 1.$$

Let L be a densely defined closed linear operator acting in X satisfying:

$$\Re(Lu, u)_X \leq -c\|u\|_X^2, \quad u \in \mathcal{D}(L),$$
$$\Re(L^*w, w)_X \leq -c\|w\|_X^2, \quad w \in \mathcal{D}(L^*),$$

with some constant $c > 0$. Then the assumptions of Example 5.1 are fulfilled with $b = 0$ and $m = 1$.

We remark that, if L is also a negative self adjoint operator, then the stronger condition (3.7) in Chapter III is verified with $\alpha = 1$, $\beta = \frac{1}{2}$.

EXAMPLE 5.7. Let X be a Banach space, and let $\mathcal{X} = \mathcal{C}([0, T]; X)$. In $\mathcal{X}$ consider the operator

$$\begin{cases} \mathcal{D}(\mathcal{T}) = \{u \in \mathcal{C}([0, T]; X) \cap \mathcal{C}^1((0, T]; X); t\dfrac{du}{dt} \in \mathcal{C}([0, T]; X)\} \\[2mm] (\mathcal{T}u)(t) = t\dfrac{du}{dt} + ku, \quad 0 \leq t \leq T, \end{cases}$$

with a suitable constant $k > 0$. Then it is easy to check that $\mathcal{T}$ satisfies $(\mathcal{T})$ with $b = 0$ and $p = 0$. Therefore, in order to handle an equation

$$t\frac{dMv}{dt} - Lv = f(t), \quad 0 < t \leq T,$$

in X, we can use Theorems 5.1, 5.4 and 5.5 under the assumption that M and L are operators satisfying (M-L) in the region $\Re\lambda \geq \varepsilon$ for a certain positive number $\varepsilon < k$.

EXAMPLE 5.8. As in Example 2.2, let

$$\mathrm{rot}\,E = -\frac{\partial B}{\partial t}, \quad \mathrm{rot}\,H = \frac{\partial D}{\partial t} + J$$

be the Maxwell equations in $\mathbb{R}^3$. We maintain the same assumptions as in Example 2.2. X is taken to be $\{L^2(\mathbb{R}^3)\}^6$. L is given in the form

$$\begin{cases} \mathcal{D}(L) = \{u \in X; \displaystyle\sum_{j=1}^{3} a_j \dfrac{\partial u}{\partial x_j} \in X\} \\[3mm] Lu = \displaystyle\sum_{j=1}^{3} a_j \dfrac{\partial u}{\partial x_j} + b(x)u. \end{cases}$$

M is given as K^2, where K is the multiplication operator on X by some matrix $c(x) \geq 0$. As verified by (2.28), we have:

$$\Re(Lu, u)_X \leq -\delta\|u\|_X^2 + \lambda_0\|Ku\|_X^2, \quad u \in \mathcal{D}(L),$$

with $\lambda_0 \geq \max\{\gamma, 1\}$. This reads also

(5.11) $$\Re e(Lu - \lambda_0 Mu, u)_X \leq -\delta\|u\|_X^2, \quad u \in \mathcal{D}(L).$$

Fix $\lambda_0 \geq \max\{\gamma, 1\}$. The change of variable $u = e^{\lambda_0 t}v$ transforms the equation to

$$\frac{dMv}{dt} = (L - \lambda_0 M)v + e^{-\lambda_0 t}f(t), \quad 0 \leq t \leq T.$$

Taking into account (5.11), we deduce that

$$\Re e(\lambda Mv + \lambda_0 Mv - Lv, v)_X \geq \delta\|v\|_X^2, \quad v \in \mathcal{D}(L), \ \Re e\lambda \geq 0,$$

and $\|((\lambda + \lambda_0)M - L)v\|_X \geq \delta\|v\|_X$ for all $v \in \mathcal{D}(L)$ and $\Re e\lambda$ large. Applying the argument at the end of Example 2.2, we see that, if $((\lambda+\lambda_0)M - L)^*w = ((\bar{\lambda}+\lambda_0)M - L^*)w = 0$, then $w \in \mathcal{D}(L)$ and

$$\Re e(L^*w, w)_X = \Re e(\lambda_0+\lambda)\|Kw\|_X^2 = \Re e(w, Lw)_X \leq -\delta\|w\|_X^2 + \lambda_0\|Kw\|_X^2.$$

Hence it follows that

$$\Re e\lambda\|Kw\|_X^2 \leq -\delta\|w\|_X^2.$$

Since $\Re e\lambda \geq a > 0$, we conclude $Kw = w = 0$. Hence (M-L) has been shown to be valid with $b = 0$ and $m = 1$.

EXAMPLE 5.9. Let $a_j(x)$ $(j = 1, ...n)$, $b(x)$, $m(x)$ be $N \times N$ matrices defined for $x \in \mathbb{R}^n$, $a_j(x)$ being hermitian. Following the regular case $(m(x) = I)$, cf. for example H. Tanabe [167; pp. 73-75], we shall treat the Cauchy problems
(5.12)
$$\begin{cases} \dfrac{\partial m(x)v}{\partial t} = \displaystyle\sum_{j=1}^{n} a_j(x)\dfrac{\partial v}{\partial x_j} + (b(x) - \alpha I)v + f(x,t), & (x,t) \in \mathbb{R}^n \times [0,T], \\ m(x)v(x,0) = u_0(x), & x \in \mathbb{R}^n, \end{cases}$$

where α is a sufficiently large real number and $v = (v_1, ..., v_N)$ is the unknown function, and
(5.13)
$$\begin{cases} m(x)\dfrac{\partial u}{\partial t} = \displaystyle\sum_{j=1}^{n} a_j(x)\dfrac{\partial u}{\partial x_j} + (b(x) - \alpha I)u + f(x,t), & (x,t) \in \mathbb{R}^n \times [0,T], \\ u(x,0) = u_0(x), & x \in \mathbb{R}^n, \end{cases}$$

where $u = (u_1, ..., u_N)$ is the unknown function.

Let us first introduce the space $\mathcal{B}^0(\mathbb{R}^n)$ (resp. $\mathcal{B}^1(\mathbb{R}^n)$) consisting of all continuous and bounded functions (resp. with continuous and bounded first order derivatives on $\mathbb{R}^n$). We assume:

(5.14) $m(x)$, $x \in \mathbb{R}^n$, are diagonal matrices whose components belong to $\mathcal{B}^0(\mathbb{R}^n)$ and are non negative.

(5.15) The components of $a_j(x)$ (resp. $b(x)$) belong to $\mathcal{B}^1(\mathbb{R}^n)$ (resp. $\mathcal{B}^0(\mathbb{R}^n)$).

Let $X = L^2(\mathbb{R}^n)^N$. Define a linear operator L by

$$
\begin{cases}
\mathcal{D}(L) = \{u \in X; \sum_{j=1}^{n} a_j(x)\dfrac{\partial u}{\partial x_j} + b(x)u \in X\} \\[2mm]
Lu = \sum_{j=1}^{n} a_j(x)\dfrac{\partial u}{\partial x_j} + b(x)u - \alpha u
\end{cases}
$$

with a sufficiently large α. Let M be the multiplication operator by the matrices $m(x)$ in X. For $\Re\lambda \geq 0$,

$$
\Re((\lambda M - L)u, u)_X \geq -\Re(Lu, u)_X
$$

$$
= \tfrac{1}{2}(u, \sum_{j=1}^{n}\frac{\partial a_j}{\partial x_j}u) - \Re(bu, u)_X + \alpha\|u\|_X^2
$$

$$
\geq (-\tfrac{c}{2} - \beta + \alpha)\|u\|_X^2, \quad u \in H^1(\mathbb{R}^n)^N,
$$

where

$$
\beta = \|b(\cdot)\|_{\mathcal{B}^0}, \quad \tilde{c} = \max\left\{\left\|\frac{\partial a_j(\cdot)}{\partial x_j}\right\|_{\mathcal{B}^0}; j = 1, \ldots, n\right\}, \quad c = n\tilde{c},
$$

and $\|\cdot\|_{\mathcal{B}^0}$ has an obvious meaning by means of the components of the matrices involved. Therefore, if $\alpha > \beta + \frac{c}{2}$, then we obtain that

$$
\|(\lambda M - L)u\|_X \geq c_1\|u\|_X, \quad u \in H^1(\mathbb{R}^n),
$$

with some constant $c_1 > 0$. If a mollifier argument is used for an arbitrary functions $u \in \mathcal{D}(L)$, the preceding estimate is easily extended for all $u \in \mathcal{D}(L)$.

Moreover, by introducing the formal adjoint of $\sum_{j=1}^{n} a_j(x)\frac{\partial}{\partial x_j} + b(x)$, we recognize that, for $\Re\lambda \geq 0$, $\lambda M - L$ has a bounded inverse and the norm $\|(\lambda M - L)^{-1}\|_{\mathcal{L}(X)}$ is uniformly bounded. Since M is bounded, this in turn implies that (M-L) is valid with $b = 0$ and $m = 1$.

Theorem 5.7 then yields the following result: If $f \in C^3([0, T]; L^2(\mathbb{R}^n)^N)$ and if u_0 satisfies the relation

$$
u_0 = -\sum_{j=0}^{2}(ML^{-1})^{j+1}f^{(j)}(0) + u_1 \text{ with some } u_1 \in \mathcal{R}((ML^{-1})^4),
$$

then (5.12) possesses a unique solution v such that

$$Mv \in \mathcal{C}^1([0,T]; L^2(\mathbb{R}^n)^N), \; Lv \in \mathcal{C}([0,T]; L^2(\mathbb{R}^n)^N).$$

On the other hand, Theorem 5.10 yields: If $f \in \mathcal{C}^4([0,T]; L^2(\mathbb{R}^n)^N)$ and if u_0 satisfies the relation

$$Lu_0 + \sum_{j=0}^{3}(ML^{-1})^j f^{(j)}(0) \in \mathcal{R}((ML^{-1})^4),$$

then (5.13) possesses a unique solution u such that

$$u \in \mathcal{C}^1([0,T]; L^2(\mathbb{R}^n)^N), \; Lu \in \mathcal{C}([0,T]; L^2(\mathbb{R}^n)^N).$$

EXAMPLE 5.10. Consider the Cauchy problem for the degenerate equation

$$(5.16) \quad \begin{cases} \dfrac{\partial m(x)v}{\partial t} = L(x; D)v + f(x,t), & (x,t) \in \Omega \times (0,T], \\ v(x,t) = 0, & (x,t) \in \partial\Omega \times (0,T], \\ m(x)v(x,0) = m(x)v_0(x), & x \in \Omega, \end{cases}$$

in a bounded domain $\Omega \subset \mathbb{R}^n$ with a smooth boundary $\partial\Omega$. Here, $m(x) \geq 0$ is a non negative function in $\mathcal{C}(\overline{\Omega})$. $L(x; D)$ is a second order differential operator

$$L(x; D)v = \sum_{i,j=1}^{n} \frac{\partial}{\partial x_i}\left(a_{ij}(x)\frac{\partial v}{\partial x_j}\right) - a_0 v,$$

where $a_{ij} \in \mathcal{C}(\overline{\Omega})$ and a_0 is a positive number. $L(x; D)$ is elliptic in the usual sense that

$$\sum_{i,j=1}^{n} a_{ij}(x)\xi_i\xi_j \geq c_0|\xi|^2, \quad \xi = (\xi_1, ..., \xi_n) \in \mathbb{R}^n, \; x \in \Omega,$$

with some constant $c_0 > 0$.

Such a problem has already been handled in Examples 3.3 and 3.6 in the underlying space $L^p(\Omega)$, where $1 < p < \infty$. In Examples 5.10 and 5.11 we shall then be interested in the cases when $p = \infty$ and 1. Let us consider the case $p = \infty$. More precisely, in this example we take $X = \mathcal{C}(\overline{\Omega})$.

As it is well known by H. B. Stewart [161], the linear operator defined by

$$\begin{cases} \mathcal{D}(L) = \{u \in \bigcap_{q>n} W_q^2(\Omega); L(x; D)u \in \mathcal{C}(\overline{\Omega}) \text{ and } u = 0 \text{ on } \partial\Omega\} \\ Lu = L(x; D)u \end{cases}$$

is the generator of an analytic semigroup on X with non dense domain $\mathcal{D}(L)$. Let M be the multiplication operator by the function $m(x)$ in X.

For $\Re\lambda \geq 0$, let $\lambda Mu - Lu = f$. We notice as in Example 3.6 that

$$\Re \int_\Omega L(x; D)u \cdot \overline{u}|u|^{p-2}dx \leq 0$$

for every $1 < p < \infty$. Then by the same argument as in Example 3.6 it is observed that

$$\Re\lambda \int_\Omega m(x)|u(x)|^p dx \leq \|f\|_{L^p}\|u\|_{L^p}^{p-1} \text{ and } a_0\|u\|_{L^p} \leq \|f\|_{L^p}.$$

Therefore

$$\Re\lambda \int_\Omega m(x)|u(x)|^p dx \leq a_0^{1-p}\|f\|_{L^p}^p,$$

and hence

$$\Re\lambda \int_\Omega m(x)^p |u(x)|^p dx$$

$$\leq \|m\|_{\mathcal{C}}^{p-1}\Re\lambda \int_\Omega m(x)|u(x)|^p dx \leq \left(\frac{\|m\|_{\mathcal{C}}}{a_0}\right)^{p-1} \|f\|_{L^p}^p.$$

We have thus obtained that

(5.17) $$(\Re\lambda)^{\frac{1}{p}}\|Mu\|_{L^p} \leq \left(\frac{\|m\|_{\mathcal{C}}}{a_0}\right)^{\frac{p-1}{p}} \|f\|_{L^p}$$

for every $p \in (1, \infty)$. As a consequence, letting $p \to \infty$, we arrive at

$$\|Mu\|_{\mathcal{C}} \leq \frac{\|m\|_{\mathcal{C}}}{a_0}\|f\|_{\mathcal{C}}.$$

On the other hand, by using the arguments as in H. B. Stewart [161] it is shown that $\mathcal{R}(\lambda M - L) = X$. In this way (M-L) is fulfilled with $b = 0$ and $m = 1$.

Applying Theorems 5.7 and 5.10, we obtain the following existence and uniqueness result for (5.16). If $f \in \mathcal{C}^3([0,T]; \mathcal{C}(\overline{\Omega}))$ and if v_0 satisfies the relation

$$L(x; D)v_0 + \sum_{j=0}^{2}(-1)^j(ML^{-1})^j\frac{\partial^j f}{\partial t^j}(\cdot, 0) = L(x; D)v_1$$

with some $v_1 \in \mathcal{D}(L)$ such that $Lv_1 \in \mathcal{R}((ML^{-1})^3)$, then (5.16) possesses a unique solution in $\mathcal{C}([0,T]; \mathcal{C}(\overline{\Omega}))$. In particular, if $\frac{\partial^j f}{\partial t^j}(\cdot, 0) = 0$, $j = 0, 1, 2$, then it needs to assume that $Lv_0 \in \mathcal{R}((ML^{-1})^3)$.

EXAMPLE 5.11. Let us consider the problem (5.16) in the space $X = L^1(\Omega)$. $m(x)$ and $L(x; D)$ are the same ones as in Example 5.10 except that we need more restrictive regularity assumption $a_{i,j} \in \mathcal{C}^1(\overline{\Omega})$ $(i, j = 1, ..., n)$ on the coefficients of $L(x; D)$.

According to H. Brezis and W. Strauss [37], L is defined as

$$\begin{cases} \mathcal{D}(L) = \{u \in \mathring{W}_1^1(\Omega); L(x; D)u \in L^1(\Omega)\} \\ Lu = L(x; D)u. \end{cases}$$

Here $L(x; D)$ is of course viewed in the sense of distributions so that $Lu = f \in L^1(\Omega)$ if and only if $u \in \mathring{W}_1^1(\Omega)$ and

$$-\sum_{i,j=1}^n \left(a_{ij}\frac{\partial u}{\partial x_i}, \frac{\partial \varphi}{\partial x_j}\right)_{L^2} - (a_0 u, \varphi)_{L^2} = (f, \varphi)_{L^2}, \quad \text{for all } \varphi \in \mathring{W}_\infty^1(\Omega).$$

Then it is known by Theorem 8 of [37], that $\mathcal{D}(L) \subset \mathring{W}_q^1(\Omega)$ for all $1 \le q < \frac{n}{n-1}$ with

$$k\|u\|_{W_q^1} \le \|Lu\|_{L^1}, \quad u \in \mathcal{D}(L),$$

where $k > 0$ is a constant determined by q. In addition, L is the closure in $L^1(\Omega)$ of the operator L_2 in $L^2(\Omega)$ introduced above, namely $\mathcal{D}(L_2) = H^2(\Omega) \cap H_0^1(\Omega)$, $L_2 u = L(x; D)u$. M is the multiplication operator by the function $m(x)$ in X.

Let $\Re\lambda > 0$. Given $f \in X$, let f_ℓ be a sequence of functions $f_\ell \in C_0^\infty(\Omega)$ such that $f_\ell \to f$ in X. Put $u_\ell = (\lambda M_2 - L_2)^{-1} f_\ell$, where M_2 is the multiplication operator by $m(x)$ in $L^2(\Omega)$. From the estimate verified above,

$$a_0\|u_\ell - u_m\|_{L^q} \le \|f_\ell - f_m\|_{L^q}$$

for any $1 < q < \frac{n}{n-1}$. Since $\|\cdot\|_{L^q}$ is continuous for $q \ge 1$, it follows that

$$a_0\|u_\ell - u_m\|_{L^1} \le \|f_\ell - f_m\|_{L^1}.$$

This then shows that u_ℓ has a limit u in X with

(5.18) $a_0\|u\|_{L^1} \le \|f\|_{L^1}.$

Clearly, also $m(x)u_\ell$ converges to $m(x)u$ in X. Similarly, in view of (5.17),

$$(\Re\lambda)^{\frac{1}{q}}\|mu_\ell\|_{L^q} \le \left(\frac{\|m\|_c}{a_0}\right)^{1-\frac{1}{q}} \|f_\ell\|_{L^q}.$$

Therefore,

$$\|Mu\|_{L^1} \le (\Re\lambda)^{-1}\|f\|_{L^1}.$$

Since $L_2 u_\ell = \lambda m u_\ell - f_\ell$ is convergent in X, it follows that $u \in \mathcal{D}(L)$ with $Lu = \lambda M u - f$. From (5.18) the range $\mathcal{R}(\lambda M - L)$ is a closed subspace of X; on the other hand, since $\mathcal{R}(\lambda M - L) \supset \mathcal{R}(\lambda M_2 - L_2) = L^2(\Omega)$, the range is dense in X; hence, $\mathcal{R}(\lambda M - L) = X$. Thus we have verified that $\lambda \in \rho_M(L)$ if $\Re\lambda > 0$ and that the m. l. operator LM^{-1} fulfills the Hille-Yosida condition (H-Y) in Chapter II. Applying Theorem 2.6, we then obtain the following result. If $f \in \mathcal{C}^1([0,T]; L^1(\Omega))$ and if $v_0 \in \mathcal{D}(L)$ satisfies $f(\cdot, 0) - Lv_0 \in \overline{M(\mathcal{D}(L))}$ (closure in $L^1(\Omega)$), then (5.16) possesses a unique solution v such that $Mv \in \mathcal{C}^1([0,T]; L^1(\Omega))$, $Lv \in \mathcal{C}([0,T]; L^1(\Omega))$.

EXAMPLE 5.12. In handling (5.16) in $L^1(\Omega)$, of course we could make use of the results of section 5.1, too, since assumption (M-L) is verified with $b = 0$ and $m = 1$, but a bit more regularity and compatibility relations between $f(x,0)$ and v_0 should be required. In fact, ML^{-1} is an abstract potential operator too; to these operators we shall devote a special paragraph in the sequel.

Moreover, the hypotheses on $m(x)$ could be weakened, as we show in a moment, even if, for sake of simplicity, we shall confine ourselves to the case $L(x; D) = \Delta$. This approach is due to V. Barbu and refers to real valued functions.

Let $\Omega \subset \mathbb{R}^n$ be a bounded open subset with a smooth boundary $\partial\Omega$. Let $m \in L^1(\Omega)$ be such that $m(x) \geq 0$ almost everywhere $x \in \Omega$. Consider the operator A in $X = L^1(\Omega)$ defined by

$$\begin{cases} \mathcal{D}(A) = \{y \in L^1(\Omega); m^{-1}y \in \mathring{W}_1^1(\Omega), \Delta(m^{-1}y) \in L^1(\Omega)\} \\ Ay = \Delta(m^{-1}y). \end{cases}$$

Our goal is then to prove that A is m-dissipative in X with

$$\mathcal{D}(A) \subset \{y \in L^1(\Omega); y = mz, z \in \bigcap_{1 \leq q < n/(n-1)} \mathring{W}_q^1(\Omega)\}.$$

In fact, according to H. Brezis and W. Strauss [37], if $\Delta z \in L^1(\Omega)$ with $z \in \mathring{W}_1^1(\Omega)$, then $z \in \bigcap_{1 \leq q < n/(n-1)} \mathring{W}_q^1(\Omega)$ with the estimate

(5.19) $$\|z\|_{W_q^1} \leq C\|\Delta z\|_{L^1}, \quad z \in \mathcal{D}(A),$$

and this implies the second part of the statement. The dissipativity of A follows immediately from the obvious inequality

$$\int_\Omega Ay \cdot \operatorname{sgn} y \, dx \geq 0, \quad y \in \mathcal{D}(A).$$

(Indeed, to verify this we multiply Ay by $\phi(y)$, where ϕ is a smooth monotone approximation of sgn and let $\phi \to \operatorname{sgn}$.)

Next we have to prove that $\mathcal{R}(1 - A) = X$. Let $f \in X$ and consider $(1 - A)y = f$. Obviously, this is equivalent to the equation

$$\begin{cases} mz - \Delta z = f, & x \in \Omega, \\ z = 0, & x \in \partial\Omega, \end{cases}$$

in X. Introducing a sequence $\{m_\varepsilon\} \subset L^\infty(\Omega)$ such that $m_\varepsilon \geq 0$ and that $m_\varepsilon \to m$ in $L^1(\Omega)$ as $\varepsilon \to 0$, we consider

(5.20)
$$\begin{cases} (m_\varepsilon + \varepsilon)z_\varepsilon - \Delta z_\varepsilon = f, & x \in \Omega, \\ z_\varepsilon = 0, & x \in \partial\Omega. \end{cases}$$

Since the operator $z \mapsto (m_\varepsilon + \varepsilon)z$ is continuous and m-accretive, and since $z \mapsto -\Delta z$ is m-accretive, (5.20) possesses a unique solution $z_\varepsilon \in \overset{\circ}{W}{}^1_q(\Omega)$ with $\Delta z_\varepsilon \in X$. Moreover, a uniform estimate

$$\|(m_\varepsilon + \varepsilon)z_\varepsilon\|_{L^1} + \|\Delta z_\varepsilon\|_{L^1} \leq C,$$

is obtained, C being independent of ε. Furthermore, by (5.19), $\|z_\varepsilon\|_{W^1_q} \leq C$; hence there is a subsequence $\{m_{\varepsilon'}\}$ such that $z_{\varepsilon'} \to z$ strongly in X and weakly in $\overset{\circ}{W}{}^1_q(\Omega)$; in particular, $(m_{\varepsilon'} + \varepsilon')z_{\varepsilon'} \to mz$ at almost every point of Ω and $\Delta z_{\varepsilon'} \to \Delta z$ in distribution sense. Here we used that $\{m_\varepsilon z_\varepsilon\}$ is weakly compact in X, which is verified by the same technique as in T. Gallouët and J. -M.Morel [107] by noticing that

$$\int_{(m_\varepsilon + \varepsilon)z_\varepsilon \geq b} (m_\varepsilon + \varepsilon)z_\varepsilon dx \leq \int_{(m_\varepsilon + \varepsilon)z_\varepsilon \geq b} f(x)dx.$$

5.4 THE GEVREY CASE

In this section we shall consider the same operational equation

$(\mathcal{E})$ $$\mathcal{T}Mv - Lv = f$$

in a Banach space $\mathcal{X}$ as in Section 5.1, but we shall focus our attention on the Gevrey case, i.e. when the closed linear operators M and L satisfy

(5.21) $$\|M(\lambda M - L)^{-1}\|_{\mathcal{L}(X)} \leq \frac{C}{(1 + |\lambda|)^\alpha}, \quad \lambda \in \Sigma,$$

where $\Sigma = \{\lambda \in \mathbb{C}; \Re\mathrm{e}\lambda \geq -c(1 + |\Im\mathrm{m}\lambda|)^\alpha\}$ with some $0 < \alpha < \frac{1}{2}$. Notice that here with notation of Section 3.5, $\alpha = \beta < \frac{1}{2}$, but $2\alpha + \beta > 2$ fails. As before, $\mathcal{D}(L) \subset \mathcal{D}(M)$ and $L^{-1} \in \mathcal{L}(X)$ are assumed. B denotes $B = -ML^{-1} \in \mathcal{L}(X)$.

On the other hand, $\mathcal{T}$ is supposed to satisfy

(5.22) $$\|(\lambda - \mathcal{T})^{-1}\|_{\mathcal{L}(X)} \leq \frac{L}{1 + |\Re e\lambda|}, \quad \lambda \in \Pi,$$

where $\Pi = \{\lambda \in \mathbb{C}; \Re e\lambda \leq 0 \text{ or } |\lambda| \leq \varepsilon\}$. Finally, $\mathcal{T}^{-1}$ commutes with B.
Putting $w = -Lv$, $(\mathcal{E})$ is written in the form

$(\mathcal{E}')$ $$\mathcal{T}Bw + w = f.$$

If $\Gamma = \{\lambda = a - c(1 + |\eta|)^{\alpha} + i\eta, -\infty < \eta < \infty\}$, $a > c$, being chosen so that $\Gamma \subset \Sigma \cap \Pi$, even if $f \in \mathcal{D}(\mathcal{T})$, the integral

$$w = \tfrac{1}{2\pi i} \int_{\Gamma} \lambda^{-1}(\lambda B + 1)^{-1}(\mathcal{T} - \lambda)^{-1}\mathcal{T}f d\lambda$$

may not be convergent, since on Γ we have only that

$$|\lambda|^{-1}\|(\lambda B + 1)^{-1}\|_{\mathcal{L}(X)}\|(\mathcal{T} - \lambda)^{-1}\|_{\mathcal{L}(X)} \leq C(1 + |\eta|)^{-2\alpha}.$$

Our candidate for the solution w is obviously written as

$$w = -\tfrac{1}{2\pi i} \int_{\Gamma} \lambda^{-2}(\lambda B + 1)^{-1}\mathcal{T}f d\lambda + \tfrac{1}{2\pi i} \int_{\Gamma} \lambda^{-2}(\lambda B + 1)^{-1}\mathcal{T}(\mathcal{T} - \lambda)^{-1}\mathcal{T}f d\lambda.$$

Now the first integral converges because of the estimate

$$|\lambda|^{-2}\|(\lambda B + 1)^{-1}\|_{\mathcal{L}(X)} \leq C(1 + |\eta|)^{-(1+\alpha)},$$

and vanishes. So that,

$$w = \tfrac{1}{2\pi i} \int_{\Gamma} \lambda^{-2}(\lambda B + 1)^{-1}\mathcal{T}(\mathcal{T} - \lambda)^{-1}\mathcal{T}f d\lambda.$$

We have to give sense to this expression. If $\mathcal{T}f \in (X, \mathcal{D}(\mathcal{T}))_{\theta,\infty}$, then $|\lambda|^{\theta} \times \|\mathcal{T}(\mathcal{T} - \lambda)^{-1}\mathcal{T}f\|_X \leq C(1 + |\eta|)^{(1-\alpha)(1+\theta)}\|\mathcal{T}f\|_{(X,\mathcal{D}(\mathcal{T}))_{\theta,\infty}}$. Therefore,

$$\|\lambda^{-2}(\lambda B + 1)^{-1}\mathcal{T}(\mathcal{T} - \lambda)^{-1}\mathcal{T}f\|_X \leq C(1 + |\eta|)^{-\alpha(2+\theta)}\|\mathcal{T}f\|_{(X,\mathcal{D}(\mathcal{T}))_{\theta,\infty}}.$$

This shows that, if $\frac{1}{\alpha} - 2 < \theta < 1$, then $w \in X$ is well defined.
It is easily seen that

$$Bw = \tfrac{1}{2\pi i} \int_{\Gamma} \lambda^{-3}(\lambda B + 1 - 1)(\lambda B + 1)^{-1}\mathcal{T}(\mathcal{T} - \lambda)^{-1}\mathcal{T}f d\lambda$$

$$= \tfrac{1}{2\pi i} \int_{\Gamma} \lambda^{-3}\mathcal{T}(\mathcal{T} - \lambda)^{-1}\mathcal{T}f d\lambda - \tfrac{1}{2\pi i} \int_{\Gamma} \lambda^{-3}(\lambda B + 1)^{-1}\mathcal{T}(\mathcal{T} - \lambda)^{-1}\mathcal{T}f d\lambda$$

$$= \tfrac{1}{2\pi i} \int_{\Gamma} \lambda^{-2}(\mathcal{T} - \lambda)^{-1}\mathcal{T}f d\lambda - \tfrac{1}{2\pi i} \int_{\Gamma} \lambda^{-3}(\lambda B + 1)^{-1}\mathcal{T}(\mathcal{T} - \lambda)^{-1}\mathcal{T}f d\lambda$$

$$= (i) + (ii).$$

Clearly, $(i) = T^{-1}f$. Moreover,

$$(ii) = -\tfrac{1}{2\pi i}\int_\Gamma \lambda^{-3}(\lambda B+1)^{-1}Tfd\lambda - \tfrac{1}{2\pi i}\int_\Gamma \lambda^{-2}(\lambda B+1)^{-1}(T-\lambda)^{-1}Tfd\lambda.$$

Since the first integrand is estimated by $C(1+|\eta|)^{-(2+\alpha)}$, the first integral vanishes. Similarly, since the second integrand is estimated by $C(1+|\eta|)^{-1-2\alpha}$, the second integral is indeed convergent. Therefore, it follows that $Bw \in \mathcal{D}(T)$ with

$$TBw = f - \tfrac{1}{2\pi i}\int_\Gamma \lambda^{-2}(\lambda B+1)^{-1}T(T-\lambda)^{-1}Tfd\lambda = f - w.$$

Thus, for $f \in \mathcal{D}(T)$ with $Tf \in (\mathcal{X},\mathcal{D}(T))_{\theta,\infty}$, $\tfrac{1}{\alpha}-2 < \theta < 1$ (hence α must of course be larger than $\tfrac{1}{3}$), there exists a solution w to $(\mathcal{E}')$, and therefore a solution v to $(\mathcal{E})$. The uniqueness of solution to $(\mathcal{E})$ is already known.

Next, let us study regularity of solutions. Let $\omega \in (0,1)$ and consider $\xi^\omega T(\xi + T)^{-1}w$, $\xi > 0$. Using Theorem 1.8, we obtain that

$$\xi^\omega T(\xi + T)^{-1}w = \xi^\omega \tfrac{1}{2\pi i}\int_\Gamma \lambda^{-2}(\lambda B+1)^{-1}T(T-\lambda)^{-1}Tfd\lambda$$

$$+ \xi^{\omega+1}\tfrac{1}{2\pi i}\int_\Gamma \lambda^{-2}(\xi+\lambda)^{-1}(\lambda B+1)^{-1}T(\xi+T)^{-1}Tfd\lambda$$

$$- \xi^{\omega+1}\tfrac{1}{2\pi i}\int_\Gamma \lambda^{-2}(\xi+\lambda)^{-1}(\lambda B+1)^{-1}T(T-\lambda)^{-1}Tfd\lambda.$$

The second integral obviously vanishes. So that, after some calculation, it follows that

$$\xi^\omega T(\xi + T)^{-1}w = \xi^\omega \tfrac{1}{2\pi i}\int_\Gamma \lambda^{-1}(\xi+\lambda)^{-1}(\lambda B+1)^{-1}T(T-\lambda)^{-1}Tfd\lambda.$$

Hence,

$$\xi^\omega \|T(\xi+T)^{-1}w\|_\mathcal{X}$$
$$\leq C\xi^\omega \int_\Gamma |\lambda|^{-(1+\theta)}|\xi+\lambda|^{-1}(1+|\Im m\lambda|)^{(1-\alpha)(2+\theta)}|d\lambda|\|Tf\|_{(\mathcal{X},\mathcal{D}(T))_{\theta,\infty}}.$$

Here we observe that

$$\int_0^\infty |\xi + a + (1+\eta)^\alpha + i\eta|^{-1}(1+\eta)^{(1-\alpha)(2+\theta)-(1+\theta)}d\eta$$

$$\leq C\int_1^\infty (\xi+\eta)^{-1}\eta^{(1-\alpha)(2+\theta)-(1+\theta)}d\eta$$

$$\leq C\xi^{(1-\alpha)(2-\theta)-(1+\theta)}\int_0^\infty (1+\eta)^{-1}\eta^{(1-\alpha)(2+\theta)-(1+\theta)}d\eta,$$

and that the last integral is convergent if and only if $\theta+1+(\alpha-1)(2+\theta) < 1$ and $\theta + 2 + (\alpha-1)(2+\theta) > 1$. Thus we conclude that, if $\theta > \tfrac{1}{\alpha} - 2$, then

$$\xi^\omega \|T(\xi+T)^{-1}w\|_\mathcal{X} \leq C\|Tf\|_{(\mathcal{X},\mathcal{D}(T))_{\theta,\infty}},$$

where $\omega = (\alpha - 1)(2 + \theta) + 1 + \theta = \alpha\theta + 2\alpha - 1$, which means that $w \in (\mathcal{X},\mathcal{D}(T))_{\omega,\infty}$.

THEOREM 5.12. *Let (5.21) and (5.22) be satisfied with $\frac{1}{3} < \alpha < \frac{1}{2}$.*
Then, for any $f \in \mathcal{D}(T)$ such that $Tf \in (\mathcal{X}, \mathcal{D}(T))_{\theta,\infty}$ with $\frac{1}{\alpha} - 2 < \theta < 1$,
($\mathcal{E}$) possesses a unique solution v such that $TMv, Lv \in (\mathcal{X}, \mathcal{D}(T))_{\omega,\infty}$,
where $\omega = \alpha\theta + 2\alpha - 1$.

As an example, we consider the case where $\mathcal{X} = \mathcal{C}([0,\tau]; X)$, X being a Hilbert space, and L is an operator of the form

$$L = -S - iT,$$

where S is a positive selfadjoint operator in X and T is a selfadjoint operator in X with $\mathcal{D}(T) \subset \mathcal{D}(S^\gamma)$, $\gamma > 1$. Since

$$\|Su\|_X \le \varepsilon \|S^\gamma u\|_X + C_\varepsilon \|u\|_X, \quad u \in \mathcal{D}(S^\gamma),$$

the operator $S + iT$ is closed and its adjoint coincides with $S - iT$. Let $M \in \mathcal{L}(X)$ be a non negative selfadjoint operator in X.

For $\Re\lambda \ge 0$, let $u \in \mathcal{D}(T)$ and $(\lambda M + S + iT)u = f$. Taking the inner product by u, we obtain that

$$\|S^{\frac{1}{2}}u\|_X \le C\|f\|_X.$$

Hence, by $(\lambda M + S + iT)^* = \bar{\lambda}M + S + iT$ it is easily seen that

$$\|S^{\frac{1}{2}}u\|_X \le C\|g\|_X \text{ if } (\bar{\lambda}M + S + iT)^*u = g.$$

These inequalities then show that $\lambda M + S + iT$ has a bounded inverse for all $\Re\lambda \ge 0$ with an estimate

$$\|S^{\frac{1}{2}}(\lambda M + S + iT)^{-1}f\|_X \le C\|f\|_X, \quad f \in X.$$

Moreover, for $\Re\lambda \ge 0$,

$$\Re\lambda\|M^{\frac{1}{2}}(\lambda M + S + iT)^{-1}f\|_X \le C\|f\|_X\|S^{\frac{1}{2}}(\lambda M + S + iT)^{-1}f\|_X$$

and

$$\|\lambda M(\lambda M + S + iT)^{-1}f\|_X \le C(1 + |\lambda|)\|f\|_X, \quad f \in \mathcal{X},$$

or else

$$\|(S + iT)(\lambda M + S + iT)^{-1}f\|_X \le C(1 + |\lambda|)\|f\|_X, \quad f \in X.$$

Therefore M and $L = S + iT$ satisfy (5.2) with $m = 1$ and $b = 0$.

It may be observed that if in addition $\|M\|_{\mathcal{L}(X)} \le 1$ and $\|u\| \le \|S^{\frac{1}{2}}u\|_X$ for all $u \in \mathcal{D}(S^{\frac{1}{2}})$, then

$$\Re\lambda\|M(\lambda M + S + iT)^{-1}f\|_X \le \|f\|_X, \quad \Re\lambda > 0, f \in X,$$

and thus Theorem 2.9 applies too.

We particularize furthermore the example by taking $M = I$. This situation appears in recent advances in control theory when one tries to stabilize a dynamical system like $\frac{du}{dt} = -Su + f(t)$ by a feedback $f(t) = -iTu(t)$, where $\mathcal{D}(T) \subset \mathcal{D}(S)$. We refer to A. Favini and R. Triggiani [95].

Under this hypothesis let us assume in addition that $\mathcal{D}(T) = \mathcal{D}(S^\gamma)$, $\gamma > 1$. Taking the inner product of $(\lambda + S + iT)u = f$ and u, we recognize that $-(S + iT)$ generates a contraction semigroup on $\mathcal{X}$. We then verify that the Gevrey condition holds. To this end we need a lemma as follows.

LEMMA 5.13. *Let* $S > 0$ *and* T *be self adjoint operators with* $\mathcal{D}(T) = \mathcal{D}(S^\gamma)$, $\gamma > 1$. *Assume that*

$$\|S^\sigma(\lambda + S + iT)^{-1}\|_{\mathcal{L}(X)} \leq C, \quad \Re\lambda > 0.$$

with some $\sigma > 0$. *Then,*

$$|\lambda|^{\frac{\sigma}{\gamma}}\|(\lambda + S + iT)^{-1}\|_{\mathcal{L}(X)} \leq C, \quad \Re\lambda > 0, |\lambda| \text{ large.}$$

PROOF. We first show that

$$(5.23) \qquad \overline{S^{-(\gamma-\nu)}TS^{-\nu}} \in \mathcal{L}(X), \quad 0 < \nu \leq \gamma.$$

In fact, if $T \geq 0$ the Heinz-Kato inequality or interpolation (see Section 0.3) implies that $\|T^\theta S^{-\gamma\theta} f\|_X \leq C_\theta \|f\|_X$, $0 \leq \theta \leq 1$. Write $S^{\nu-\gamma}TS^{-\nu} = S^{-\gamma(1-\frac{\nu}{\gamma})}T^{1-\frac{\nu}{\gamma}}T^{\frac{\nu}{\gamma}}S^{-\nu}$; then, $T^{\frac{\nu}{\gamma}}S^{-\nu} \in \mathcal{L}(X)$ and $S^{-\gamma(1-\frac{\nu}{\gamma})}T^{1-\frac{\nu}{\gamma}} \in \mathcal{L}(X)$ by taking its adjoint. Therefore, (5.23) is verified. In the general case (5.23) it is proven by using the representation that $T = T^+ - T^-$ with $T^+, T^- \geq 0$.

Let us next observe that, for $\Re\lambda > 0$,

$$(\lambda + S + iT)^{-1} - (\lambda + S^\gamma)^{-1}$$

$$= (\lambda + S^\gamma)^{-1}S^{\gamma-\sigma}\left(I - S^{1-\gamma} - i\overline{S^{\sigma-\gamma}TS^{-\sigma}}\right)S^\sigma(\lambda + S + iT)^{-1}$$

$$= S^{\gamma-\sigma}(\lambda + S^\gamma)^{-1}\left(I - S^{1-\gamma} - i\overline{S^{\sigma-\gamma}TS^{-\sigma}}\right)S^\sigma(\lambda + S + iT)^{-1}.$$

Now it is seen by interpolation (notice that $\mathcal{D}(S^{\gamma-\sigma}) = [X, \mathcal{D}(S^\gamma)]_{1-\frac{\sigma}{\gamma}}$, see Section 0.3 and [172; p. 103]) that

$$\|S^{\gamma-\sigma}(\lambda + S^\gamma)^{-1}\|_{\mathcal{L}(X)} \leq \frac{C}{(|\lambda| + 1)^{\frac{\sigma}{\gamma}}}, \quad \Re\lambda > 0.$$

This combined with the assumption then yields that

$$\|(\lambda + S + iT)^{-1}\|_{\mathcal{L}(X)} \leq \|(\lambda + S^\gamma)^{-1}\|_{\mathcal{L}(X)} + C(|\lambda| + 1)^{-\frac{\sigma}{\gamma}}.$$

Therefore the proof of Lemma 5.13 is complete.

As a consequence we establish the following result.

THEOREM 5.14. *Let S, T be two self adjoint operators in X with $S > 0$ and $\mathcal{D}(T) = \mathcal{D}(S^\gamma)$, $\gamma > 1$. Then,*

$$|\lambda|^{\frac{1}{2\gamma}} \|(\lambda + S + iT)^{-1}\|_{\mathcal{L}(X)} \leq C, \quad \Re\lambda > 0.$$

COROLLARY 5.15. *Let S, T satisfy, in addition, $(Su, Tu)_X \in \mathbb{R}$ for all $u \in \mathcal{D}(T)$. Then,*

$$|\lambda|^{\frac{1}{\gamma}} \|(\lambda + S + iT)^{-1}\|_{\mathcal{L}(X)} \leq C, \quad \Re\lambda > 0.$$

PROOF. Take the inner product of $(\lambda + S + iT)u = f$ and Su. Then,

$$\Re\lambda \|S^{\frac{1}{2}}u\|_X^2 + \|Su\|_X^2 \leq \|f\|_X \|Su\|_X.$$

So that, $\|Su\|_X \leq \|f\|_X$, from which the result follows.

Finally, let us consider the Cauchy problem

(E)
$$\begin{cases} \dfrac{du}{dt} + (S + iT)u = f(t), & 0 \leq t \leq \tau, \\ u(0) = u_0, \end{cases}$$

in a Hilbert space X. Here, S and T are the same operators as in Theorem 5.14.

THEOREM 5.16. *Let $1 < \gamma < \frac{3}{2}$. Then, if $f \in \mathcal{C}^{1+\theta}([0,\tau]; X)$, $2(\gamma-1) < \theta < 1$, and if $u_0 \in \mathcal{D}((S + iT)^2)$ satisfies the relations*

$$f(0) - (S+iT)u_0 \in \mathcal{D}(S^\gamma), \quad f'(0) - (S+iT)\{f(0) - (S+iT)u_0\} \in \mathcal{D}(S^\gamma),$$

then the solution u to (E) enjoys the maximal regularity $\frac{du}{dt}$, $(S + iT)u \in \mathcal{C}^\omega([0,\tau]; X)$ with $\omega = \frac{\theta}{2\gamma} + \gamma - 1$.

The proof of this theorem is now quite analogous to that of Theorem 5.7, and we may leave it to the reader.

REMARK. We recall that a strongly continuous semigroup e^{tA} on a Banach space X is called of Gevrey class $\delta > 1$ for $t > t_0$ if e^{tA} is infinitely differentiable for $t \in (t_0, \infty)$ and if, for any compact $K \subset (t_0, \infty)$ and each $k > 0$, there exists a constant $c = c(k, K)$ such that $\left\|\frac{d^n}{dt^n} e^{tA}\right\|_{\mathcal{L}(X)} \leq ck^n (n!)^\delta$ for all $t \in K$ and $n = 0, 1, 2, \ldots$. We refer to S. Taylor [170] for the theory of such semigroups. In particular, in that paper it is shown that, if the generator A of a strongly continuous semigroup satisfies

$$|\eta|^{\frac{1}{\sigma}} \|(i\eta - A)^{-1}\|_{\mathcal{L}(X)} \leq C, \quad \eta \in \mathbb{R},$$

with some $\sigma > 1$, then e^{tA} is of Gevrey class $> \sigma$. The results obtained in this section furnish conditions (even easy to check) under which an operator generates this type of semigroups.

5.5 ABSTRACT POTENTIAL OPERATORS

This section is closely related to certain basic results contained in Chapter II and furnishes an alternative approach to them. Its main difference consists in singling out the bounded linear operator $B = ML^{-1}$ rather than the multivalued linear operator $A = LM^{-1}$ in order to treat the degenerate differential equation $\frac{d}{dt}Mv - Lv = f(t)$.

According to K. Yosida [187; p. 412], a closed linear operator B from a Banach space X into itself is called an abstract potential operator if $B = A^{-1}$, where A is the generator of a bounded semigroup of class $\mathcal{C}_0$ on X.

We have the following characterization of abstract potential operators.

LEMMA 5.17. *A closed linear operator B is an abstract potential operator in X if and only if $\overline{\mathcal{R}(B)} = X$ and there is a positive constant M such that*

$$(5.24) \qquad \|\{\xi B(\xi B - 1)^{-1}\}^m\|_{\mathcal{L}(X)} \leq M, \quad \xi > 0,$$

for all $m = 0, 1, 2, ...$, M being independent of m.

PROOF. We need only to prove the sufficient part, since the necessary one is trivial. If (5.24) holds, then the linear operators $J_\xi = B(\xi B - 1)^{-1}$ are bounded in X and satisfy the resolvent equation $J_\xi - J_\eta = (\eta - \xi)J_\xi J_\eta$ for any $\xi, \eta > 0$. (5.24) $(m = 1)$ implies that $\mathcal{K}(J_\xi) = \{0\}$, in view of the fact that $\mathcal{K}(J_\xi) \cap \overline{\mathcal{R}(J_\xi)} = \{0\}$ (see K. Yosida [187; pp. 216-217]) and $\mathcal{K}(J_\xi) = \mathcal{K}(B), \mathcal{R}(J_\xi) = \mathcal{R}(B)$. Therefore, there is a linear operator A such that $J_\xi = (\xi - A)^{-1}$; in fact, $A = \xi - J_\xi^{-1} = B^{-1}$. Moreover, $\mathcal{D}(A)$ is everywhere dense in X and $\|\{\xi(\xi - A)^{-1}\}^m\|_{\mathcal{L}(X)} \leq M, m = 0, 1, 2, \ldots$.

REMARK. If B generates a bounded analytic semigroup on X with a strongly dense range, then B is an abstract potential operator in X, for we then have the estimates $\|\lambda B(\lambda B - 1)^{-1}\|_{\mathcal{L}(X)} = \|B(B - \lambda^{-1})^{-1}\|_{\mathcal{L}(X)} \leq M$ for all $\Re\lambda > 0$, and thus by Theorem 1 of K. Yosida [187; Chapter VIII, Section 3], $\mathcal{K}(J_\xi) = \mathcal{K}(B) = \{0\}$. Therefore, B is invertible and its inverse B^{-1} verifies $\|\lambda(\lambda - B^{-1})^{-1}\|_{\mathcal{L}(X)} \leq M$ for $\Re\lambda > 0$, so implying the result.

Our next aim is to overcome the restrictive assumption $\overline{\mathcal{R}(B)} = X$.

THEOREM 5.18. *In addition to (5.24) assume that $\mathcal{D}(B)$ is dense in X. Then B_1, the part of B in $\overline{\mathcal{R}(B)}$, is an abstract potential operator in $\overline{\mathcal{R}(B)}$.*

PROOF. It is well known that $\mathcal{D}(B_1)$ is dense in $\overline{\mathcal{R}(B)}$. Moreover, since $\|f\|_X \leq M\|(\eta - B)f\|_X$ for all $f \in \mathcal{D}(B)$ and $\eta > 0$, we deduce that $\eta - B_1, \eta > 0$, has a bounded inverse in $\mathcal{L}(\overline{\mathcal{R}(B)})$ and $(\eta - B_1)^{-1}$ coincides with the part of $(\eta - B)^{-1}$ in $\overline{\mathcal{R}(B)}$. This implies that we can apply to B_1 the same argument as in the proof of Lemma 5.17. It then follows that

$f \in \overline{\mathcal{R}(B)}$ belongs to $\overline{\mathcal{R}(B_1)}$ if and only if $\lim_{\xi \to \infty} \|(\xi B_1 - 1)^{-1} f\|_X = 0$. On the other hand, if $u = Bf \in \mathcal{R}(B)$, we have: $(\xi B_1 - 1)^{-1} u = \xi^{-1} \{f + (\xi B - 1)^{-1} f\}$, and thus $\lim_{\xi \to \infty} \|(\xi B_1 - 1)^{-1} u\|_X = 0$.

It follows that $\mathcal{R}(B) \subset \overline{\mathcal{R}(B_1)} \subset \overline{\mathcal{R}(B)}$, and this together with the uniform boundedness of $\|(\xi B + 1)^{-1}\|_{\mathcal{L}(X)}$ yields the density of $\mathcal{R}(B_1)$ in $\overline{\mathcal{R}(B)}$. We then conclude that $\mathcal{K}(B_1) = \{0\}$ and so the proof is complete.

COROLLARY 5.19. *If B generates a bounded analytic semigroup on X, then B_1 is an abstract potential operator in $\overline{\mathcal{R}(B)}$.*

PROOF. The same argument as in the first half of the proof of Theorem 5.18 yields that $\|\lambda(\lambda - B_1)^{-1}\|_{\mathcal{L}(\overline{\mathcal{R}(B)})} \leq M$ for all $\Re\lambda > 0$. But, since $(\lambda - B_1)^{-1} = B_1^{-1}(\lambda B_1^{-1} - 1)^{-1}$, it is straightforward to see that $\|(\lambda B_1^{-1} - 1)^{-1}\|_{\mathcal{L}(\overline{\mathcal{R}(B)})} \leq M'$. This implies that B_1^{-1} generates an analytic semigroup in $\overline{\mathcal{R}(B)}$.

When X is a Hilbert space, it is known as Lumer-Phillips' theorem that, if $\Re e(Bu, u)_X \leq 0$ for all $u \in \mathcal{D}(B)$ and if $\xi_0 - B$ has a bounded inverse for some $\xi_0 > 0$, then B generates a contraction semigroup on X. In this case, for $\xi > 0$,

$$\|u\|_X^2 - \|\xi B(\xi B - 1)^{-1} u\|_X^2$$
$$= -2\Re e(u, (\xi B - 1)^{-1} u)_X - \|(\xi B - 1)^{-1} u\|_X^2, \quad u \in \mathcal{D}(B).$$

Hence, $\|\xi B(\xi B - 1)^{-1}\|_{\mathcal{L}(X)} \leq 1$, $\xi > 0$, is equivalent to $\|(\xi B - 1)^{-1} f\|_X^2 \leq -2\Re e(f, (\xi B - 1)^{-1} f)_X$, $f \in X$, $\xi > 0$. Furthermore, such a condition is satisfied if and only if $-\|v\|_X^2 \leq -2\xi \Re e(Bv, v)_X$ for all $\xi > 0$ and $v \in \mathcal{D}(B)$, that is, if and only if $\Re e(Bv, v)_X \leq 0$ for all $v \in \mathcal{D}(B)$. Hence, B_1 is an abstract potential operator in $\overline{\mathcal{R}(B)}$.

Let us apply this remark to operator pencils in a complex Hilbert space X. Let M and L be two closed linear operators in X such that $M \in \mathcal{L}(X)$ and L is densely defined with $L^{-1} \in \mathcal{L}(X)$. Assume that

(5.25) $\Re e(Mu, Lu)_X \leq 0, \quad u \in \mathcal{D}(L),$

(5.26) $\Re e(M^*v, L^*v)_X \leq 0, \quad v \in \mathcal{D}(L^*).$

Then (5.25) implies that $B = ML^{-1}$ satisfies $\Re e(Bu, u) \leq 0$ for all $u \in \mathcal{D}(B)$. Moreover, since

$$\|(\xi M - L)u\|_X^2 = \|Lu\|_X^2 + \xi^2 \|Mu\|_X^2 - 2\xi \Re e(Mu, Lu)_X$$
$$\geq \|Lu\|_X^2, \quad u \in \mathcal{D}(L),$$

for all $\xi > 0$, $\xi M - L$ has a closed range. We show that it coincides with the entire space. Otherwise, there would exist an element $h \neq 0$ orthogonal to

$\mathcal{R}(\xi M - L)$. This yields $|(Lu, h)_X| \leq \xi \|M\|_{\mathcal{L}(X)} \|u\|_X \|h\|_X$ and therefore $h \in \mathcal{D}(L^*)$. Moreover, we observe that $(\xi M^* - L^*)h = 0$. Hence, $0 = ((\xi M^* - L^*)h, M^*h)_X = \xi \|M^*h\|_X^2 - (L^*h, M^*h)_X$. This combined with (5.26) yields that $\xi \|M^*h\|_X^2 = 0$; hence, $M^*h = L^*h = h = 0$, which is a contradiction. Thus, $\mathcal{R}(\xi M - L) = X$ for all $\xi > 0$. We arrive at the following result.

PROPOSITION 5.20. *Let X be a Hilbert space. Let $M \in \mathcal{L}(X)$ and let L be densely defined, $L^{-1} \in \mathcal{L}(X)$. Assume either (5.25) and (5.26) or (5.25) and $\mathcal{R}(\xi_0 M - L) = X$ with some $\xi_0 > 0$. Then B_1, the restriction of B to $\overline{\mathcal{R}(B)}$, is an abstract potential operator in $\overline{\mathcal{R}(B)}$.*

PROOF. We need only to observe that in the second case $\xi_0 - B$ has a bounded inverse.

Our next step is to consider the Cauchy problem

(D-E.1)
$$\begin{cases} \dfrac{dMv}{dt} = Lv + f(t), & 0 \leq t \leq T, \\ Mv(0) = u_0, \end{cases}$$

where the operator $B = ML^{-1}$ satisfies an assumption like (5.24). Before establishing our result, we must recall that, if X is a reflexive Banach space and if B is a closed linear linear operator satisfing $\|\eta(\eta - B)^{-1}\|_{\mathcal{L}(X)} \leq C$ for all $\eta > 0$, then X is the direct sum of $\mathcal{K}(B)$ and $\overline{\mathcal{R}(B)}$, see K. Yosida [187; pp. 216-218]. Moreover, $\mathcal{R}(B) = \mathcal{R}(B_1)$. Let P be the corresponding projection operator from X onto $\mathcal{K}(B)$. We have the following theorem.

THEOREM 5.21. *Let X be reflexive and let M and L be two closed linear operators from X into itself, $0 \in \rho(L)$, $\mathcal{D}(L) \subset \mathcal{D}(M)$, with*

(5.27)
$$\|\{\xi M(\xi M - L)^{-1}\}^m\|_{\mathcal{L}(X)} \leq C, \quad \xi > 0,$$

for all $m = 0, 1, 2, ...$, C being a constant independent of m. Then, for any $f \in C^1([0,T]; X)$ and any $u_0 \in \mathcal{R}(ML^{-1}) = M(\mathcal{D}(L))$, (D-E.1) possesses a unique solution v such that $Mv \in C^1([0,T]; X)$, $Lv \in C([0,T]; X)$. It is given by

$$v(t) = -L^{-1}Pf(t) + L^{-1}B_1^{-1}e^{tB_1^{-1}}u_0$$
$$+ L^{-1}B_1^{-1} \int_0^t e^{(t-\tau)B_1^{-1}}(1 - P)f(\tau)d\tau, \quad 0 \leq t \leq T,$$

where B_1 denotes the part of $B = ML^{-1}$ in $\overline{\mathcal{R}(B)}$ and P is the projection from X onto $\mathcal{K}(B)$.

PROOF. Denoting $Lv(t)$ by $w(t)$, the equation (D-E.1) is written as $\frac{d}{dt}Bw - w = f(t)$. In addition, this equation splits into

$$Pw(t) = -Pf(t), \quad 0 \leq t \leq T,$$

and

$$\frac{d}{dt}B_1(1-P)w - (1-P)w = (1-P)f(t), \quad 0 \le t \le T.$$

On the other hand, the initial condition is written as $B_1(1-P)w(0) = u_0$. Consequently, this problem has a unique solution which is given by

$$(1-P)w(t) = B_1^{-1}e^{tB_1^{-1}}u_0 + B_1^{-1}\int_0^t e^{(t-\tau)B_1^{-1}}(1-P)f(\tau)d\tau, \quad 0 \le t \le T.$$

Hence the desired result is proved.

COROLLARY 5.22. *If X is reflexive and if M and L are two closed linear operators in X satisfying $L^{-1} \in \mathcal{L}(X)$ and*

$$\|\lambda M(\lambda M - L)^{-1}\|_{\mathcal{L}(X)} \le C, \quad \Re\lambda > 0,$$

then (D-E.1) possesses a unique solution v such that $Mv \in \mathcal{C}([0,T];X) \cap \mathcal{C}^1((0,T];X)$ for any $f \in \mathcal{C}^\theta([0,T];X)$, $0 < \theta \le 1$, and any $u_0 \in \overline{\mathcal{R}(ML^{-1})}$.

When X is not reflexive, these results are obtained provided that X is replaced by $\mathcal{K}(B) \oplus \overline{\mathcal{R}(B)}$ (the sum in every case is a direct one).

As verified above, if either (5.25) and (5.26) or (5.25) and $\mathcal{R}(\xi_0 M - L) = X$ with some $\xi_0 > 0$ are satisfied, then (5.27) is valid with $C = 1$ and Theorem 5.21 is applicable.

EXAMPLE 5.13. We note that (5.25) and (5.26) seem particularly satisfactory when M and L are defined by means of differential operators, but they can be verified in other interesting cases. For example, let $X = L^2(0,1)$ and $L = \frac{d^2}{dx^2}$: $H^2(0,1) \cap H_0^1(0,1) \to L^2(0,1)$, while M is the multiplication operator by a function $m(x) \ge 0$, $m \in \mathcal{C}^2([0,1])$, which is concave. Then,

$$\int_0^1 m(x)u(x)\overline{u}''(x)dx = -\int_0^1 m'(x)u(x)\overline{u}'(x)dx - \int_0^1 m(x)|u'(x)|^2dx,$$

while, since

$$\int_0^1 m'(x)u(x)\overline{u}'(x)dx$$

$$= \int_0^1 m'(x)\{a(x)a'(x) + b(x)b'(x) + i(b(x)a'(x) - a(x)b'(x))\}dx,$$

where $u(x) = a(x) + ib(x)$, it follows that

$$\Re\int_0^1 m'(x)u(x)\overline{u}'(x)dx = -\tfrac{1}{2}\int_0^1 m''(x)(|a(x)|^2 + |b(x)|^2)dx \ge 0.$$

Hence, M and L fulfil (5.25).

For other interesting cases where our assumptions are fulfilled, we refer to Ph. Clément and J. Prüss [52].

5.6 THE CASE OF $\lambda = 0$ AN ISOLATED
SINGULAR POINT OF $(\lambda L - M)^{-1}$

In this section we shall discuss the abstract version of the differential equation of Sobolev type which was studied in Example 3.10. We shall devote our attention to the equation

$$(5.28) \qquad \frac{dMv}{dt} = Lv + f(t), \quad 0 < t \leq T,$$

in a Banach space X, where M and L are two closed linear operators in X with $\mathcal{D}(L) \subset \mathcal{D}(M)$, L being invertible, with the property that $\lambda = 0$ is a pole of order $k+1$, $k = 0, 1, 2, ...$, of the bounded operator $L(\lambda L - M)^{-1} = (\lambda - B)^{-1}$, $B = ML^{-1} \in \mathcal{L}(X)$.

Notice that in the quoted example, $k = 0$, so that the situation is the best possible one, that is, $\lambda = 0$ is a simple pole for $(\lambda - B)^{-1}$.

We recall that if B is a closed linear operator in X which has $\lambda = 0$ as pole of $(\lambda - B)^{-1}$ of order $k + 1$, then the decomposition $X = \mathcal{K}(B^m) \oplus \mathcal{R}(B^m)$ holds for every $m \geq k + 1$. Moreover, $\mathcal{R}(B^m) = \overline{\mathcal{R}(B^m)} = \mathcal{R}(B^{k+1})$, $\mathcal{K}(B^m) = \mathcal{K}(B^{k+1})$ (see for example K. Yosida [187; p. 229]). If $\gamma:|\lambda| = \varepsilon$ is a circumference of sufficiently small radius with $|\lambda| \leq \varepsilon$ not containing singularities other than $\lambda = 0$ and if

$$P = \tfrac{1}{2\pi i} \int_\gamma (\lambda - B)^{-1} d\lambda,$$

then P is a projection onto $\mathcal{K}(B^{k+1})$ and $\mathcal{R}(1 - P) = \mathcal{R}(B^{k+1})$. Furthermore, it is easy to verify that $\|\lambda^{k+1}(\lambda - B)^{-1}\|_{\mathcal{L}(X)} \leq C$ for any λ such that $0 < |\lambda| \leq \varepsilon$. Hence if the equivalent Cauchy problem

$$(5.29) \qquad \begin{cases} \dfrac{d}{dt}Bw - w = f(t), \quad 0 < t \leq T, \\[2mm] Bw(0) = u_0 \end{cases}$$

has a solution, it is unique. The argument is analogous to the one used previously in this chapter, see Theorem 5.5. It can be shown in a more direct way, exploiting the particular case $X = \mathcal{K}(B^{k+1}) \oplus \mathcal{R}(B^{k+1})$. We refer to A. Favini [81].

Concerning existence, since $\mathcal{K}(B^k) \subset \mathcal{K}(B^{k+1})$, we observe that, if $f \in \mathcal{C}([0, T]; X)$, then the equation in (5.29) is equivalent to the system

$$(5.30) \qquad \frac{d}{dt}B_1(1 - P)w - (1 - P)w = (1 - P)f(t),$$

$$(5.31) \qquad \frac{d}{dt}B_2 Pw - Pw = Pf(t),$$

where B_1 and B_2 denote the parts of B in $\mathcal{R}(B^{k+1})$ and in $\mathcal{K}(B^{k+1})$ respectively. It is also known by T. Kato [117; p. 178] that the spectra of B_1 and B_2 coincide with the spectrum of B minus $\{0\}$ and with $\{0\}$ respectively. Hence, $B_2 \in \mathcal{L}(\mathcal{K}(B^{k+1}))$. On the other hand, B_1 is a closed operator in $\mathcal{R}(B^{k+1})$, mapping $\mathcal{D}(B_1) = \mathcal{D}(B) \cap \mathcal{R}(B^{k+1})$ onto $\mathcal{R}(B^{k+1})$ in a one-to-one fashion. This implies that (5.30) with assigned initial condition $B_1(1 - P)w(0) = (1 - P)u_0$ possesses a unique solution given by

$$(1 - P)w(t) = B_1^{-1}e^{tB_1^{-1}}(1 - P)u_0$$
$$+ \int_0^t B_1^{-1}e^{(t-\tau)B_1^{-1}}(1 - P)f(\tau)d\tau, \quad 0 \leq t \leq T.$$

As regards (5.31), we point out by K. Yosida [187; p. 328] that

$$(\lambda B_2 + 1)^{-1} = \sum_{i=0}^{\infty}(-1)^i\lambda^i B_2^i, \quad \lambda \neq 0,$$

and hence all the powers B_2^m vanish provided that $m \geq k + 1$. This yields the following result.

LEMMA 5.23. *Assume that* $B \in \mathcal{L}(X)$ *has* $\lambda = 0$ *as the unique singularity of* $(\lambda - B)^{-1}$ *with a pole of order* $k + 1$. *If* $f \in \mathcal{C}^k([0,T];X)$, *then the equation in (5.29) possesses the unique solution*

$$w(t) = \sum_{j=0}^{k} B^j f^{(j)}(t), \quad 0 < t \leq T.$$

PROOF. Since $Bw(t) = \sum_{j=1}^{k} B^j f^{(j-1)}(t)$,

$$\frac{d}{dt}Bw(t) = \sum_{j=1}^{k} B^j f^{(j)}(t) = w(t) + f(t).$$

As a consequence of this lemma, we obtain the following proposition.

PROPOSITION 5.24. *Assume that* B *has* $\lambda = 0$ *as a pole of* $(\lambda - B)^{-1}$ *of order* $k + 1$. *If* $f \in \mathcal{C}^k([0,T];X)$, *then the equation in (5.29) possesses the general solution*

$$w(t) = \sum_{j=0}^{k} B_2^j P f^{(j)}(t) + B_1^{-1}e^{tB_1^{-1}}(1 - P)u_1$$
$$+ \int_0^t B_1^{-1}e^{(t-\tau)B_1^{-1}}(1 - P)f(\tau)d\tau, \quad 0 \leq t \leq T,$$

where $u_1 \in X$.

REMARK. We note that no initial condition can be assigned arbitrarily to $Bw(t)$ at $t = 0$. In fact, all admissible initial values u_0 must be of the form

$$u_0 = \overline{w} + \sum_{j=0}^{k-1} B^{j+1} P f^{(j)}(0),$$

where $\overline{w} = (1 - P)u_1 \in \mathcal{R}(B^{k+1})$. On the other hand, if $B \in \mathcal{L}(X)$, taking $u_1 = (B_1^k)^{-1}w_0$, for every $w_0 \in \mathcal{R}(B^{k+1})$, there exists a solution w to (5.29) such that $\|B^{k+1}w(t) - w_0\|_X \to 0$ as $t \to 0$.

COROLLARY 5.25. *Under the assumption in Proposition 5.24, if $f \in C^k([0,T]; X)$ with $f^{(j)}(0) = 0$ for $j = 0, 1, ..., k - 1$, and if $u_0 \in \mathcal{R}(B^{k+1})$, then (5.29) possesses a unique solution w.*

As applications, we verify the following interesting cases:

If X is a Hilbert space and B_0 is a compact linear operator from X into itself such that $\lambda_0 \neq 0$ is an element of $\sigma(B_0)$, then Proposition 5.24 holds true with $B = \lambda_0 - B_0$.

Let B be a closed linear operator with a compact resolvent. If $0 \in \sigma(B)$, then it is an isolated eigenvalue with finite multiplicity (cf. T. Kato [117; p. 181]), and therefore Proposition 2.24 applies again.

Let B be a closed Fredholm operator in X such that

$$r(B) = \lim_{n \to \infty} \alpha(B^n) < \infty, \ r^*(B) = \lim_{n \to \infty} \beta(B^n) < \infty,$$

where, for any closed linear operator in X, $\alpha(S)$ denotes the dimension of $\mathcal{K}(S)$ and $\beta(S)$ is the codimension of $\mathcal{R}(S)$ in X. Then there is a positive integer m such that $X = \mathcal{K}(B^m) \oplus \mathcal{R}(B^m)$ (see M. Schechter [151; pp. 274-275]) and $\lambda = 0$ is an isolated singularity of $(\lambda - B)^{-1}$.

Our next step is to consider the general case that not necessarily $\mathcal{D}(L) \subset \mathcal{D}(M)$.

More precisely, let M and L be two closed linear operators in X such that $\lambda = 0$ is a pole of $L(\lambda L - M)^{-1}$ of order $k+1$ and hence there is $r > 0$ such that $(\lambda L - M)^{-1} \in \mathcal{L}(X)$ for all $0 < |\lambda| < r$. Let us fix $|k_0| > \frac{1}{r}$. We then observe that, if $B = M(k_0 M - L)^{-1}$, then $\lambda = 0$ is a pole of $(\lambda - B)^{-1}$ of order $k + 1$.

Indeed,

$$\lambda - B = (\lambda k_0 - 1)(\lambda(\lambda k_0 - 1)^{-1}L - M)(L - k_0 M)^{-1}.$$

Moreover, it is observed that if $|\lambda| \leq \frac{1}{2|k_0|}$ then $|\lambda k_0 - 1| \geq ||\lambda||k_0| - 1| \geq \frac{1}{2}$ and hence $|\lambda(\lambda k_0 - 1)^{-1}| \leq 2|\lambda|$. Let $\varepsilon = \min\{\frac{r}{2}, \frac{1}{2|k_0|}\}$. Then, for all

$0 < |\lambda| \leq \varepsilon$ we have $|\lambda(\lambda k_0 - 1)^{-1}| \leq r$, so that for these λ's, $\lambda - B$ has a bounded inverse

$$(\lambda - B)^{-1} = (\lambda k_0 - 1)^{-1}(L - k_0 M)(\lambda(\lambda k_0 - 1)^{-1}L - M)^{-1}.$$

Since $|\lambda k_0 - 1| \geq \frac{1}{2}$ and

$$M(\mu L - M)^{-1} = -1 + \mu L(\mu L - M)^{-1}, \quad \mu \in \rho_L(M),$$

we deduce that for $0 < |\lambda| \leq \varepsilon$

$$\|M(\lambda(\lambda k_0 - 1)^{-1}L - M)^{-1}\|_{\mathcal{L}(X)} \leq C|\lambda|^{-(k+1)}$$

as well. Therefore the estimate

$$\|(\lambda - B)^{-1}\|_{\mathcal{L}(X)} \leq C'|\lambda|^{-(k+1)}, \quad 0 < |\lambda| \leq \varepsilon,$$

is obtained, as affirmed.

Application of Proposition 5.24, after a usual change of variable argument, yields the following theorem.

THEOREM 5.26. *Let* $\lambda = 0$ *be a pole of* $L(\lambda M - L)^{-1}$ *of order* $k + 1$ *and let* $B = M(L - k_0 M)^{-1}$ *for* $|k_0|$ *suitably large. Let* P *be the corresponding projection onto* $\mathcal{K}(B^{k+1})$. *If* $f \in C^k([0, T]; X)$ *and* $u_1 \in \mathcal{R}(B^{k+1})$, *then (5.28) possesses the general solution*

$$v(t) = e^{k_0 t} \sum_{j=0}^{k} (L - k_0 M)^{-1} B_2^j P \frac{d^j}{dt^j} \{e^{-k_0 t} f(t)\}$$

$$+ (L - k_0 M)^{-1} B_1^{-1} e^{t(B_1^{-1} + k_0)} u_1$$

$$+ \int_0^t (L - k_0 M)^{-1} B_1^{-1} e^{(t-s)(B_1^{-1} + k_0)} (1 - P) f(s) ds.$$

As before, B_1 *and* B_2 *denote the restrictions of* B *to* $\mathcal{R}(B^{k+1})$ *and* $\mathcal{K}(B^{k+1})$, *respectively.*

Clearly this result improves Theorem 5.7 in virtue of the particular assumption on the resolvents.

Let us mention again some applications.

Let L be a bounded operator in $\mathcal{L}(X)$ and M be a densely defined closed operator in X such that $(\lambda_0 M - L)^{-1}$ is a compact operator in X. Since $\lambda_0 M(\lambda_0 M - L)^{-1} = 1 + L(\lambda_0 M - L)^{-1}$, we deduce that $S = \lambda_0 M(\lambda_0 M - L)^{-1}$ is a Fredholm operator with $r(S) < \infty$, $r^*(S) < \infty$. Moreover, $\mathcal{R}(S) = \mathcal{R}(M)$ is closed in X and this implies (M. Schechter

[152; pp. 150-152]) that $\lambda = 0$ is a pole of $(\lambda - S)^{-1}$ with $X = \mathcal{K}(S^m) \oplus \mathcal{R}(S^m)$ for a certain positive integer m. Therefore, Theorem 5.26 applies.

If L is a compact operator from $\mathcal{D}(M) \to X$ and $(\lambda_0 L + M)^{-1} \in \mathcal{L}(X)$, then a similar result as above is true again.

Let $\mathcal{D}(M) \subset \mathcal{D}(L)$, M be a Fredholm operator in X with $\alpha(M) = \beta(M) < \infty$, and let $(\lambda_0 M - L)^{-1} \in \mathcal{L}(X)$. Then it is well known by Lemma 1.5 of M. Schechter [152; p. 165] that $B = M(\lambda_0 M - L)^{-1}$ is also Fredholm with $\alpha(B) = \beta(B) = \alpha(M)$. If $r(B) < \infty$, we can conclude by M. Schechter [151; p. 151] that $\lambda = 0$ is a pole of $(\lambda - B)^{-1}$.

These results extend to infinite dimensional Banach spaces many results obtained by several authors in various papers and monographs. We refer to F. R. Gantmacher [108], S. L. Campbell [40,41], S. L. Campbell, C. D. Meyer Jr. and N. J. Rose [42].

The representation of X under the form $X = \mathcal{K}(M^{k+1}) \oplus \mathcal{R}(M^{k+1})$ can furnish a very useful tool to treat the resolvent equation $(\lambda M - L)u = f$ where L is a polynomial in the operator M, both in the parabolic case and the hyperbolic one. One example is described below.

EXAMPLE 5.14. Let us consider the abstract equation

$$\frac{d}{dt}\{(s - K)v\} = aKv - bK^2v, \quad 0 < t \leq T,$$

in a Banach space X. Here, K is the generator of an analytic semigroup on X, $0 \in \rho(K)$, $s < 0$ is a simple pole of $(\lambda - K)^{-1}$ so that $X = \mathcal{K}(s - K) \oplus \mathcal{R}(s - K)$, a and b are two positive constants.

Put $M = s - K$ and $L = aK - bK^2 = -bM^2 + (2bs - a)M - bs^2 + as$. Let $P:X \to \mathcal{K}(M)$ be the projection onto $\mathcal{K}(M)$. Let M_1 be the restriction of M to $\mathcal{R}(M) = \overline{\mathcal{R}(M)}$. We observe that $(\lambda M - L)u = f$ if and only if

(5.32) $$s(a - bs)Pu = Pf,$$

(5.33) $$(\lambda + a - 2\beta s)M_1(1 - P)u + bM_1^2(1 - P)u + s(a - bs)(1 - P)u$$
$$= (1 - P)f.$$

Since a, $b > 0$ and $s < 0$, the first equation is uniquely solvable. On the other hand, the second one reads equivalently, using $v = M_1(1 - P)u$,

$$(\lambda + a - 2bs)v + bM_1v + s(a - bs)M_1^{-1}v = (1 - P)f.$$

Notice that $M_1^{-1} \in \mathcal{L}(\mathcal{R}(M))$ and $-bM_1$ generates an analytic semigroup in $\mathcal{R}(M)$. Indeed, since $(\mu + bM)u = f$ has a unique solution for all $f \in X$ and $\Re e\mu \geq 0$, $\mu(1 - P)u + bM_1(1 - P)u = (1 - P)f$ has a unique solution $(1 - P)u \in \mathcal{D}(M_1)$ with $\|(\mu + bM_1)^{-1}\|_{\mathcal{L}(\mathcal{R}(M))} \leq \frac{C}{1+|\mu|}$. Furthermore

$\mathcal{D}(M_1)$ is dense in $\mathcal{R}(M)$ because $\mathcal{D}(M)$ is dense in X so that, given $x \in X$, for all $\varepsilon > 0$ there exists $y \in \mathcal{D}(M)$ with

$$\|x - y\|_X < \varepsilon(\|1 - P\|_{\mathcal{L}(X)} + 1)^{-1}.$$

Hence, $\|(1 - P)(x - y)\|_{\mathcal{R}(M)} < \varepsilon$ and $(1 - P)y = y - Py \in \mathcal{D}(M_1)$. It follows that $-bM_1 - s(a - bs)M_1^{-1}$ generates an analytic semigroup in $\mathcal{R}(M)$. This implies that the equation (5.33) is uniquely solvable for any $\Re\lambda$ sufficiently large and its solution is $(1 - P)u = M_1^{-1}v$. We thus deduce from (5.32) and (5.33) that

$$\|\lambda M u\|_X = |\lambda|\|M_1(1 - P)u\|_X = |\lambda|\|v\|_{\mathcal{R}(M)}$$
$$\leq C\|(1 - P)f\|_{\mathcal{R}(M)} \leq C\|f\|_X,$$

and hence we are in a position to invoke the methods and results in Chapter IV. If K generates a $\mathcal{C}_0$ semigroup on X, a corresonding reasoning works and yields that there exists $C > 0$ such that, possibly after a change of variable argument, for $\Re\lambda$ large, one has the estimate

$$(\Re\lambda + c)\|Mu\|_X \leq C_1\|f\|_X.$$

Therefore we conclude that in this case (5.2) is verified with $m = 1$ and $b = 0$.

As a particular interesting example, let $X = L^2(\Omega)$ and $K = \Delta$ with $\mathcal{D}(K) = H^2(\Omega) \cap H_0^1(\Omega)$. Then the equation above is just the Sobolev equation. For a different approach we refer to G. A. Sviridyuk [163].

5.7 EQUATIONS OF HIGHER ORDER IN TIME

In this section we shall show that the Cauchy problem for the higher order differential equation

$$(5.34) \quad \begin{cases} A_n u^{(n)} + \cdots + A_j u^{(j)} + \cdots + A_1 u' + A_0 u = f(t), & 0 \leq t \leq T, \\ u(0) = u_0, \ u'(0) = u_1, \ ..., \ u^{(n-1)}(0) = u_{n-1} \end{cases}$$

in a Banach space X can be reduced to a first order system in a suitable product space E to which the technique developed in Section 5.2 applies provided that, for example, the operator pencil $P(\lambda) = \sum_{j=0}^{n} \lambda^j A_j$ have an inverse $P(\lambda)^{-1}$ whose norm in $\mathcal{L}(X, \mathcal{D}(A_j))$ is polynomially bounded for $j = 0, 1, 2, ..., n$, and for any $\lambda \in \mathbb{C}$ in the halfplane $\Re\lambda \geq const.$ Here, A_j, $j = 0, 1, 2, ..., n$, are closed linear operators in X, $f(t)$ is a given function in $\mathcal{C}([0, T]; X)$ and $u_j \in X$ for $j = 0, 1, 2, ..., n - 1$.

In the next chapter we shall indicate a number of conditions on the single operators A_j implying that our actual assumptions on the operator pencil $P(\lambda)$ are verified, pointing out in the meantime the deep difference between the parabolic case and the (highly degenerate) hyperbolic case.

To limit the extent of the paragraph we shall confine our attention to strict solutions of (5.34), that is, functions $u \in \mathcal{C}^n([0,T]; X)$ such that $u^{(j)}(t) = \frac{d^j u}{dt^j}(t) \in \mathcal{D}(A_j)$ for all $0 \leq t \leq T$ and every $j = 0, 1, ..., n$, $A_j u^{(j)} \in \mathcal{C}([0,T]; X)$, and (5.34) holds. Our main assumption on the operators A_j reads:

(5.35) A_0 and $P(\lambda) = \sum_{j=0}^n \lambda^j A_j$ have a bounded inverse for all λ in the logarithmic region

$$\Lambda = \{\lambda \in \mathbb{C}; \Re\lambda \geq \tilde{a} + b\log(1 + |\lambda|)\},$$

where $\tilde{a}$, $b \geq 0$, and there exist $m_j \geq 0$, $j = 1, 2, ..., n$, such that

$$\|A_j P(\lambda)^{-1}\|_{\mathcal{L}(X)} \leq C(1 + |\lambda|)^{m_j}, \quad \lambda \in \Lambda, \ j = 1, 2, ..., n,$$

with some constant C.

Define $\nu = \max\{m_j + j; j = 1, 2, ..., n\}$ and $X_j = \bigcap_{k=j}^n \mathcal{D}(A_k)$, $j = 1, 2, ..., n - 1$, where it is obviously supposed that $X_j \neq \{0\}$ and X_j is endowed with the norm $\|u\|_{X_j} = \max\{\|u\|_{\mathcal{D}(A_k)}; k = j, j + 1, ..., n\}$. Let

$$X_n = X, \quad Z = X_1 \times X_2 \times \cdots \times X_n,$$

$$C_j(\lambda) = \sum_{k=j}^n \lambda^{k-j} A_k, \quad \mathcal{D}(C_j(\lambda)) = X_j, \ j = 1, 2, ..., n - 1,$$

$$C_n(\lambda) = I = \text{(the identity operator in } X).$$

Let us set

$$u^{(j)}(t) = v_j(t), \quad j = 0, 1, ..., n - 1,$$

$$v(t) = (v_0(t), v_1(t), ..., v_{n-1}(t)),$$

$$v_0 = (u_0, ..., u_{n-1}), \ X^n = X \times X \times \cdots \times X,$$

$$F(t) = (0, 0, ..., 0, f(t))$$

and introduce two operators L and M in X^n by

$$\mathcal{D}(L) = \mathcal{D}(A_0) \times \mathcal{D}(A_1) \times \cdots \times \mathcal{D}(A_{n-1}),$$

$$Lv = (v_1, v_2, ..., v_{n-1}, -\sum_{j=0}^{n-1} A_j v_j),$$

$$\mathcal{D}(M) = \underbrace{X \times X \times \cdots \times X}_{n-1 \text{ times}} \times \mathcal{D}(A_n), \quad Mv = (v_0, v_1, ..., v_{n-2}, A_n v_{n-1}).$$

Then (5.34) assumes the form

$$\begin{cases} M\dfrac{dv}{dt} = Lv + F(t), & 0 \leq t \leq T, \\ v(0) = v_0. \end{cases}$$

In order to find a space in which the arguments in Section 5.2 hold, a careful description of the solution v to $(\lambda M - L)v = h$ is needed. To this end we observe that $(\lambda M - L)v = h = (h_0, h_1, ..., h_{n-1})$ is equivalent to the system

$$\lambda v_j - v_{j+1} = h_j, \quad j = 0, 1, 2, ..., n-2,$$

$$\sum_{j=0}^{n-2} A_j v_j + (\lambda A_n + A_{n-1})v_{n-1} = h_{n-1}.$$

In view of assumption (5.35), we deduce that, for all $\lambda \in \Lambda$,

$$v_0 = P(\lambda)^{-1} \sum_{j=0}^{n-1} C_{j+1}(\lambda)h_j,$$

$$v_j = \lambda^j v_0 - \sum_{k=0}^{j-1} \lambda^{j-1-k} h_k, \quad j = 1, 2, ..., n-1.$$

It follows that

$$v_j = \sum_{k=0}^{j-1} \{\lambda^j P(\lambda)^{-1} C_{k+1}(\lambda) - \lambda^{j-1-k}\} h_k$$

$$+ \sum_{k=j}^{n-1} \lambda^j P(\lambda)^{-1} C_{k+1}(\lambda)h_k, \quad j = 1, 2, ..., n-1,$$

and hence, by some cumbersome calculations we conclude that $(\lambda M - L)u = f = (f_1, f_2, ..., f_n) \in Z$ has the unique solution $u = (u_1, u_2, ..., u_n)$, where

$$u_1 = \sum_{k=1}^{n} P(\lambda)^{-1} C_k(\lambda) f_k,$$

$$u_j = \sum_{k=1}^{j-1} \lambda^{j-1-k} \{\lambda^k P(\lambda)^{-1} C_k(\lambda) - 1\} f_k$$

$$+ \sum_{k=j}^{n} \lambda^{j-1} P(\lambda)^{-1} C_k(\lambda) f_k, \quad j = 2, 3, ..., n.$$

We in fact show that Z is just a good choice for the ambient space to our equation, because $\|M(\lambda M - L)^{-1}\|_{\mathcal{L}(Z)}$ is polynomially bounded on Λ. We first observe that (5.35) yields

$$\|A_0 P(\lambda)^{-1}\|_{\mathcal{L}(X)} \leq 1 + C \sum_{j=1}^{n} |\lambda|^j (1 + |\lambda|)^{m_j} \leq C'(1 + |\lambda|)^\nu, \quad \lambda \in \Lambda.$$

We estimate $\|u_1\|_{X_1} = \|u_1\|_X + \sum_{j=1}^{n} \|A_j u_1\|_X$. It is seen that, for $j = 1, 2, ..., n$,

$$\|u_1\|_X \leq C(1 + |\lambda|)^{\nu+n} \sum_{k=1}^{n} (1 + |\lambda|)^{-k} \|f_k\|_{X_k} \leq C(1 + |\lambda|)^{\nu+n-1} \|f\|_Z,$$

$$\|A_j u_1\|_X \leq C \sum_{k=1}^{n} (1 + |\lambda|)^{m_j+n} (1 + |\lambda|)^{-k} \|f_k\|_{X_k}$$

$$\leq C(1 + |\lambda|)^{\nu+n-1-j} \|f\|_Z;$$

and therefore,

$$\|u_1\|_{X_1} \leq C(1 + |\lambda|)^{\nu+n-1} \|f\|_Z, \quad f \in Z, \ \lambda \in \Lambda.$$

Suppose now $j \in \{2, 3, ..., n-1\}$. Then,

$$\|u_j\|_X \leq C\left\{ \sum_{k=1}^{j-1} |\lambda|^{j-1-k} \{ |\lambda|^k (1 + |\lambda|)^\nu (1 + |\lambda|)^{n-k} + 1 \} \|f_k\|_{X_k} \right.$$

$$\left. + \sum_{k=j}^{n} |\lambda|^{j-1} (1 + |\lambda|)^{\nu+n-k} \|f_k\|_{X_k} \right\} \leq C(1 + |\lambda|)^{\nu+n+\max\{n-3,-1\}} \|f\|_Z.$$

Let $i \geq j$. Then,

$$\|A_i u_j\|_X \leq C\left\{ \sum_{k=1}^{j-1} |\lambda|^{j-1} \|A_i P(\lambda)^{-1} C_k(\lambda) f_k\|_X + \sum_{k=1}^{j-1} |\lambda|^{j-1-k} \|A_i f_k\|_X \right.$$

$$\left. + \sum_{k=j}^{n} |\lambda|^{j-1} \|A_i P(\lambda)^{-1} C_k(\lambda) f_k\|_X \right\}$$

$$\leq C\left\{ (1 + |\lambda|)^{j-1+m_i+n} \sum_{k=1}^{n} (1 + |\lambda|)^{-k} \|f_k\|_{X_k} \right.$$

$$\left. + |\lambda|^{j-1} \sum_{k=1}^{j-1} |\lambda|^{-k} \|f\|_Z \right\} \leq C\left\{ (1 + |\lambda|)^{\nu+n-2} + (1 + |\lambda|)^{j-2} \right\} \|f\|_Z.$$

Here, $2 \leq j \leq n - 1$ implies $\nu + n - 2 \geq j - 2$. Thus we conclude that

$$\|A_i u_j\|_X \leq C(1 + |\lambda|)^{\nu+n-2}\|f\|_Z,$$

and

$$\|u_j\|_X + \sum_{i=j}^{n} \|A_i u_j\|_X \leq C\Big\{(1 + |\lambda|)^{\nu+n+\max\{n-3,-1\}}$$
$$+ \sum_{i=j}^{n}(1 + |\lambda|)^{\nu+n-2}\Big\}\|f\|_Z \leq C(1 + |\lambda|)^{\nu+n+\max\{n-3,-1\}}\|f\|_Z.$$

It remains to estimate $\|A_n u_n\|_X$. Since

$$u_n = \sum_{j=1}^{n} \lambda^{n-1} P(\lambda)^{-1} C_j(\lambda) f_j - \sum_{j=1}^{n-1} \lambda^{n-1-j} f_j,$$

it is bounded by

$$\|A_n u_n\|_X$$
$$\leq \sum_{j=1}^{n} |\lambda|^{n-1}\|A_n P(\lambda)^{-1}\|_{\mathcal{L}(X)}\|C_j(\lambda) f_j\|_X + \sum_{j=1}^{n-1} |\lambda|^{n-1-j}\|A_n f_j\|_X$$
$$\leq C\Big\{\sum_{j=1}^{n} |\lambda|^{n-1}(1 + |\lambda|)^{m_n+n-j}\|f_j\|_{X_j} + \sum_{j=1}^{n-1} |\lambda|^{n-1-j}\|f_j\|_{X_j}\Big\}$$
$$\leq C(1 + |\lambda|)^{\nu+n-2}\|f\|_Z.$$

This together with the estimates obtained above yields that

$$\|M(\lambda M - L)^{-1}\|_{\mathcal{L}(X)}$$
$$\leq C\{(1 + |\lambda|)^{\nu+n-1} + (1 + |\lambda|)^{\nu+n+\max\{n-3,-1\}} + (1 + |\lambda|)^{\nu+n-2}\},$$
$$\leq C(1 + |\lambda|)^{\nu+n+\max\{n-3,-1\}}.$$

In this way, Theorem 5.10 applies with $m = \nu + n + \max\{n - 2, 0\}$. The function f shall be required to belong to $\mathcal{C}^h([0, T]; X)$, where $h > m+bT+2$. Moreover the initial values u_j must satisfy the compatibility relations with $f(0), f'(0), \ldots, f^{(h-1)}(0)$ that are deduced from the general link detailed in the quoted theorem.

In this case they reduce formally to operate $L^{-1}M\frac{d}{dt}$ $(h - 1)$-times to both the nembers of $L^{-1}Mu' + u = L^{-1}F$ and evaluate the obtained identity at $t = 0$.

As already said, in the next chapter we shall describe conditions on the operators A_j implying assumption like (5.35), even with better estimates (the parabolic situation).

5.8 INTEGRATED SOLUTIONS TO
IMPLICIT DIFFERENTIAL EQUATIONS

In this section we show that the Cauchy problem (D-E.1) in a Banach space X, under the assumption (M-L) with $b = 0$, has always a "solution" provided that it is understood as an integrated solution even if (D-E.1) may not be solvable in strict sense, that is, the initial condition and the equation in (D-E.1) are not compatible. The main results are contained in W. Arendt and A. Favini [8] and extend well known properties of k-integrated solutions of the problem (D-E.1) in the non degenerate case ($M = I$); see a.e. W. Arendt [9], H. Kellermann and M. Hieber [119], and the even more general theory of regularized semigroup in M. Hieber, A. Holderrieth and F. Neubrander [116].

We have seen that in general both regularity (for arbitrary f) and compatibility relations between $f^{(j)}$ and v_0, for some $j \in \mathbb{N}$, are necessary to have a solution, even if we require the equation in (D-E.1) to be satisfied only for $t > 0$. Nevertheless it is well known in control theory that also a not compatible system (D-E.1) allows a corresponding "answer" v, in some weak sense, to an "arbitrary" input f. Our aim here is to clarify the meaning of this answer. In particular, we show that after definition of k-integrated solution to (D-E.1), which reduces to the one relative to the non degenerate system when $M = I$, it always exists, provided that k is sufficiently large, for any $v_0 \in \mathcal{D}(M)$ and $f \in \mathcal{C}([0,T]; X)$.

In order to motivate our notion of integrated solution, we observe that, if v is a solution of $\mathcal{C}^1$ class, the position

$$v_1(t) = \int_0^t v(\tau)d\tau,$$

$$v_k(t) = \int_0^t v_{k-1}(\tau)d\tau, \quad k = 2, 3, 4, ...,$$

leads to the relation

$$(5.36) \qquad Mv_k(t) = \int_0^t Lv_k(\tau)d\tau + \frac{t^k}{k!}Mv_0 + f_{k+1}(t), \quad k = 1, 2, 3, ...,$$

where $f_{k+1}(t) = \int_0^t \frac{(t-\tau)^k}{k!}f(\tau)d\tau$. On the other hand, introducing $w_k(t) = Lv_k(t)$, $B = ML^{-1}$, $u_0 = Mv_0$, (5.36) is written as

$$(5.37) \qquad Bw_k(t) = \int_0^t w_k(\tau)d\tau + \frac{t^k}{k!}u_0 + f_{k+1}(t), \quad k = 1, 2, 3, ...$$

We are then ready to the definition as follows.

DEFINITION. A function $v \in \mathcal{C}([0, T]; \mathcal{D}(L))$ such that (5.36) holds (with v instead of v_k) is *a k-integrated solution* to (D-E.1).

Equivalently we could say that v is a k-integrated solution to (D-E.1) if $w = Lv$ is a k-integrated solution to

(5.38)
$$\begin{cases} \dfrac{d}{dt}Bw - w = f(t), & 0 \le t \le T, \\ Bw(0) = u_0. \end{cases}$$

REMARK. We could think of an alternative definition of integrated solution to (D-E.1), in some strict analogy to the regular case $M = I$, weaking the (apparently more restrictive) assumption that the function v in the above definition belongs to $\mathcal{C}([0, T]; \mathcal{D}(L))$. Precisely, we could require that $v \in \mathcal{C}([0, T]; X)$, $v(t) \in \mathcal{D}(M)$ for all $t \in [0, T]$, $Mv \in \mathcal{C}([0, T]; X)$ and

$$Mv(t) = L \int_0^t v(\tau)d\tau + \frac{t^k}{k!}u_0 + f_{k+1}(t).$$

But if $S = L^{-1}M (\in \mathcal{L}(\mathcal{D}(L)))$, this reads

$$Sv(t) = \int_0^t v(\tau)d\tau + \frac{t^k}{k!}Sv_0 + L^{-1}f_{k+1}(t).$$

Notice that the integral $\int_0^t v(\tau)d\tau \in \mathcal{D}(L)$. Hence this formula is considered in the space $\mathcal{D}(L)$, endowed with the graph norm. As an operator from $\mathcal{D}(L)$ to itself, the operator S satisfies precisely assumption (M-L), since

$$\|(\lambda S + 1)^{-1}f\|_{\mathcal{D}(L)} = \|LP(\lambda)^{-1}Lf\|_X \le C(1 + |\lambda|)^m\|f\|_{\mathcal{D}(L)},$$

for all $f \in \mathcal{D}(L)$ and $\Re\lambda \ge 0$. Thus it follows that the two definitions are in fact equivalent.

THEOREM 5.27. *Let us assume (M-L) with $b = 0$. If $k \ge m + 3$, then (D-E.1) in Section 5.2 possesses a unique k-integrated solution v for any $f \in \mathcal{C}([0, T]; X)$ and any $v_0 \in \mathcal{D}(M)$.*

PROOF. We need to show that (5.38) has a unique k-integrated solution. To accomplish this, we remark that if $w = w(t)$ is a $\mathcal{C}^1$ solution to

(5.39)
$$\begin{cases} \dfrac{d}{dt}Bw - w = \dfrac{t^{k-1}}{(k-1)!}u_0 + f_k(t), & 0 \le t \le T, \\ Bw(0) = 0 \end{cases}$$

then such a solution satisfies the equation (5.37) and conversely. To solve (5.39) it suffices to apply Theorem 5.7 with the nonhomogeneous function $F(t)$ given by

$$F(t) = \frac{t^{k-1}}{(k-1)!}u_0 + f_k(t), \quad 0 \le t \le T.$$

Since, for each $j = 0, 1, 2, ..., m+1$,

$$F^{(j)}(t) = \frac{t^{k-1-j}}{(k-1-j)!}u_0 + \int_0^t \frac{(t-\tau)^{k-1-j}}{(k-1-j)!} f(\tau)d\tau,$$

assumption $k \geq m+3$ yields $F^{(j)}(0) = 0$, $j = 0, 1, 2, ..., m+1$, and Theorem 5.7 in fact applies.

We remark that if the function w satisfying (5.39) is k-times continuously differentiable on $[0, T]$, then $w^{(k)}(t) = z(t)$ fulfils (5.38), for, deriving both the members of (5.37), we infer

$$Bw^{(j)}(t) = w^{(j-1)}(t) + \frac{t^{k-j}}{(k-j)!}u_0 + \int_0^t \frac{(t-\tau)^{k-j}}{(k-j)!} f(\tau)d\tau,$$

$$j = 1, 2, 3, ..., k.$$

Hence a further derivation furnishes the result. However, in general, no initial condition to $z(t)$ can be specified.

On the other hand, even if (D-E.1) could have no C^1 solution, according to Theorem 5.27 all possible solutions to the equation in (D-E.1) are to be sought among its integrated solutions. More precisely, if we denote by v_k the k-integrated solution to (D-E.1) with $k \geq m+3$, then by the definition it follows that, whenever a C^1 solution v exists, we have: $v = v_k^{(k)}$. However, in contrast to the regular case (see W. Arendt [9]), it can happen that v_{m+3} is infinitely differentiable from $[0, T]$ into X, but there does not exist any C^1 solution (see Example 5.15 below).

We illustrate the preceding remarks by means of two trivial examples.

EXAMPLE 5.14. Consider the algebraic-differential system

$$\begin{cases} \dfrac{d}{dt}(u+v) + u = f(t), & 0 \leq t \leq T, \\ \quad\quad\quad 0 + v = g(t), & 0 \leq t \leq T, \end{cases}$$

where $f, g \in C([0, T]; \mathbb{C})$. Obviously the solution is given by

$$u(t) = e^{-t}(u(0) + g(0)) - g(t) + \int_0^t e^{-(t-\tau)}(f(\tau) + g(\tau))d\tau,$$

$$v(t) = g(t).$$

The assumption (M-L) holds with $m = 0$ (and hence $k = 3$). In this case the 3-integrated solution $x(t)$, $y(t)$ to the problem satisfies

$$y(t) = \int_0^t \frac{(t-\tau)^2}{2} g(\tau)d\tau,$$

while

$$(x(t) + y(t))' = -x(t) + \frac{t^2}{2}(x(0) + y(0)) + \int_0^t \frac{(t-\tau)^2}{2} f(\tau) d\tau.$$

implies

$$\frac{d}{dt}(x(t) + y(t))^{(3)} = -(x(t) + y(t))^{(3)} + f(t) + g(t).$$

Hence we have:

$$\frac{d^3}{dt^3}(x(t) + y(t)) = e^{-t}(x + y)^{(3)}(0) + \int_0^t e^{-(t-\tau)}(f(\tau) + g(\tau)) d\tau.$$

Therefore, $u(t) = \frac{d^3}{dt^3} x(t)$, $v(t) = \frac{d^3}{dt^3} y(t)$ solves uniquely the Cauchy problem with given $u(0)$.

EXAMPLE 5.15. Consider

$$\begin{cases} \dfrac{d}{dt} v + u = f(t), & 0 \le t \le T, \\ v = g(t), & 0 \le t \le T, \\ v(0) = v_0, \end{cases}$$

where f, $g \in \mathcal{C}([0,T]; \mathbb{C})$ and $v_0 \in \mathbb{C}$. (M-L) holds with $m = 1$ and hence $k = 4$. The 4-integrated solution x, y to the problem is characterized by the relations

$$\frac{dy}{dt}(t) = -x(t) + \frac{t^3}{6} v_0 + \int_0^t \frac{(t-\tau)^3}{6} f(\tau) d\tau,$$

$$y(t) = \int_0^t \frac{(t-\tau)^3}{6} g(\tau) d\tau$$

so that the unique $\mathcal{C}^1$ solution to the problem exists if and only if $v_0 = g(0)$ and $g \in \mathcal{C}^1([0,T]; \mathbb{C})$; it in fact coincides with $(x^{(4)}(t), y^{(4)}(t))$.

In order to find the k-integrated solution to (D-E.1) we have used Theorem 5.7 which was derived by a very general result like Theorem 5.1. We can expect that an "ad hoc" technique allows to obtain a slightly more general version of Theorem 5.27 in which mainly L is not necessarily invertible.

THEOREM 5.28. *Suppose that there exist $k > 0$, $a \ge 0$, $0 \le \beta < 1$ such that $P(\lambda)$ is invertible for $\Re \lambda \ge a$ and*

$$\|P(\lambda)^{-1}\|_{\mathcal{L}(X, \mathcal{D}(L))} \le k(1 + |\lambda|)^{m+\beta}, \qquad \Re \lambda \ge a,$$

where $m \in \{-1, 0, 1, ...\}$. Then for any $v_0 \in \mathcal{D}(M)$, (D-E.1) possesses a unique $m + 2$-times integrated solution.

Before beginning the proof of the theorem, we point out that one integration has been gained with respect to Theorem 5.27. We shall need the following lemma from Proposition 3.1 of W. Arendt [9].

LEMMA 5.29. *Let $a \geq 0$, $0 \leq \beta < 1$, $k \geq 0$, $m \in \{-1, 0, 1, ...\}$ and let Z be a Banach space. Assume that $Q:\{\lambda \in \mathbb{C}; \Re e \lambda > a\} \to Z$ is holomorphic and*

$$\|Q(\lambda)\|_Z \leq C|\lambda|^{m+\beta}, \quad \Re e \lambda > a.$$

Then there exists a continuous function $S:[0, \infty) \to Z$ with $S(0) = 0$ which satisfies $\sup_{t \geq 0} \|e^{-\omega t} S(t)\|_Z < \infty$ for all $\omega > a$, and such that

$$Q(\lambda) = \lambda^{m+2} \int_0^\infty e^{-\lambda t} S(t) dt, \quad \Re e \lambda > a.$$

PROOF OF THEOREM 5.28. By applying the lemma to $Q(\lambda) = P(\lambda)^{-1} \in Z = \mathcal{L}(X, \mathcal{D}(L))$, there exists a continuous function $S:[0, \infty) \to Z$ such that $S(0) = 0$, $\sup_{t \geq 0} \|e^{-\omega t} S(t)\|_{\mathcal{L}(X, \mathcal{D}(L))} < \infty$, $\omega > a$, and that

$$P(\lambda)^{-1} = \lambda^{m+2} \int_0^\infty e^{-\lambda t} S(t) dt, \quad \Re e \lambda > a.$$

Hence $t \to LS(t)$ and $t \to MS(t)$ are continuous from $[0, \infty)$ into $\mathcal{L}(X)$ with

$$\sup_{t \geq 0} e^{-\omega t} \|LS(t)\|_{\mathcal{L}(X)} < \infty, \quad \sup_{t \geq 0} e^{-\omega t} \|MS(t)\|_{\mathcal{L}(X)} < \infty, \quad \omega > a.$$

Let $\xi > a$. Then,

$$\xi^{m+3} \int_0^\infty e^{-\xi t} \frac{t^{m+2}}{(m+2)!} dt = 1 = (\xi M - L) P(\xi)^{-1}$$

$$= \xi^{m+3} \int_0^\infty e^{-\xi t} \{MS(t) - \int_0^t LS(\tau) d\tau\} dt.$$

This yields

$$\int_0^\infty e^{-\xi t} \left\{ \frac{t^{m+2}}{(m+2)!} - MS(t) + \int_0^t LS(\tau) d\tau \right\} dt = 0$$

for all $\xi > a$. From the uniqueness theorem for Laplace transforms we deduce that

(5.40) $$\int_0^t LS(\tau) d\tau = MS(t) - \frac{t^{m+2}}{(m+2)!}, \quad t \geq 0.$$

Let $f \in \mathcal{C}([0, T]; X)$, $v_0 \in \mathcal{D}(M)$, and define

(5.41) $$v(t) = S(t) M v_0 + \int_0^t S(\tau) f(t - \tau) d\tau, \quad 0 \leq t \leq T.$$

We show that v is an $m + 2$-integrated solution, i.e. v satisfies (5.36) for $k = m + 2$. Indeed, since $v \in \mathcal{C}([0,T]; \mathcal{D}(L))$,

$$
\begin{aligned}
Mv(t) &= MS(t)Mv_0 + \int_0^t MS(\tau)f(t-\tau)d\tau \\
&= \int_0^t LS(\tau)Mv_0 d\tau + \frac{t^{m+2}}{(m+2)!}Mv_0 \\
&\qquad + \int_0^t \int_0^\tau LS(\sigma)f(t-\tau)d\sigma d\tau + f_{m+3}(t).
\end{aligned}
$$

Therefore, in order to show (5.36), it suffices to show that

$$
(5.42) \qquad \int_0^t Lv(\tau)d\tau = \int_0^t LS(\tau)Mv_0 d\tau + \int_0^t \int_0^\tau LS(\sigma)f(t-\tau)d\sigma d\tau.
$$

We have:

$$
\int_0^t Lv(\tau)d\tau = \int_0^t LS(\tau)Mv_0 d\tau + \int_0^t L\int_0^\tau S(\sigma)f(\tau-\sigma)d\sigma d\tau,
$$

and then, by Fubini's theorem,

$$
\begin{aligned}
\int_0^t L\int_0^\tau S(\sigma)f(\tau-\sigma)d\sigma d\tau &= \int_0^t \int_\sigma^t LS(\sigma)f(\tau-\sigma)d\tau d\sigma \\
&= \int_0^t \int_0^\sigma LS(t-\sigma)f(\tau)d\tau d\sigma = \int_0^t \int_\tau^t LS(t-\sigma)f(\tau)d\sigma d\tau \\
&= \int_0^t \int_0^\tau LS(\sigma)f(t-\tau)d\sigma d\tau.
\end{aligned}
$$

Thus (5.42) is verified. This finishes the proof of existence.

In order to prove the uniqueness we need to establish some commutation properties. We observe that, for $\xi, \eta > a$, $\xi \neq \eta$, we have:

$$
\frac{P(\xi)^{-1} - P(\eta)^{-1}}{\eta - \xi} = P(\xi)^{-1}MP(\eta)^{-1},
$$

and consequently $P(\xi)^{-1}MP(\eta)^{-1} = P(\eta)^{-1}MP(\xi)^{-1}$, $\xi, \eta > a$. Moreover, if we fix $\eta > a$, then, for all $\xi > a$,

$$
\xi^{m+2} \int_0^\infty e^{-\xi t}S(t)MP(\eta)^{-1}dt = P(\xi)^{-1}MP(\eta)^{-1}
$$

$$
= P(\eta)^{-1}MP(\xi)^{-1} = \xi^{m+2}\int_0^\infty e^{-\xi t}P(\eta)^{-1}MS(t)dt,
$$

and thus the uniqueness theorem gives another commutativity relation $S(t)MP(\eta)^{-1} = P(\eta)^{-1}MS(t)$, $\eta > a$. Let $y \in \mathcal{D}(L)$, $x = P(\eta)y$, with $\eta > a$. Then this commutativity implies

$$(\eta M - L)S(t)My = (\mu M - L)S(t)MP(\eta)^{-1}x$$
$$= MS(t)x = MS(t)(\eta M - L)y.$$

Hence, $LS(t)My = MS(t)Ly$, $y \in \mathcal{D}(L)$. Moreover, it follows from (5.40) that

$$(5.43) \qquad \frac{d}{dt}MS(t) = LS(t) + \frac{t^{m+1}}{(m+1)!}, \quad t > 0.$$

Let $v \in \mathcal{C}([0,T]; \mathcal{D}(L))$ be an $m+2$-integrated solution. Then by the definition,

$$\frac{d}{dt}Mv(t) = Lv(t) + \frac{t^{m+1}}{(m+1)!}Mv_0 + f_{m+2}(t), \quad 0 \le t \le T.$$

Fix $0 < t \le T$. If $s \in [0,t]$, let $w(s) = MS(t-s)Mv(s)$. Then, by (5.43),

$$w'(s) = -LS(t-s)Mv(s) - \frac{(t-s)^{m+1}}{(m+1)!}Mv(s) + MS(t-s)\frac{d}{ds}Mv(s),$$

so that it is easily seen that

$$w'(s) = -\frac{(t-s)^{m+1}}{(m+1)!}Mv(s) + \frac{s^{m+1}}{(m+1)!}MS(t-s)Mv_0 + MS(t-s)f_{m+2}(s).$$

Hence,

$$0 = w(t) - w(0) = \int_0^t w'(\tau)d\tau = -\int_0^t \frac{(t-\tau)^{m+1}}{(m+1)!}Mv(\tau)d\tau$$
$$+ \int_0^t \frac{\tau^{m+1}}{(m+1)!}MS(t-\tau)Mv_0 d\tau + \int_0^t MS(t-\tau)f_{m+2}(\tau)d\tau.$$

Differentiating $m+2$-times yields

$$(5.44) \qquad Mv(t) = MS(t)Mv_0 + \int_0^t MS(\tau)f(t-\tau)d\tau.$$

If M is injective, this implies that v is given in the form (5.41), and uniqueness follows. In the general case, let v_1, v_2 be two solutions for some $k \in \{0,1,2,...\}$. Let $v = v_1 - v_2$. Then $v \in \mathcal{C}([0,T]; \mathcal{D}(L))$ and $Mv(t) = \int_0^t Lv(\tau)d\tau$, $0 \le t \le T$. Moreover, (5.44) implies that $Mv(t) = 0$ identically, and thus $Lv(t) = 0$ identically. Since $\xi M + L$, $\xi > a$, is invertible, we conclude from $(\xi M - L)v(t) = 0$ that $v(t) = 0$ identically.

We want to recall that the Laplace tramsform technique similar to ours has been also used by F. Neubrander and A. Rhandi [142; pp. 91-92].

5.9 SOME FURTHER RESULTS ON
CONVERGENCE OF SOLUTIONS

Our main purpose in this section is to establish convergence results on solutions v_n of the abstract equations

$$(\mathcal{E}_n) \qquad\qquad\qquad TM_n v_n - Lv_n = f_n$$

to the solution v of

$$(\mathcal{E}) \qquad\qquad\qquad TMv - Lv = f$$

under the assumptions made in this chapter, and then to apply them to degenerate equations.

This problem has been already considered in literature by means of techniques completely different from ours. We in fact refer to X. Xu [181], where the case of maximal monotone operators is investigated, and to R. K. Lamm and I. G. Rosen [124] relative to the linear case, where a large literature on the subject is provided, too. The treatment of these authors is heavily based on the methods by R. W. Carroll and R. E. Showalter [48], and R. E. Showalter [155] and hence only weak solutions are concerned therein.

For sake of simplicity and for the reader's convenience we immediately list below the assumptions on the operators involved.

(5.45) The closed linear operators M_n, L_n ($n = 1, 2, 3, ...$) and M, L act from the Banach space $\mathcal{X}$ into itself and satisfy $\mathcal{D}(L_n) \subset \mathcal{D}(M_n)$, $\mathcal{D}(L) \subset \mathcal{D}(M)$, $P_n(\lambda) = \lambda M_n - L_n$ and $P(\lambda) = \lambda M - L$ have bounded inverses for all $\lambda \in \Pi_\delta$, where $\Pi_\delta = \{\lambda \in \mathbb{C}; \Re\lambda \geq -\delta\}$, $\delta > 0$, and there exists an integer $m \geq 1$ such that

$$\|L_n P_n(\lambda)^{-1}\|_{\mathcal{L}(\mathcal{X})} \leq C(1 + |\lambda|)^m, \quad \lambda \in \Pi_\delta,$$
$$\|LP(\lambda)^{-1}\|_{\mathcal{L}(\mathcal{X})} \leq C(1 + |\lambda|)^m, \quad \lambda \in \Pi_\delta,$$

with some constant C independent of n.

(5.46) $L_n P_n(\lambda)^{-1} \to LP(\lambda)^{-1}$ strongly in $\mathcal{L}(\mathcal{X})$ as $n \to \infty$ for all $\lambda \in \Pi_\delta$.

As it was observed previously in proving Theorem 3.31, if (5.45) holds, then (5.46) reduces to assume that $B_n = M_n L_n^{-1}$ is strongly convergent in $\mathcal{L}(\mathcal{X})$ to the operator $B = ML^{-1}$ as $n \to \infty$.

Concerning the closed linear operator T from $\mathcal{X}$ into itself, we assume that it fulfils $(\mathcal{T})$ in Section 5.1 with $b = 0$ and $p = 0$, that is,

$$\|(\lambda - T)^{-1}\|_{\mathcal{L}(\mathcal{X})} \leq C, \quad \Re\lambda \leq a,$$

where $a > 0$.

In view of Theorem 5.1 we know that $(\mathcal{E}_n)$ and $(\mathcal{E})$ possess the solutions

$$v_n = \tfrac{1}{2\pi i} \int_\varphi \lambda^{-(m+2)} P_n(\lambda)^{-1} (\mathcal{T} - \lambda)^{-1} \mathcal{T}^{m+2} f_n d\lambda,$$

$$v = \tfrac{1}{2\pi i} \int_\varphi \lambda^{-(m+2)} P(\lambda)^{-1} (\mathcal{T} - \lambda)^{-1} \mathcal{T}^{m+2} f d\lambda,$$

respectively, where φ is the straight line $\Re\lambda = \delta$ oriented by $\Im\lambda$ increasing from $-\infty$ to ∞, provided that $f_n,\, f \in \mathcal{D}(\mathcal{T}^{m+2})$ for all n. The arguments that have yielded Theorem 3.31, when applied to the present situation, give the following result.

THEOREM 5.29. *Under the assumptions (5.45), (5.46) and $(\mathcal{T})$ with $b = 0$ and $p = 0$, if $f_n,\, f \in \mathcal{D}(\mathcal{T}^{m+2})$ and $\mathcal{T}^{m+2} f_n \to \mathcal{T}^{m+2} f$ in $\mathcal{X}$ as $n \to \infty$, then $L_n v_n \to L v$ in $\mathcal{X}$ as $n \to \infty$.*

In order to translate last result to the convergence problem relative to the Cauchy problem for degenerate equations of type (D-E.1) in a Banach space X, where this time we have closed linear operators M_n, L_n acting in X satisfying (5.45) and (5.46) with X instead of $\mathcal{X}$, we meet the difficulty, due to the high degeneration of the operator pencil involved, that necessarily some compatibility relations between v_0, $f(0)$, $f'(0)$, $\dots$, $f^{(m+1)}(0)$, like $L v_0 = -\sum_{j=0}^{m+1} (-1)^j B^j f^{(j)}(0) + B^{m+2} \bar{v}$ for a suitable $\bar{v} \in X$, must hold, according to the theorems obtained in Section 5.2.

Here, of course, we are concerned with the convergence of solutions v_n of the approximating problems

(D-E$_n$.1)
$$\begin{cases} \dfrac{d}{dt} M_n v_n = L_n v_n + f_n(t), & 0 \le t \le T, \\ M_n v_n(0) = M_n v_{0,n}, \end{cases}$$

$n = 1, 2, 3, \dots$, to the solution v of

(D-E.1)
$$\begin{cases} \dfrac{d}{dt} M v = L v + f(t), & 0 \le t \le T, \\ M v(0) = M v_0. \end{cases}$$

In virtue of the aforementioned remark we easily deduce the following proposition.

PROPOSITION 5.30. *Let us assume that operators M_n, L_n and M, L satisfy (5.45) and (5.46) in the space X. Let $f_n,\, f \in C^{m+2}([0,T]; X)$,*

$$\|f_n - f\|_{C^{m+2}} \to 0 \quad \text{as } n \to \infty$$

with the conditions

$$L_n v_{0,n} = - \sum_{j=0}^{m+1} B_n^j f_n^{(j)}(0) + B_n^{m+2} w_{0,n},$$

$$L v_0 = - \sum_{j=0}^{m+1} B^j f^{(j)}(0) + B^{m+2} w_0,$$

where $\|w_{0,n} - w_0\|_X \to 0$ *as* $n \to \infty$, $w_{0,n}$, $w_0 \in X$. *Then*, $L_n v_n \to L v$ *in* $\mathcal{C}([0,T];X)$ *as* $n \to \infty$.

PROOF. It is an immediate consequence of the representation formula of the solutions to (D-E$_n$.1) and (D-E.1) as expressed in Theorems 5.7 and 5.29.

Clearly it could be happen that the approximating problems (D-E$_n$.1) be regular in the sense that M_n has a bounded inverse and $L_n M_n^{-1}$ generates a $\mathcal{C}_0$ semigroup or an integrated semigroup. In the first case we may weaken the strong compatibility and regularity assumptions relative to $v_{0,n}$ and f_n, but they reappear in the limit problem (D-E.1) necessarily. In the second case we already know that these conditions are necessary to guarentee the existence of a solution to (D-E$_n$.1). Proposition 5.30 then says in particular that if (5.45),(5.46) and ($\mathcal{T}$) are verified and $f_n \to f$ in $\mathcal{C}^{m+2}([0,T];X)$ as $n \to \infty$, then the solution v_n to (D-E$_n$.1) converges pointwise to the solution v to (D-E.1) provided that $L_n v_n(0) \to L v(0)$. The notion of integrated solutions that we have introduced in the preceding section seems then convenient to avoid the difficulty connected with regularity and compatibility. Indeed, we have the following proposition for integrated solutions.

PROPOSITION 5.31. *Let us suppose that (5.45) and (5.46) hold in the Banach space* X, *and let* $M_n v_{0,n} \to M v_0$ *in* X *as* $n \to \infty$, $f_n \to f$ *in* $\mathcal{C}([0,T];X)$, *where* f_n, $f \in \mathcal{C}([0,T];X)$. *If* z_n *and* z *denote the* $m+3$- *integrated solutions of problems (D-E$_n$.1) and (D-E.1) respectively, then for all* $0 \le t \le T$, $L_n z_n(t) \to L z(t)$ *in* X *as* $n \to \infty$.

PROOF. First of all, the existence of z_n and z is assured by Theorem 5.27. In virtue of the definition of an integrated solution, we have that $L_n z_n(t) = w_n(t)$ satisfies $B_n w_n(0) = 0$ and

$$\frac{d}{dt} B_n w_n - w_n = F_n(t), \quad 0 \le t \le T,$$

where

$$F_n(t) = \frac{t^{m+2}}{(m+2)!} w_{0,n} + \int_0^t \frac{(t-\tau)^{m+2}}{(m+2)!} f_n(\tau) d\tau, \ w_{0,n} = M_n v_{0,n}.$$

Our assumptions imply that F_n is $(m+2)$-times continuously differentiable and $F_n^{(j)} \to F^{(j)}$ in $\mathcal{C}([0,T];X)$, $j = 0,1,2,...,m+2$, where $F_n^{(i)}(0) = F^{(i)}(0) = 0$ for $i = 0,1,2,\ldots,m+1$, and

$$F(t) = \frac{t^{m+2}}{(m+2)!}w_0 + \int_0^t \frac{(t-\tau)^{m+2}}{(m+2)!}f(\tau)d\tau, \quad w_0 = Mv_0.$$

Hence, in virtue of Proposition 5.30 (applied to $v_{0,n} = v_0 = 0$, $f_n = F_n$, $f = F$, $w_{0,n} = w_0 = 0$), we deduce that $w_n(t)$ converges to $w(t)$ uniformly on $[0,T]$, $Bw(0) = 0$ and

$$\frac{d}{dt}Bw - w = F(t), \quad 0 \le t \le T;$$

i.e. for the $(m+3)$-integrated solution z_n of problem (D-E$_n$.1), $L_n z_n(t)$ converges uniformly on $[0,T]$ to $Lz(t)$, where z is the $(m+3)$-integrated solution to (D-E.1).

EXAMPLE 5.15. Consider the case when X is a Hilbert space. Let M and L be two closed linear linear operators in X such that

$$\Re e(Lv,v)_X \le -c_0\|v\|_X^2, \quad v \in \mathcal{D}(L),$$
$$\Re e(L^*f,f)_X \le -c_0\|f\|_X^2, \quad f \in \mathcal{D}(L^*),$$

where c_0 is a positive constant, $M \in \mathcal{L}(X)$ is selfadjoint and non negative. As appoximating operators M_n and L_n to M and L respectively, we take closed linear operators M_n and L_n, $n = 1,2,3,...$, in X satisfying

$$\Re e(L_n v,v)_X \le -c_0\|v\|_X^2, \quad v \in \mathcal{D}(L_n),$$
$$\Re e(L_n^*f,f)_X \le -c_0\|f\|_X^2, \quad f \in \mathcal{D}(L_n^*),$$

with the same positive constant $c_0 > 0$. In addition, $M_n \in \mathcal{L}(X)$ is selfadjoint and non negative, $M_n \to M$ and $L_n^{-1} \to L^{-1}$ strongly in $\mathcal{L}(X)$ as $n \to \infty$. From Example 5.1, we know that for fixed $a > 0$, $P(\lambda) = \lambda M - L$ and $P_n(\lambda) = \lambda M_n - L_n$ have bounded inverses with

$$\|P(\lambda)^{-1}\|_{\mathcal{L}(X)} \le c_0^{-1}, \|P_n(\lambda)^{-1}\|_{\mathcal{L}(X)} \le c_0^{-1}, \quad \Re e\lambda \ge a.$$

On the other hand, since $\|M_n\|_{\mathcal{L}(X)} \le K_1 < \infty$, it follows from $L_n P_n(\lambda)^{-1} = -1 + \lambda M_n P_n(\lambda)^{-1}$ that $\|L_n P_n(\lambda)^{-1}\|_{\mathcal{L}(X)} \le 1 + |\lambda|K_1 c_0^{-1}$, $\Re e\lambda \ge a$. Therefore, Propositions 5.30 and 5.31 can be applied with $m = 1$.

In order to illustrate and to clarify the use of integrated solutions and the range of application of the last affirmations, we treat the very simple (but lighting, to our opinion) equation (without fixed initial conditions)

$$\begin{cases} \dfrac{dv}{dt} = -u + 1, & t \ge 0, \\ 0 = -v + t, & t \ge 0. \end{cases}$$

Let $u(0) = u_0$, $v(0) = v_0$. We approximate this problem (notice that the operators M and L there satisfy all our assumptions in the space $\mathbb{C}^2$ with $m = 1$) by means of solutions to

$$\begin{cases} \dfrac{d}{dt}\left(\dfrac{1}{n}u_n + v_n\right) = -u_n + 1, & t \geq 0, \\[2mm] \dfrac{d}{dt}\left(\dfrac{1}{n}v_n\right) = -v_n + t, & t \geq 0, \end{cases}$$

where $\frac{1}{n}u_n(0) + v_n(0) = \frac{1}{n}u_0 + v_0$, $\frac{1}{n}v_n(0) = \frac{1}{n}v_0$. Instead of solving these equations directly, according to Proposition 5.31, we consider the 4-integrated solution z_n, w_n, i. e.

$$\frac{d}{dt}\left(\frac{1}{n}z_n + w_n\right) = -z_n + \frac{t^3}{6}\left(\frac{1}{n}u_0 + v_0\right) + \int_0^t \frac{(t-\tau)^3}{6}\,d\tau,$$

$$\frac{d}{dt}\left(\frac{1}{n}w_n\right) = -w_n + \frac{t^3}{6n}v_0 + \int_0^t \frac{(t-\tau)^3}{6}\tau\,d\tau,$$

$$w_n(0) = z_n(0) = 0.$$

Letting to the limit as $n \to \infty$, we recognize that $w_n(t)$ converges to $\frac{t^5}{5!}$, independently of any initial conditions. Analogously we see that $z_n^{(4)}(t)$ converges to 0 pointwise, since $z_n^{(4)}(t) = e^{-nt}u_0 + n^2 e^{-nt}t(v_0 + \frac{1}{n})$, $t > 0$. Hence, $(z_n^{(4)}(t), w_n^{(4)}(t)) \to (0, t)$, $t > 0$. Notice that in fact $(0, t)$ is the unique true solution to the original degenerate problem.

EXAMPLE 5.16. Let $m(x)$, $a_j(x)$, $b(x)$, $x \in \mathbb{R}^n$, $j = 1, 2, ..., n$, be the square $N \times N$ matrices introduced in Example 5.9, where α is a sufficiently large real number, and let us suppose that all the assumptions indicated there are satisfied, so that in the space $X = L^2(\mathbb{R}^n)^N$, endowed with the usual inner product, we have:

$$\|\lambda M(\lambda M - L)^{-1}\|_{\mathcal{L}(X)} \leq C(1 + |\lambda|), \quad \Re e \geq -a_0, \quad a_0 > 0.$$

Here, M is multiplication by the diagonal matrix $m(\cdot)$ in X, and $L = A - \alpha I$, where $Au = \mathcal{A}u$ with $\mathcal{A}u = \sum_{j=1}^n a_j(\cdot)\frac{\partial u}{\partial x_j} + b(\cdot)u$ and $u \in \mathcal{D}(A) = \{u \in X; \mathcal{A}u \in X\}$.

We consider a sequence $a_j^s(x)$, $b_s(x)$, $m_s(x)$, $s \in \mathbb{N}$, of matrices fulfilling the same assumptions as $a_j(x)$, $b(x)$, $m(x)$ and, moreover,

$$|m_i^s(x)| \leq K, \quad s \in \mathbb{N}, \, i = 1, 2, ..., N,$$

$$m_i^s(x) \to m_i(x) \text{ as } s \to \infty, \quad i = 1, 2, ..., N, \, x \in \mathbb{R}^n.$$

Since we want to apply the representation $L_p^{-1} - L^{-1} = L_p^{-1}(L - L_p)L^{-1}$, we shall assume $\mathcal{D}(L) \subset \mathcal{D}(L_p)$, $p \in \mathbb{N}$, all components $a_{j,s}^{ik}(x)$, $b_{ik}^s(x)$ of

$a_j^s(x)$, $b_s(x)$ converge pointwise to the corresponding components of $a_j(x)$ and $b(x)$ respectively, are bounded uniformly in $s \in \mathbb{N}$, and the sequences $\{\beta_s\}$, $\{c_s\}$, where

$$\beta_s = \sup_{x \in \mathbb{R}^n} \sup_{i,k=1,2,...,N} |b_{ik}^s(x)|, \quad c_s = n \sup_{x \in \mathbb{R}^n} \sup_{j=1,2,...,n} \sup_{i,k=1,2,...,N} \left| \frac{\partial a_{j,s}^{ik}(x)}{\partial x_j} \right|,$$

are bounded. If $\overline{\beta} = \max\{\beta, \sup_s \beta_s\}$, $\overline{C} = \max\{C, \sup_s C_s\}$, where

$$\beta = \sup_{x \in \mathbb{R}^n} \sup_{i,k=1,2,...,n} |b_{ik}(x)|, \quad C = n \sup_{x \in \mathbb{R}^n} \sup_{j=1,2,...,n} \sup_{i,k=1,2,...,N} \left| \frac{\partial a_j^{ik}(x)}{\partial x_j} \right|,$$

taking $\alpha > \overline{\beta} + \frac{\overline{C}}{2}$, we easily get the uniform estimate for the modified resolvents that is needed in (5.45), namely with $m = 1$. Here, Propositions 5.30 and 5.31 give a (eventually non degenerate) sequence of problems approximating (D-E.1) in this case.

DEGENERATE DIFFERENTIAL
EQUATIONS OF THE SECOND ORDER

The ideas and methods of Chapters III-V will be applied to solve abstract degenerate differential equations of the second order in time. In a particularly interesting situation the treatment extends to arbitrary order n.

More precisely, the first section is devoted to the equation of parabolic type $\frac{d}{dt}(Cu') + Bu' + Au = f(t)$ and it is shown that suitable assumptions allow to handle it by means of the tools described in Chapters III and IV. The time-dependent operator coefficients is examined, too. Moreover, it is seen in Theorem 6.7 that the theory can be extended to treat the equation of higher order $\frac{d}{dt}(A_n u^{(n-1)}) + \sum_{j=0}^{n-1} A_j u^{(j)} = f(t)$. Section 6.2 lists a number of partial differential equations of interest in applied mathematics that satisfy the conditions in Section 6.1.

In the next Section 6.3 we examine the equation $C\frac{d^2u}{dt^2} + B\frac{du}{dt} + Au = f(t)$ in a Hilbert space H, $0 \leq C \in \mathcal{L}(H)$, having in mind above all its application to partial differential equations of hyperbolic-parabolic type like $m_1(x)\frac{\partial^2 u}{\partial t^2} + m_2(x)\frac{\partial u}{\partial t} - \Delta u = f(x,t)$, where $m_1(x)$, $m_2(x) \geq 0$, see Example 6.18.

In Section 6.4 we prove that the theory developed in Chapter V for highly degenerate equations permits to treat the equation $C\frac{d^2u}{dt^2} + B\frac{du}{dt} + Au = f(t)$ when $C = C^*$ is an unbounded operator. Some examples are also given. In Section 6.5 it is proven that the operational approach works for some degenerate abstract elliptic equations.

Last section contains a treatment of elliptic hyperbolic equations and it is shown that the famous Tricomi equations do satisfy our assumptions.

6.1 EQUATIONS OF PARABOLIC TYPE

Let A, B, C be three closed linear operators acting in a Banach space X such that

(6.1) $\mathcal{D}(B) \subset \mathcal{D}(A)$ and B has a bounded inverse.

Given $f \in \mathcal{C}([0,T]; X)$, where $T > 0$ is fixed, we shall study existence, uniqueness and regularity (in time) of the solution (in some senses to be specified in a moment) to the initial value problem

(6.2)
$$\begin{cases} \dfrac{d}{dt}(Cu') + Bu' + Au = f(t), & 0 < t \leq T, \\ u(0) = u_0, \ Cu'(0) = Cu_1. \end{cases}$$

Here, $u_0 \in X$ and $u_1 \in \mathcal{D}(C)$ are given initial values.

In order to justify all of our future treatment of (6.2), we quickly outline the heuristic reasoning basic (and natural) for it. Putting $u'(t) = \frac{du}{dt}(t) = v(t)$, (6.2) becomes the problem (D-E.1) in Chapter IV, that we write here for reader's convenience

(6.3)
$$\begin{cases} \dfrac{d}{dt}Mz(t) - Lz(t) = F(t), & 0 < t \leq T, \\ Mz(0) = Mz_0, \end{cases}$$

where $z(t) = (u(t), v(t))$, $z_0 = (u_0, u_1)$, $F(t) = (0, f(t))$, $0 < t \leq T$, and

$$M = \begin{pmatrix} I & 0 \\ 0 & C \end{pmatrix}, \quad L = \begin{pmatrix} 0 & I \\ -A & -B \end{pmatrix}.$$

The problem is to single out a product space and corresponding domains for the operators M and L such that our previous results (at first of parabolic type, and then of hyperbolic type) can be applied. This question was considered in the preceding chapter and we saw that a distinguishing role is played by the resolvent $P(\lambda)^{-1} = (\lambda^2 C + \lambda B + A)^{-1}$ and the estimates of its norm as an operator acting between different spaces.

Here we shall discuss in some details the situation where (6.3) has properties of parabolic type, guaranteed by "good" spectral features of the pairs (B, C), (A, B), together with suitable relations between them. First we shall examine the easiest case where $\mathcal{D}(B) \subset \mathcal{D}(A)$; if $\mathcal{D}(A) \subset \mathcal{D}(B)$, we shall meet with some more difficulties, but the introduction of multivalued operators shall permit us to overcome them, generalizing the results to be found in S. G. Krein [121; Chapter 3] to the degenerate equations.

To begin with, we observe that in both the cases, the pivot space $\mathcal{D}(B) \times X$ works, where $\mathcal{D}(B)$ (in virtue of (6.1)) is endowed with the graph norm $\|u\|_{\mathcal{D}(B)} = \|Bu\|_X$, $u \in \mathcal{D}(B)$. Moreover, we must define precisely what we mean by a solution to (6.2).

DEFINITION. A function $u:(0,T] \to X$ is a solution to (6.2), if $u(t) \in \mathcal{D}(A)$, $0 < t \leq T$,

$$u \in \mathcal{C}^1((0,T]; X), \ u'(t) \in \mathcal{D}(B) \cap \mathcal{D}(C),$$
$$Cu' \in \mathcal{C}^1((0,T]; X), \ Bu' \in \mathcal{C}((0,T]; X),$$

and if u satisfies the equation in (6.2) for each $0 < t \leq T$ together with the initial conditions in the X-norm sense that $\lim_{t \to 0} u(t) = u_0$, $\lim_{t \to 0} Cu'(t) = Cu_1$.

Besides such a solution, we are also concerned with seeking a solution regular at $t = 0$, according that $Cu' \in C^1([0,T]; X)$, $Bu' \in C([0,T]; X)$, $u \in C^1([0,T]; X)$. Moreover it is required that $u_0 \in \mathcal{D}(A)$ and $u_1 \in \mathcal{D}(B) \cap \mathcal{D}(C)$.

We shall make the following assumptions:

(6.4) The C modified resolvent set $\rho_C(B)$ contains the region

$$\Sigma_\alpha = \{\lambda \in \mathbb{C}; \, \Re e \lambda \leq c(1 + |\Im m \lambda|)^\alpha\}$$

with $c > 0$.

(6.5) $\qquad \|C(\lambda C + B)^{-1}\|_{\mathcal{L}(X)} \leq C(1 + |\lambda|)^{-\beta}, \quad \lambda \in -\Sigma_\alpha,$

where $0 < \beta \leq \alpha \leq 1$.

Our first result then reads as follows.

THEOREM 6.1. *Let us assume (6.1),(6.4) and (6.5). If $u_0, u_1 \in \mathcal{D}(B)$ and $f \in C^\sigma([0,T]; X)$ with $2\alpha + \beta > 2$, $\frac{2-\alpha-\beta}{\alpha} < \sigma \leq 1$, then problem (6.2) has a unique solution.*

PROOF. Let us consider the operators M and L in the space $\mathcal{D}(B) \times X$, endowed with the product topology, with the domains $\mathcal{D}(M) = \mathcal{D}(B) \times (\mathcal{D}(B) \cap \mathcal{D}(C))$ and $\mathcal{D}(L) = \mathcal{D}(B) \times \mathcal{D}(B)$. For $-\lambda \in \Sigma_\alpha$, we define

$$P(\lambda) = \lambda^2 C + \lambda B + A, \quad \mathcal{D}(P(\lambda)) = \mathcal{D}(B) \cap \mathcal{D}(C).$$

Then, $P(\lambda) = (\lambda + AB^{-1})(\lambda C + B) - \lambda AB^{-1}C$. So that, if $-\lambda \in \Sigma_\alpha$ and $|\lambda|$ is sufficiently large, then $P(\lambda) = (\lambda + AB^{-1})Q(\lambda)(\lambda C + B)$, where

$$Q(\lambda) = 1 - \lambda(\lambda + AB^{-1})^{-1}AB^{-1}C(\lambda C + B)^{-1}.$$

Moreover, since $AB^{-1} \in \mathcal{L}(X)$,

$$\|\lambda(\lambda + AB^{-1})^{-1}AB^{-1}C(\lambda C + B)^{-1}\|_{\mathcal{L}(X)} \leq C(1 + |\lambda|)^{-\beta}.$$

This implies that for all $-\lambda \in \Sigma_\alpha$, $|\lambda|$ large, the operator $P(\lambda)$ has a bounded inverse. On the other hand, it is not restrictive to suppose, under the actual assumptions, that this holds for any $-\lambda \in \Sigma_\alpha$ and $\mathcal{D}(B) \subset \mathcal{D}(C)$, because otherwise the change of variable $u(t) = e^{kt}w(t)$, $k > 0$, transforms the equation into

$$\frac{d}{dt}(Cw') + (B + 2kC)w' + (A + kB + k^2 C)w = e^{-kt} f(t)$$

and $\mathcal{D}(B + 2kC) \subset \mathcal{D}(C)$.

Next we show that equation $(\lambda M - L)(u, v) = (f_1, f_2) \in \mathcal{D}(B) \times X$ has a unique solution for $-\lambda \in \Sigma_\alpha$ with an estimate

$$\|L(u, v)\|_{\mathcal{D}(B) \times X} \leq C(1 + |\lambda|)^{1-\beta}\|(f_1, f_2)\|_{\mathcal{D}(B) \times X}.$$

To this end, we have to solve

$$\begin{cases} \lambda u - v = f_1, \\ Au + \lambda(\lambda C + B)u = (\lambda C + B)f_1 + f_2. \end{cases}$$

It has the unique solution

$$u = P(\lambda)^{-1}(\lambda C + B)f_1 + P(\lambda)^{-1}f_2,$$
$$v = -P(\lambda)^{-1}Af_1 + \lambda P(\lambda)^{-1}f_2.$$

Hence,

$$(\lambda M - L)^{-1} = [A_{ij}(\lambda)], \quad M(\lambda M - L)^{-1} = [B_{ij}(\lambda)], \quad i, j = 1, 2,$$

where

$$A_{11}(\lambda) = \lambda^{-1} - \lambda^{-1}P(\lambda)^{-1}, \quad A_{12}(\lambda) = P(\lambda)^{-1},$$
$$A_{21}(\lambda) = -P(\lambda)^{-1}A, \quad A_{22}(\lambda) = \lambda P(\lambda)^{-1},$$
$$B_{11}(\lambda) = \lambda^{-1} - \lambda^{-1}P(\lambda)^{-1}A, \quad B_{12}(\lambda) = P(\lambda)^{-1},$$
$$B_{21}(\lambda) = -CP(\lambda)^{-1}A, \quad B_{22}(\lambda) = \lambda CP(\lambda)^{-1}.$$

Furthermore, from

$$\|BP(\lambda)^{-1}\|_{\mathcal{L}(X)} \leq C(1 + |\lambda|)^{-\beta}, \quad \|CP(\lambda)^{-1}\|_{\mathcal{L}(X)} \leq C(1 + |\lambda|)^{-\beta-1},$$

it is easily observed that

$$\|\lambda^{-1}B(f_1 - P(\lambda)^{-1}Af_1) + BP(\lambda)^{-1}f_2\|_X \leq C(1 + |\lambda|)^{-\beta}\|(f_1, f_2)\|_{\mathcal{D}(B) \times X}.$$

Analogously, we verify that

$$\|-CP(\lambda)^{-1}Af_1 + \lambda CP(\lambda)^{-1}f_2\|_X \leq C(1 + |\lambda|)^{-\beta}\|(f_1, f_2)\|_{\mathcal{D}(B) \times X}.$$

Then the assertion of the theorem is an immediate consequence of Theorems 3.8 and 3.9.

THEOREM 6.2. *Let us assume (6.4),(6.5) and $\mathcal{D}(B) \subset \mathcal{D}(A) \cap \mathcal{D}(C)$. Moreover, let $u_0, u_1 \in \mathcal{D}(B)$ and $f \in C^\theta([0,T]; X)$, $2\alpha + \beta > 1$, $\frac{2-\alpha-\beta}{\alpha} < \theta \leq 1$, satisfy $f(0) - Au_0 - Bu_1 = Cw$ with a suitable $w \in \mathcal{D}(B)$. Then, problem (6.2) has a solution u such that*

$$Bu', \ (Cu')' \in C^\omega([0,T]; X), \quad \omega = \alpha\theta + \alpha + \beta - 2.$$

PROOF. Our goal is to apply Theorem 3.26 and to this end we need to reduce (6.2) to a first order system. On the other hand, using the notation of Theorem 3.26, it seems to be too complicated to interpret and characterize X_A^φ in this context and hence we shall substitute it by means of a smaller but simple space. First of all, we observe that the change of unknown function $z(t) = e^{kt}w(t), k > 0$, transforms (6.3) into

(6.6)
$$\begin{cases} \dfrac{d}{dt}(Mw) - L_1 w = e^{-kt} F(t), & 0 < t \leq T, \\ Mw(0) = Mz_0, \end{cases}$$

where

$$L_1 = \begin{pmatrix} -k & I \\ -A & -kC - B \end{pmatrix}$$

acts in $\mathcal{D}(B) \times X$ with domain $\mathcal{D}(B) \times \mathcal{D}(B)$. It is easily verified that, taking k suitably large, the following representation holds

$$M(\lambda M - L_1)^{-1} = \begin{pmatrix} P(\lambda + k)^{-1}((\lambda + k)C + B) & P(\lambda + k)^{-1} \\ -CP(\lambda + k)^{-1}A & (\lambda + k)CP(\lambda + k)^{-1} \end{pmatrix}$$

for all $-\lambda \in \Sigma_\alpha$; so that, denoting $\mathcal{D}(B) \times X$ by Y, in view of (6.4) and (6.5),

$$\|M(\lambda M - L_1)^{-1}\|_{\mathcal{L}(Y)} \leq C(1 + |\lambda|)^{-\beta}, \quad -\lambda \in \Sigma_\alpha.$$

Since $\mathcal{D}(L_1 M^{-1}) = M(\mathcal{D}(L)) = \{(x, Cy); (x,y) \in \mathcal{D}(B) \times \mathcal{D}(B)\}$, the condition

$$F(0) - L_1 z_0 \in \mathcal{D}(L_1 M^{-1}) \subset X_{L_1 M^{-1}}^\varphi$$

in Theorem 3.26 reads

$$-ku_0 + u_1 = x \in \mathcal{D}(B),$$
$$f(0) - Au_0 - (kC + B)u_1 = Cy, \ y \in \mathcal{D}(B),$$

and this is ensured by $u_0, u_1 \in \mathcal{D}(B)$, $f(0) - Au_0 - Bu_1 = C(y + ku_1) = Cw$, where $w = y + ku_1 \in \mathcal{D}(B)$. The affirmation then is an immediate consequence of Theorem 3.26 when applied to (6.6) if we take into account the definition of the space Y.

Let us address our attention to the case when $\mathcal{D}(A)$ is smaller than $\mathcal{D}(B)$. It is also not restrictive to assume that in its turn $\mathcal{D}(B) \subset \mathcal{D}(C)$ and this hypothesis shall be maintained in the sequel. The essential step is to invert $P(\lambda) = \lambda^2 C + \lambda B + A$ in the presence of the difficulty that AB^{-1} is not a bounded operator. To overcome this we introduce the multivalued linear operator $V = BC^{-1}$, $\mathcal{D}(V) = \{Cx; x \in \mathcal{D}(B)\}$ and the closed linear operator $U = AB^{-1}$ (The hypothesis that $0 \in \rho(A)$ guarantees that this is true). We also suppose that (6.4) and (6.5) hold for B, C. Moreover,

$$(6.7) \qquad \|B(\lambda B + A)^{-1}\|_{\mathcal{L}(X)} \le C(1 + |\lambda|)^{-\gamma}, \quad -\lambda \in \Sigma_\alpha,$$

where $0 < \gamma \le 1$.

(6.8) $\mathcal{D}(V) \subset \mathcal{D}(U)$ and there is $\delta > 0$ such that

$$\|U(\lambda + V)^{-1}\|_{\mathcal{L}(X)} \le C(1 + |\lambda|)^{-\delta}, \quad -\lambda \in \Sigma_\alpha, |\lambda| \text{ large.}$$

Assumption (6.8) is crucial and we shall indicate some concrete examples for which it is fulfilled.

Let us first note that under (6.4),(6.5),(6.7) and (6.8) we have:

$$P(\lambda) = (\lambda + A(\lambda C + B)^{-1})(\lambda C + B)$$
$$= (\lambda + UB(\lambda C + B)^{-1})(\lambda C + B) = (\lambda + U[1 - \lambda C(\lambda C + B)^{-1}])(\lambda C + B).$$

Since $(\lambda C + B)(\mathcal{D}(A)) \subset \mathcal{D}(U)$, guaranteed by $B(\mathcal{D}(A)) \subset \mathcal{D}(AB^{-1})$ and $C(\mathcal{D}(A)) = CB^{-1}B(\mathcal{D}(A)) = V^{-1}B(\mathcal{D}(A)) \subset \mathcal{D}(V) \subset \mathcal{D}(U)$ (by (6.8)), we have:

$$P(\lambda) = (\lambda + U)(1 - \lambda U(\lambda + U)^{-1}(\lambda + V)^{-1})(\lambda C + B).$$

We should now assume that

$$Q(\lambda) = 1 - \lambda U(\lambda + U)^{-1}(\lambda + V)^{-1}$$

has a bounded inverse. Notice, however, that

$$\|\lambda(\lambda + U)^{-1}U(\lambda + V)^{-1}\|_{\mathcal{L}(X)} \le C(1 + |\lambda|)^{1-(\delta+\gamma)}$$

in virtue of (6.7),(6.8), and therefore $\delta + \gamma > 1$ yields that $Q(\lambda)$ has a bounded inverse for all $-\lambda \in \Sigma_\alpha$, $|\lambda|$ large. We are now ready to prove the theorem.

THEOREM 6.3. *Assume (6.4),(6.5),(6.7) and (6.8) with* $0 < \beta, \gamma \le \alpha \le 1$, $\gamma + \beta, \gamma + \delta > 1$, $2\alpha + \beta + \gamma > 3$, *and let* $\frac{3 - \alpha - \beta - \gamma}{\alpha} < \sigma \le 1$. *Then for any*

$u_0 \in \mathcal{D}(A)$, $u_1 \in \mathcal{D}(B)$ and $f \in C^\sigma([0,T]; X)$, the problem (6.2) possesses a unique solution.

PROOF. As in the proof of Theorem 6.1, it is not restrictive to assume $\mathcal{D}(B) \subset \mathcal{D}(C)$. Moreover, the operators M and L, as defined by $\mathcal{D}(M) = \mathcal{D}(B) \times \mathcal{D}(B)$, $\mathcal{D}(L) = \mathcal{D}(A) \times \mathcal{D}(B)$, $M(u,v) = (u, Cv)$, $L(u,v) = (v, -Au - Bv)$ act in the product space $\mathcal{D}(B) \times X$. Then it is a simple matter to recogunize that

$$M(\lambda M - L)^{-1} = \begin{pmatrix} \lambda^{-1} + \lambda^{-1} R(\lambda) & P(\lambda)^{-1} \\ CR(\lambda) & \lambda CP(\lambda)^{-1} \end{pmatrix},$$

with $R(\lambda) = \lambda P(\lambda)^{-1}(\lambda C + B) - 1$ (which is equal to $-P(\lambda)^{-1}A$ on the domain $\mathcal{D}(A)$). This is a consequence of the analogous formula used in Theorem 6.1, when we observe that now $-\lambda^{-1} P(\lambda)^{-1} A$ is restriction of $\lambda^{-1}\{\lambda P(\lambda)^{-1}(\lambda C + B) - 1\}$ to $\mathcal{D}(A)$.

Next step is the following representation of $R(\lambda)$, precisely

$$R(\lambda) = (\lambda C + B)^{-1} Q(\lambda)^{-1}\{\lambda(\lambda + U)^{-1}(\lambda C + B) - Q(\lambda)(\lambda C + B)\}$$
$$= (\lambda C + B)^{-1} Q(\lambda)^{-1}\{\lambda(\lambda + U)^{-1}(\lambda C + B)$$
$$- [1 - \lambda U(\lambda + U)^{-1} C(\lambda C + B)^{-1}](\lambda C + B)\}$$
$$= (\lambda C + B)^{-1} Q(\lambda)^{-1}\{\lambda(\lambda + U)^{-1}(\lambda C + B) - (\lambda C + B) + \lambda U(\lambda + U)^{-1} C\}$$
$$= (\lambda C + B)^{-1} Q(\lambda)^{-1}[\lambda(\lambda + U)^{-1} - 1]B.$$

In addition, if $f = (f_1, f_2) \in \mathcal{D}(B) \times X$, then

$$\|BR(\lambda)f_1\|_X \le \|B(\lambda C + B)^{-1}\|_{\mathcal{L}(X)} \|Q(\lambda)^{-1}\|_{\mathcal{L}(X)}$$
$$\times \|\lambda(\lambda + U)^{-1} - 1\|_{\mathcal{L}(X)} \|Bf_1\|_X$$
$$\le C|\lambda|^{2-(\gamma+\beta)} \|f_1\|_{\mathcal{D}(B)},$$

$$\|CR(\lambda)f_1\|_X \le \|C(\lambda C + B)^{-1}\|_{\mathcal{L}(X)} \|Q(\lambda)^{-1}\|_{\mathcal{L}(X)}$$
$$\times \|\lambda(\lambda + U)^{-1} - 1\|_{\mathcal{L}(X)} \|Bf_1\|_X$$
$$\le C|\lambda|^{1-(\gamma+\beta)} \|f_1\|_{\mathcal{D}(B)}.$$

By similar calculations,

$$\|BP(\lambda)^{-1}f_2\|_X \le C|\lambda|^{1-(\gamma+\beta)} \|f_2\|_X,$$
$$\|CP(\lambda)^{-1}f_2\|_X \le C|\lambda|^{-(\gamma+\beta)} \|f_2\|_X.$$

In virtue of the expression of $M(\lambda M - L)^{-1}$ written above we conclude that

$$\|M(\lambda M - L)^{-1}\|_{\mathcal{L}(\mathcal{D}(B) \times X)} \le C|\lambda|^{1-(\gamma+\beta)}, \quad \lambda \in -\Sigma_\alpha, \lambda \ne 0.$$

We again make use of Theorems 3.8 and 3.9 to obtain the desired results.

THEOREM 6.4. *Under the same assumptions as in Theorem 6.3, if, in addition,* $f(0) - Au_0 - Bu_1 = Cw$, *where* $w \in \mathcal{D}(B)$, *then the solution* u *to (6.2) obtained in Theorem 6.3 enjoys the regularity* Au, Bu', $\frac{d}{dt}(Cu') \in C^\omega([0,T]; X)$, $\omega = \alpha\sigma + \alpha + \beta + \gamma - 3$.

PROOF. It suffices to apply Theorem 3.26, by observing that under the present hypothesis

$$
\begin{pmatrix} 0 \\ f(0) \end{pmatrix} - \begin{pmatrix} 0 & -I \\ A & B \end{pmatrix} \begin{pmatrix} u_0 \\ u_1 \end{pmatrix} = \begin{pmatrix} I & 0 \\ 0 & C \end{pmatrix} \begin{pmatrix} u_0 \\ w \end{pmatrix},
$$

and $(u_0, w) \in \mathcal{D}(L) = \mathcal{D}(A) \times \mathcal{D}(B)$.

Obviously, in the best possible situation that $\alpha = \beta = \gamma = 1$, then σ can vary arbitrarily in $(0,1)$ and we obtain precisely the maximal regularity property Au, $Bu' \in C^\sigma([0,T]; X)$.

We also observe that in the conditions listed in Theorem 6.3, the one really restrictive reduces to (6.8), since it imposes very strong links between U and V, that is, between A, B and C. It should be very important for applications to substitute (6.8) with another hypothesis where the multi-valued linear operator $V = BC^{-1}$ does not appear. As a result in this direction we shall prove the following one.

THEOREM 6.5. *Let us assume (6.1),(6.4),(6.5),(6.7), and*

(6.9) $\mathcal{D}(B)$ *is invariant under* C;

(6.10) $\|BCu\|_X \le m\|Bu\|_X$ *for all* $u \in \mathcal{D}(B)$;

(6.11) $\mathcal{D}(AB^{-1}) \supset (X, \mathcal{D}(B))_{\tau,1}$, $0 < \tau < \beta$, *where* $(X, \mathcal{D}(B))_{\tau,1}$ *denotes the real interpolation space.*

Then the Condition (6.8) holds.

PROOF. In virtue of (6.9) and (6.10), we have, for $\lambda \in -\Sigma_\alpha$:

$$
\|BC(\lambda C + B)^{-1}u\|_X = \|B(\lambda + BC^{-1})^{-1}u\|_X
$$
$$
\le m\|B(\lambda C + B)^{-1}u\|_X \le m'(1 + |\lambda|)^{1-\beta}\|u\|_X.
$$

Let us recall that by (6.5), $\|C(\lambda C + B)^{-1}u\|_X \le m''(1 + |\lambda|)^{-\beta}\|u\|_X$, and hence by interpolation

$$
\|C(\lambda C + B)^{-1}u\|_{(X,\mathcal{D}(B))_{\tau,1}} \le m'''(1 + |\lambda|)^{\tau-\beta}\|u\|_X, \quad \lambda \in \Sigma_\alpha, u \in X.
$$

Then, (6.11) yields the result with $\delta = \beta - \tau$.

REMARK. If B is a positive operator with bounded imaginary powers $B^{i\eta}$, $\eta \in \mathbb{R}$, and $\|B^{i\eta}\|_{\mathcal{L}(X)} \le$ const. for all η, $|\eta| \le \rho$, with $\mathcal{D}(B^{1+\tau}) \subset \mathcal{D}(A)$, then (6.11) is verified, since by [172; p. 103] $\mathcal{D}(B^\tau)$ coincides with the complex interpolation space $[X, \mathcal{D}(B)]_\tau \supset (X, \mathcal{D}(B))_{\tau,1}$ and $\mathcal{D}(B^\tau) \subset \mathcal{D}(AB^{-1})$ from the assumption.

Our next goal is to extend Theorems 6.3 and 6.4 to the general equation of arbitrary order $n \geq 2$ in time, that is, to consider the initial value problem

(6.12)
$$
\begin{cases}
\dfrac{d}{dt}(A_n u^{(n-1)}) + \displaystyle\sum_{j=0}^{n-1} A_j u^{(j)} = f(t), \quad 0 < t \leq T, \\[2mm]
u^{(j)}(0) = u_j, \ j = 0, 1, ..., n-2, \\[2mm]
A_n u^{(n-1)}(0) = A_n u_{n-1}.
\end{cases}
$$

In this way we shall be able to treat the degenerate case corresponding to the regular situation $A_n = I$ investigated in A. Favini and H. Tanabe [91].

It seems, however, a very hard task to reduce immediately (6.12) to a first order system in a suitable product space and thus we shall exploit the device of M. K. Balaev [13]. To this end we shall assume that the closed linear operators A_j, $j = 0, 1, 2, ..., n-1$, have bounded inverse so that the problem to find a solution to (6.12), which is defined as a function $u \in C^{(n-1)}((0, T]; X)$, $u^{(j)}(t) \in \mathcal{D}(A_j)$ $(0 \leq j \leq n-1)$, $u^{(n-1)}(t) \in \mathcal{D}(A_n)$, $A_j u^{(j)} \in C((0, T]; X)$ $(0 \leq j \leq n-1)$, $A_n u^{(n-1)} \in C^1((0, T]; X)$ satisfying the equation in (6.12) for every $0 < t \leq T$ and the initial conditions in the sense that $\lim_{t \to 0} u^{(j)}(t) = u_j$ in $\mathcal{D}(A_{j+1})$ $(0 \leq j \leq n-2)$ and $\lim_{t \to 0} A_n u^{(n-1)}(t) = A_n u_{n-1}$ in X, is clearly equivalent to solve the corresponding problem (6.3) in which

$$
M = \begin{pmatrix}
A_1 & A_2 & A_3 & \cdots & A_{n-1} & A_n \\
0 & A_2 & A_3 & \cdots & A_{n-1} & A_n \\
0 & 0 & A_3 & \cdots & A_{n-1} & A_n \\
\vdots & \vdots & \vdots & \cdots & \vdots & \vdots \\
0 & 0 & 0 & \cdots & A_{n-1} & A_n \\
0 & 0 & 0 & \cdots & 0 & A_n
\end{pmatrix},
$$

$$
L = - \begin{pmatrix}
A_0 & 0 & 0 & \cdots & 0 & 0 \\
A_0 & A_1 & 0 & \cdots & 0 & 0 \\
A_0 & A_1 & A_2 & \cdots & 0 & 0 \\
\vdots & \vdots & \vdots & \cdots & \vdots & \vdots \\
A_0 & A_1 & A_2 & \cdots & A_{n-2} & 0 \\
A_0 & A_1 & A_2 & \cdots & A_{n-2} & A_{n-1}
\end{pmatrix},
$$

are linear operators acting in $X^n = X \times \cdots \times X$ with $\mathcal{D}(M) = \mathcal{D}(A_1) \times \cdots \times \mathcal{D}(A_n)$ and $\mathcal{D}(L) = \mathcal{D}(A_0) \times \cdots \times \mathcal{D}(A_{n-1})$, $z(t) = (u(t), u'(t), ..., u^{(n-1)}(t))$, $F(t) = (f(t), ..., f(t))$, and $z_0 = (u_0, u_1, ..., u_{n-1})$.

In order to express $M(\lambda M - L)^{-1}$ in term of A_k, we prefer to observe that it coincides with $(\lambda - LM^{-1})^{-1}$, where M^{-1} is to be viewed as a multivalued linear operator in general, because A_n is not supposed to have a single valued inverse operator. Now,

$$M^{-1} = \begin{pmatrix} A_1^{-1} & -A_1^{-1} & 0 & 0 & \cdots & 0 & 0 & 0 \\ 0 & A_2^{-1} & -A_2^{-1} & 0 & \cdots & 0 & 0 & 0 \\ \vdots & \vdots & \vdots & \vdots & \cdots & \vdots & \vdots & \vdots \\ 0 & 0 & 0 & 0 & \cdots & 0 & A_{n-1}^{-1} & -A_{n-1}^{-1} \\ 0 & 0 & 0 & 0 & \cdots & 0 & 0 & A_n^{-1} \end{pmatrix},$$

and hence

$$-LM^{-1}$$

$$= \begin{pmatrix} U_1 & -U_1 & 0 & 0 & \cdots & 0 \\ U_1 & -U_1 + U_2 & -U_2 & 0 & \cdots & 0 \\ U_1 & -U_1 + U_2 & -U_2 + U_3 & -U_3 & \cdots & 0 \\ \vdots & \vdots & \vdots & \vdots & \cdots & \vdots \\ U_1 & -U_1 + U_2 & -U_2 + U_3 & -U_3 + U_4 & \cdots & -U_{n-1} \\ U_1 & -U_1 + U_2 & -U_2 + U_3 & -U_3 + U_4 & \cdots & -U_{n-1} + U_n \end{pmatrix}$$

$$= \mathfrak{A}_1 + \mathfrak{A}_2,$$

where

$$\mathfrak{A}_1 = \begin{pmatrix} U_1 & 0 & 0 & \cdots & 0 & 0 \\ U_1 & U_2 & 0 & \cdots & 0 & 0 \\ \vdots & \vdots & \vdots & \cdots & \vdots & \vdots \\ U_1 & U_2 & U_3 & \cdots & U_{n-1} & 0 \\ U_1 & U_2 & U_3 & \cdots & U_{n-1} & U_n \end{pmatrix},$$

$$\mathfrak{A}_2 = \begin{pmatrix} 0 & -U_1 & 0 & \cdots & 0 & 0 \\ 0 & -U_1 & -U_2 & \cdots & 0 & 0 \\ \vdots & \vdots & \vdots & \cdots & \vdots & \vdots \\ 0 & -U_1 & -U_2 & \cdots & -U_{n-2} & 0 \\ 0 & -U_1 & -U_2 & \cdots & -U_{n-2} & -U_{n-1} \end{pmatrix},$$

and $U_k = A_{k-1} A_k^{-1}$, $k = 1, 2, \ldots, n$. It follows that our resolvent $(\lambda - LM^{-1})^{-1}$ coincides with

$$(\lambda + \mathfrak{A}_1)^{-1} (\mathfrak{J} + \mathfrak{A}_2 (\lambda + \mathfrak{A}_1)^{-1})^{-1},$$

provided that the inverses exist as single valued operators, where $\mathfrak{J}$ denotes the identity matrix on X^n.

On the other hand, a very long (but trivial) computation leads to the following expression for $(\lambda + \mathfrak{A}_1)^{-1}$, precisely

$$(\lambda + \mathfrak{A}_1)^{-1} = (A_{ij}(\lambda))$$

with

$$\begin{cases} A_{ii}(\lambda) = (\lambda + U_i)^{-1}, & 1 \le i \le n, \\ A_{ij}(\lambda) = 0, & 1 \le i < j \le n, \\ A_{ij}(\lambda) = -\lambda^{i-j-1}(\lambda + U_{j+i-j})^{-1} \cdots (\lambda + U_{j+1})^{-1} U_j (\lambda + U_j)^{-1}, \\ & 1 \le j < i \le n. \end{cases}$$

Moreover, $\mathfrak{A}_2(\lambda + \mathfrak{A}_1)^{-1}$ is expressed by means of operator matrices whose terms are the sum of elements like

$$\lambda^{j-i} U_j (\lambda + U_{j+1})^{-1} (\lambda + U_j)^{-1} \cdots (\lambda + U_{i+1})^{-1} U_i (\lambda + U_i)^{-1}, \quad i \le j.$$

The following assumptions seem then suitable in order to obtain a generalization of Theorem 6.3:

(6.13) There exist $\alpha, \beta_1, \dots, \beta_n \in (0, 1]$ such that

$$\|(\lambda + U_j)^{-1}\|_{\mathcal{L}(X)} \le \frac{m}{(1 + |\lambda|)^{\beta_j}}, \quad -\lambda \in \Sigma_\alpha,$$

for each $j = 1, \dots, n$;

(6.14) $\mathcal{D}(U_{j+1}) \subset \mathcal{D}(U_j)$, $j = 1, \dots, n-1$;

(6.15) There is $\delta > 0$ such that

$$\|U_j(\lambda + U_{j+1})^{-1}\|_{\mathcal{L}(X)} \le m(1 + |\lambda|)^{-\delta}, \quad -\lambda \in \Sigma_\alpha, |\lambda| \text{ large};$$

(6.16) $\nu = \min_{1 \le j < i \le n} \left\{ \sum_{k=j}^{i} \beta_k - i + j \right\} > 0$, $\tau = \min\{\nu, \beta_1, \dots, \beta_n\}$, $\delta + \sum_{j=k}^{s} \beta_j > s - k + 1$, for $s = k, k+1, \dots, n-1$, $k \in \{1, \dots, n-1\}$.

Notice that if $\beta_1 = \cdots = \beta_{n-1} = 1$, (that is the best situation) then $\tau = \beta_n$. We here verify the lemma.

LEMMA 6.6. *If (6.13)~(6.16) hold, then there exists $k > 0$ such that*

$$\|(\lambda - LM^{-1})^{-1}\|_{\mathcal{L}(X^n)} \le k(1 + |\lambda|)^{-\tau}, \quad -\lambda \in \Sigma_\alpha, |\lambda| \text{ large}.$$

PROOF. We exploit the expressions described above for the resolvents $(\lambda - LM^{-1})^{-1}$, $(\lambda + \mathfrak{A}_1)^{-1}$ and $\mathfrak{A}_2(\lambda + \mathfrak{A}_1)^{-1}$. We have: $\|A_{ij}(\lambda)\|_{\mathcal{L}(X)} = 0$

for $j > i$, $\|A_{ii}(\lambda)\|_{\mathcal{L}(X)} \le m(1 + |\lambda|)^{-\beta_i}$, $1 \le i \le n$, and by (6.13) and (6.16),

$$\|A_{ij}(\lambda)\|_{\mathcal{L}(X)} \le C(1 + |\lambda|)^{i-j-\sum_{k=j}^{i} \beta_k} \le C(1 + |\lambda|)^{-\nu}, \quad \lambda \in -\Sigma_\alpha.$$

This yields that $\|(\lambda + \mathfrak{A}_1)^{-1}\|_{\mathcal{L}(X^n)} \le C(1 + |\lambda|)^{-\tau}$, $\lambda \in -\Sigma_\alpha$. Since the terms

$$\|\lambda^{j-i} U_j (\lambda + U_{j+1})^{-1} (\lambda + U_j)^{-1} \cdots (\lambda + U_{i+1})^{-1} U_i (\lambda + U_i)^{-1}\|_{\mathcal{L}(X)}$$

are estimated by means of

$$C(1 + |\lambda|)^{j-i}(1 + |\lambda|)^{-\delta}(1 + |\lambda|)^{1-(\beta_i + \cdots + \beta_j)}, \quad i \le j,$$

in virtue of (6.15) and (6.16), we conclude that there is $\sigma > 0$ such that all the entries $B_{ij}(\lambda)$ of $\mathfrak{A}_2(\lambda + \mathfrak{A}_1)^{-1}$ satisfy

$$\|B_{ij}(\lambda)\|_{\mathcal{L}(X)} \le C(1 + |\lambda|)^{-\sigma}, \quad \lambda \in \Sigma,$$

and hence $\|B_{ij}(\lambda)\|_{\mathcal{L}(X)} \to 0$ as $|\lambda| \to \infty$, $\lambda \in -\Sigma_\alpha$. The proof of the lemma is complete.

REMARK. If $\beta_1 = \cdots = \beta_{n-1} = 1$, $0 < \beta_n \le 1$, then $\tau = \nu = \beta_n$ and the last assumption in (6.16) is verified, too.

REMARK. Under (6.13)~(6.16), the operator pencil $P(\lambda) = \sum_{j=1}^{n} \lambda^j A_j$, $\lambda \in \mathbb{C}$, with $\mathcal{D}(P(\lambda)) = \mathcal{D}(A_0)$, satisfies

(6.17) $\|A_j P(\lambda)^{-1}\| \le C(1 + |\lambda|)^{n-j-\sum_{r=1}^{n} \beta_r}$, $\quad \lambda \in \Sigma_\alpha$, $|\lambda|$ large.

We give the proof of this estimate directly for $n = 3$; the proof for arbitrary n follows the same lines (see A. Favini and H. Tanabe [91]).

We note that in the present case

$$P(\lambda) = \left(\lambda + A_0 A_1^{-1} - \lambda A_0 A_1^{-1}(\lambda A_3 + A_2)(\lambda^2 A_3 + \lambda A_2 + A_1)^{-1}\right)$$
$$\times (\lambda^2 A_3 + \lambda A_2 + A_1).$$

Observe also that

$$\sum_{j=1}^{3} \lambda^{j-1} A_j = (1 - \lambda U_2(\lambda + U_3)^{-1}(\lambda + U_2)^{-1})(\lambda + U_2)(\lambda A_3 + A_2)$$

has a bounded inverse

$$\left(\sum_{j=1}^{3} \lambda^{j-1} A_j\right)^{-1} = (\lambda A_3 + A_2)^{-1}(\lambda + U_2)^{-1}(1 - \lambda U_2(\lambda + U_3)^{-1}(\lambda + U_2)^{-1}),$$

in view of the bound

$$\|\lambda U_2(\lambda + U_3)^{-1}(\lambda + U_2)^{-1}\|_{\mathcal{L}(X)} \leq C(1 + |\lambda|)^{1-\delta-\beta_2}.$$

It then follows that

$$P(\lambda) = (1 - \lambda U_1(\lambda + U_2)^{-1}(1 - \lambda U_2(\lambda + U_3)^{-1}(\lambda + U_2)^{-1})^{-1}(\lambda + U_1)^{-1})$$
$$\times (\lambda + U_1)(\lambda^2 A_3 + \lambda A_2 + A_1).$$

If we denote the inverse

$$(1 - \lambda U_1(\lambda + U_2)^{-1}(1 - \lambda U_2(\lambda + U_3)^{-1}(\lambda + U_2)^{-1})^{-1}$$

by $R(\lambda)$ (it exists because of (6.16)), and if we observe that $\|R(\lambda)\|_{\mathcal{L}(X)} \leq$ const. for $-\lambda \in \Sigma_\alpha$, $|\lambda|$ large, then we deduce that

$$P(\lambda)^{-1} = (\lambda A_3 + A_2)^{-1}(\lambda + U_2)^{-1}(1 - \lambda U_2(\lambda + U_3)^{-1}(\lambda + U_2)^{-1})^{-1}$$
$$\times (\lambda + U_1)^{-1}R(\lambda).$$

Hence, for each $i = 0, 1, 2, 3$,

$$\|A_i P(\lambda)^{-1}\|_{\mathcal{L}(X)} \leq C(1 + |\lambda|)^{3-i-(\beta_1+\beta_2+\beta_3)}, \quad -\lambda \in \Sigma_\alpha, |\lambda| \text{ large.}$$

Concerning the arbitrary order n, we remark that if

$$Q_j(\lambda) = \lambda^j A_n + \lambda^{j-1} A_{n-1} + \cdots + \lambda A_{n-j+1} + A_{n-j}, \quad j = 1, 2, \ldots, n,$$

then $P(\lambda)$ is shown to be equal to

$$P(\lambda)^{-1} = Q_{n-1}(\lambda)^{-1}(\lambda + U_1)^{-1}S_n(\lambda),$$

where

$$S_n(\lambda) = (1 - \lambda U_1 Q_{n-2}(\lambda)Q_{n-1}(\lambda)^{-1}(\lambda + U_1)^{-1})^{-1}$$
$$= (1 - \lambda U_1(\lambda + U_2)^{-1}S_{n-1}(\lambda)(\lambda + U_1)^{-1})^{-1}.$$

After some calculations we can arrive at the expression

$$S_n(\lambda) = (1 - \lambda U_{n-2}(\lambda + U_{n-1})^{-1}(1 - \lambda U_{n-1}(\lambda + U_n)^{-1}(\lambda + U_{n-1})^{-1})^{-1}$$
$$\times (\lambda + U_{n-2})^{-1}.$$

The assumptions (6.13)$\sim$(6.16) then guarantee that $S_n(\lambda)$ is in fact well defined for $|\lambda|$ sufficiently large, $-\lambda \in \Sigma_\alpha$. The estimates for the norm in

$\mathcal{L}(X)$ of $A_i P(\lambda)$ is then obvious taking into account that $Q_i(\lambda)$ has the same structure as $P(\lambda)$ with order i.

In addition we also remark that the general results obtained in Section 5.7 apply directly to the abstract equation (6.12) by reducing it to a first order system in the space $Z = X_1 \times X_2 \times \cdots \times X_n$, where $X_j = \mathcal{D}(A_j) \cap \mathcal{D}(A_{j+1}) \cap \cdots \cap \mathcal{D}(A_n) = \mathcal{D}(A_j)$, $j = 1, \ldots, n-1$, $X_n = X$, but then we obtain the resolvent estimate

$$(6.18) \qquad \|M(\lambda M - L)^{-1}\|_{\mathcal{L}(Z)} \leq C(1 + |\lambda|)^{2n - \sum_{r=1}^{n} \beta_r + \max\{n-3,-1\}},$$

since $m_j = n - j - \sum_{r=1}^{n} \beta_r$; so that, $\nu = n - \sum_{r=1}^{n} \beta_r$ and hence $m = \nu + n + \max\{n-2, 0\}$ in that section coincides with $2n - \sum_{r=1}^{n} \beta_r + \max\{n - 2, 0\}$. Therefore, properties (6.13)~(6.16) allow us to deduce estimates much better than (6.18). On the other hand, we notice that to obtain (6.18) we in fact used only a few assumptions entering in (6.16).

The preceding lemma yields the following result.

THEPREM 6.7. *Let us assume (6.13)~(6.16) and let $0 < \tau \leq \alpha \leq 1$, $2\alpha + \tau > 2$, $\frac{2-\alpha-\tau}{\alpha} < \theta < 1$. Then, for all $f \in \mathcal{C}^\theta([0,T]; X)$ and $u_j \in \mathcal{D}(A_j)$, $j = 0, 1, \ldots, n-1$, (6.12) possesses a unique solution u such that $u^{(j)}(t) \to u_j$ as $t \to 0$ in $\mathcal{D}(A_j)$ for $j = 0, 1, \ldots, n-2$. Moreover, if $u_0 \in \mathcal{D}(A_0)$, $u_j \in \mathcal{D}(A_{j-1})(j = 1, \ldots, n)$ and the compatibility relation*

$$f(0) - \sum_{j=0}^{n-1} A_j u_j = A_n u_n$$

holds for a certain $u_n \in \mathcal{D}(A_{n-1})$, then the solution enjoyes the regularity

$$A_j u^{(j)}, \ \tfrac{d}{dt} A_n u^{(n-1)} \in \mathcal{C}^\omega([0,T]; X), \qquad \omega = \alpha\theta + \alpha + \tau - 2,$$

and the equation in (6.12) holds even at $t = 0$.

PROOF. We only need to recall the meaning of the operators M and L introduced previously to obtain the first assertion in view of Theorems 3.8 and 3.9, because $(u_0, u_1, \ldots, u_{n-1}) \in \mathcal{D}(L)$ if and only if $u_j \in \mathcal{D}(A_j)$, $j = 0, 1, \ldots, n-1$. Concerning the second assertion we apply Theorem 3.26, taking $\mathcal{D}(LM^{-1})$ instead of (the larger) interpolation space $(X \times \cdots \times X, \mathcal{D}(LM^{-1}))_\varphi$, entering into the compatibility relation on $F(0) - LV_0$, $F(0) = (f(0), \ldots, f(0))$, $V_0 = (u_0, u_1, \ldots, u_{n-1})$. This translates into $f(0) - \sum_{i=0}^{k} A_i u_i = \sum_{j=k+1}^{n} A_j v_j$, where $(v_1, \ldots, v_n) \in \mathcal{D}(L) = \mathcal{D}(A_0) \times \cdots \times \mathcal{D}(A_{n-1})$. Taking $v_j = u_j$ $(j = 1, \ldots, n)$ we complete the proof.

Let us devote now our attention to the initial value problem with time dependent operator coefficients

$$(6.19) \qquad \begin{cases} \dfrac{d}{dt}(C(t)u)' + B(t)u' + A(t)u = f(t), & 0 < t \leq T, \\ u(0) = u_0, \ C(0)u'(0) = C(0)u_1, \end{cases}$$

in a Banach space X, where $A(t)$, $B(t)$, $C(t)$ are three families of closed linear operators in X defined for $0 \leq t \leq T$. The definition of solution to (6.19) is just a natural extension of that for the time homogeneous problem (6.2).

For sake of brevity we shall confine ourselves to the case:

(6.20) $\mathcal{D}(B(t)) \subset \mathcal{D}(A(t)) \cap \mathcal{D}(C(t))$, $\mathcal{D}(B(t)) \equiv \mathcal{D}$ are independent of t;

(6.21) There exist k_1, k_2, $k_3 > 0$ such that

$$k_1\|u\|_X \leq \|B(t)u\|_X \leq k_2\|B(s)u\|_X \leq k_3\|B(t)u\|_X, \quad u \in \mathcal{D},$$

for all $0 \leq s, t \leq T$.

Introducing the operators $M(t)$, $L(t)$, $0 \leq t \leq T$, by

$$M(t)(u, v) = (u, C(t)v), \quad (u, v) \in \mathcal{D}(M(t)) = \mathcal{D} \times \mathcal{D},$$
$$L(t)(u, v) = (v, -A(t)u - B(t)v), \quad (u, v) \in \mathcal{D}(L(t)) = \mathcal{D} \times \mathcal{D},$$

we see that (6.19) becomes

(6.22)
$$\begin{cases} \dfrac{d}{dt}M(t)U - L(t)u = F(t), & 0 < t \leq T, \\ M(0)U(0) = M(0)U_0, \end{cases}$$

where $U(t) = (u(t), u'(t))$, $F(t) = (0, f(t))$, $U_0 = (u_0, u_1)$, and the ambient space is $\mathcal{D} \times X$.

Our goal is to describe some simple conditions on the single operator coefficients permitting to apply the results obtained in Chapter IV. The first step consists in the change of the dependent variable $U(t) = e^{kt}V(t)$, where $k > 0$ is fixed, so that (6.22) reads

(6.23)
$$\begin{cases} \dfrac{d}{dt}M(t)V - L_1(t)V = F_1(t), & 0 < t \leq T, \\ M(0)V(0) = M(0)U_0, \quad C(0)u'(0) = C(0)u_1, \end{cases}$$

with $L_1(t) = L(t) - kM(t)$, $\mathcal{D}(L_1(t)) = \mathcal{D} \times \mathcal{D}$, $F_1(t) = e^{-kt}F(t)$. Then the operators $B_1(t) = -M(t)L_1(t)^{-1}$, $0 \leq t \leq T$, are formally given by

$$B_1(t)(u, v) = (P(k, t)^{-1}(kC(t) + B(t))u + P(k, t)^{-1}v,$$
$$- C(t)P(k, t)^{-1}A(t)u - kC(t)P(k, t)^{-1}v), \quad (u, v) \in \mathcal{D} \times X,$$

provided that the inverse $P(k, t)^{-1}$ of $P(k, t) = k^2C(t) + kB(t) + A(t)$ exists as a bounded operator on X.

We prove the following result.

THEOREM 6.8. *Let us assume that $C(t)$ are non negative bounded linear operators on X and (6.20) and (6.21) hold. Furthermore,*

$$\|C(t)(\lambda C(t) + B(t))^{-1}\|_{\mathcal{L}(X)} \le k(1 + |\lambda|)^{-\beta}, \quad -\lambda \in \Sigma_1,$$

$$\|B(t)(\lambda B(t) + A(t))^{-1}\|_{\mathcal{L}(X)} \le k(1 + |\lambda|)^{-1}, \quad -\lambda \in \Sigma_1,$$

$$B(0)B(\cdot)^{-1}, \, A(\cdot)B(0)^{-1}, \, C(\cdot)B(0)^{-1} \in \mathcal{C}([0,T]; \mathcal{L}(X)),$$

$$B(0)B(\cdot)^{-1}f, \, A(\cdot)B(0)^{-1}f, \, C(\cdot)B(0)^{-1} \in \mathcal{C}^2([0,T]; X) \text{ for all } f \in X$$

$$\|C'(t)f\|_X \le m\|C(t)^\nu f\|_X, \quad f \in \mathcal{D},$$

m being independent of t. Suppose that $0 < \beta$, $\nu \le 1$, $\nu + 2\beta > 2$, $\beta^2 + 4 < 5\beta + \nu$ and let $1 - \beta < \theta < (\nu + 2\beta - 2)(2 - \beta)^{-1}$. Then, for all $u_0, u_1 \in \mathcal{D}$ and $f \in \mathcal{C}^\theta([0,T]; X)$ such that

$$f(0) - A(0)u_0 - B(0)u_1 - C'(0)u_1 = C(0)z$$

for a certain $z \in \mathcal{D}$, (6.19) possesses a unique solution such that

$$\tfrac{d}{dt}C(\cdot)u', \, B(\cdot)u', \, A(\cdot)u \in \mathcal{C}^\omega([0,T]; X), \quad \omega = \theta + \beta - 1.$$

Moreover, the equation in (6.19) takes place even at $t = 0$.

PROOF. This is an easy translation of Proposition 4.15 to the present problem. All the assumptions there are satisfied as soon as we observe that the compatibility relation translates into

$$u_1 - ku_0 = x \in \mathcal{D},$$

$$f(0) - A(0)u_0 - B(0)u_1 - C'(0)u_1 = C(0)y - kC(0)u_1 = C(0)z,$$

with $z = y - ku_1 \in \mathcal{D}$. Moreover, the operators $M(t)$ have the property $\|M'(t)F\|_{\mathcal{D} \times X} \le m\|M(t)^\nu F\|_{\mathcal{D} \times X}$ for all $F \in \mathcal{D} \times X$.

We remark that this theorem seems to be interesting in that it reduces solvability of (6.19) to verify that the pairs $(B(t), C(t))$ and $(B(t), A(t))$ satisfy those spectral properties studied in Chapters III and IV; hence, a wide range of application to partial differential equations is allowed. Using Propositions 4.14 and 4.15, we next prove that, when $B(t)$ and $A(t)$ do not depend on t, the regularity $C(\cdot)f \in \mathcal{C}^2([0,T]; X)$, $f \in \mathcal{D}$, can be weakened to Hölder continuity of the first derivative $C'(\cdot)f$.

THEOREM 6.9. *Let $A(t) \equiv A$, $B(t) \equiv B$, $C(t) \in \mathcal{L}(X)$ be a non negative bounded linear operator on X such that $\mathcal{D}(B) \subset \mathcal{D}(A)$,*

$$C(\cdot) \in \mathcal{C}^{1+\varphi}([0,T]; \mathcal{L}(X)), 0 < \varphi \le 1,$$

and $\|C'(t)f\|_X \leq k\|C(t)^\nu f\|_X$ for all $f \in \mathcal{D}(B)$,

$$\|C(t)(\lambda C(t) + B)^{-1}\|_{\mathcal{L}(X)} \leq k(1 + |\lambda|)^{-\beta}, \quad -\lambda \in \Sigma_1,$$

k being independent of t. If $2\beta + \nu > 2$, $(1 - \beta)(2 - \beta) < (2\beta + \nu - 2)\varphi$, let $\theta \in (1 - \beta, \frac{2\beta + \nu - 2}{2 - \beta}\varphi)$. Then, for all $f \in C^\theta([0, T]; X)$ and any u_0, $u_1 \in \mathcal{D}(B)$ such that

$$f(0) - Au_0 - Bu_1 - C'(0)u_1 = C(0)z, \quad \text{for a certain } z \in \mathcal{D}(B),$$

(6.19) possesses a unique solution such that

$$\frac{d}{dt}C(\cdot)u', \, Bu', \, Au \in C^\omega([0, T]; X), \quad \omega = \theta + \beta - 1.$$

Moreover, u satisfies the equation of (6.19) even at $t = 0$.

The proof is an easy consequence of Propositions 4.14 and 4.15. Notice that it holds in the particularly interesting case where $C(t) = m(t)C$ with $m \in C^{1+\varphi}([0, T]; \mathbb{R})$, $m(t) \geq 0$, $|m'(t)| \leq km(t)^\nu$ on $[0, T]$, C is a non negative operator in $\mathcal{L}(X)$, and where (6.4) and (6.5) hold with $\alpha = 1$. Hence, all the preceding assumptions are verified with the same indexes.

6.2 EXAMPLES

We devote this section to list a number of initial boundary value problems for partial differential equations to which the previous abstract results apply. Our aim is only to motivate the general treatment and hence we shall confine to simple cases, in order to avoid the technical complications due to not essential complexity.

EXAMPLE 6.1. Let us consider the problem
(6.24)
$$\begin{cases} \dfrac{\partial}{\partial t}\left(m(x)\dfrac{\partial u}{\partial t}\right) - \Delta\dfrac{\partial u}{\partial t} + A(x; D)u = f(x, t), & (x, t) \in \Omega \times (0, T], \\[2mm] u = \dfrac{\partial u}{\partial t} = 0, & (x, t) \in \partial\Omega \times (0, T], \\[2mm] u(x, 0) = u_0(x), & x \in \Omega, \\[2mm] m(x)\dfrac{\partial u}{\partial t}(x, 0) = m(x)u_1(x), & x \in \Omega, \end{cases}$$

where Ω is a bounded open set in $\mathbb{R}^n$, $n \geq 1$, with a smooth boundary $\partial\Omega$, $m \in L^\infty(\Omega)$, $m(x) \geq 0$ in Ω, $A(x; D)$ is a second order linear differential operator on Ω with coefficients continuous on $\overline{\Omega}$, and u_0, u_1 are given functions on Ω, $f(x, t)$ is a scalar valued continuous function on $\overline{\Omega} \times [0, T]$.

(6.24) and its generalizations have been very much treated in the literature (see e.g. R. W. Carroll and R.E. Showalter [48]). The usual approach
is to work in the negative Sobolev space $H^{-1}(\Omega)$, for then if A denotes
$A(x; D)$ with Dirichlet boundary conditions operating from $H_0^1(\Omega)$ into
$H^{-1}(\Omega)$, $B = -\Delta$ with $\mathcal{D}(B) = \mathcal{D}(A)$ and C is the multiplication operator
by the function m so that $C \in \mathcal{L}(H_0^1(\Omega), L^2(\Omega))$, then (6.4) and (6.5) are
verified with $\alpha = \beta = 1$. Therefore, Theorems 6.1 and 6.2 are applicable.

If we require more regular solutions to (6.24), we have seen (Example
3.1) that if $\mathcal{D}(B) = H_0^1(\Omega) \cap H^2(\Omega)$, $B = -\Delta$ operates in $X = L^2(\Omega)$ and
C has the meaning as above, then (6.4) and (6.5) hold with $\alpha = 1$, $\beta = \frac{1}{2}$.
Now A can be viewed as a bounded linear operator from $H^2(\Omega)$ into X
and then Theorems 6.1 and 6.2 work with $\sigma \in (\frac{1}{2}, 1)$. In particular, if

$$f(\cdot, 0) + \Delta u_1 - A(x; D)u_0 = mw,$$

with $w \in H_0^1(\Omega) \cap H^2(\Omega)$, u_0, $u_1 \in \mathcal{D}(B)$, $f \in C^\sigma([0, T]; L^2(\Omega))$, then (6.24)
possesses a unique solution u such that

$$\Delta \frac{\partial u}{\partial t} \in C^{\sigma - \frac{1}{2}}([0, T]; L^2(\Omega)), \quad \frac{\partial}{\partial t}\left(m\frac{\partial u}{\partial t}\right) \in C^{\sigma - \frac{1}{2}}([0, T]; L^2(\Omega)).$$

In view of Example 3.1, this approach extends to general $L^p(\Omega)$, $1 < p <$
∞, provided that we take $\sigma \in (\frac{1}{p}, 1)$, and u_0, $u_1 \in \mathring{W}_p^1(\Omega) \cap W_p^2(\Omega)$.

EXAMPLE 6.2. Let $\Omega = (0, 1)$ for sake of simplicity and consider
(6.25)

$$\begin{cases} \dfrac{\partial}{\partial t}\left(m(x)\dfrac{\partial u}{\partial t}\right) + (-1)^m \dfrac{\partial^{2m+1}u}{\partial x^{2m}\partial t} + A(x; D)u = f(x, t), \\ \qquad\qquad\qquad\qquad\qquad\qquad 0 < x < 1,\, 0 < t \leq T, \\ \dfrac{\partial^i u}{\partial x^i}(x, t) = \dfrac{\partial^{i+1}u}{\partial x^i \partial t}(x, t) = 0, \quad 0 \leq i \leq m-1,\, x = 0, 1,\, 0 < t \leq T, \\ u(x, 0) = u_0(x), \qquad\qquad 0 < x < 1, \\ m(x)\dfrac{\partial u}{\partial t}(x, 0) = m(x)u_1(x), \qquad 0 < x < 1, \end{cases}$$

where $m \in C([0, T])$, $m(x) \geq 0$ on $[0, T]$, $f \in C([0, T]; L^2(0, 1))$, u_0, $u_1 \in$
$L^2(0, 1)$. Let us take $\mathcal{D}(B) = H_0^m(0, 1) \cap H^{2m}(0, 1)$, $Bu = (-1)^m \frac{\partial^{2m}u}{\partial x^{2m}}$,
$u \in \mathcal{D}(B)$, and suppose $A(x; D)$ to be a linear differential operator of order
$\leq 2m$ with continuous coefficients on $[0, 1]$.

If we maintain the same meaning as in Example 6.1 to the operators A
and C, we observe that $(\lambda C + B)u = f$ implies, by repeating integration
by parts and taking in account the boundary conditions in $u \in \mathcal{D}(B)$ and
Sobolev imbedding theorem, that

$$(6.26) \qquad \lambda \int_0^1 m(x)|u(x)|^2 dx + \|u^{(m)}\|_{L^2}^2 = \int_0^1 f\bar{u}\,dx.$$

Considering the real and imaginary parts of this formula and using Poincaré inequality (eventually after adding the absolute value of the imaginary parts to the real parts as done in Example 3.1), we check that

$$\|u^{(m)}\|_{L^2} \le C\|f\|_{L^2},$$

$$(1+|\lambda|)\int_0^1 m(x)|u(x)|^2 dx \le C\|f\|_{L^2}\|u\|_{L^2}, \quad \lambda \in -\Sigma_1,$$

and this in its turn implies that

$$(1+|\lambda|)\|Cu\|_{L^2}^2 \le \|f\|_{L^2}^2, \quad \lambda \in -\Sigma_1.$$

Since B is self adjoint and C is bounded from $X = L^2(0,1)$ into itself, we deduce an analogous estimate for the adjoint operator $\overline{\lambda}C + B$, thus implying again that (6.4) and (6.5) are verified with $\alpha = 1$, $\beta = \frac{1}{2}$. It follows that all the results in Theorems 6.1 and 6.2 can be applied. Interpretations of them in this case as well as translation of the compatibility relation are straightforward.

Of course, making use of the technique delivered for Example 3.1, it should be be possible to extend these estimates to $2m$-order strongly elliptic operators $B = \Sigma_{|\alpha| \le 2m} a_\alpha D^\alpha$ both in $L^2(\Omega)$ and in $L^p(\Omega)$, $1 < p < \infty$, with suitable hypotheses on the coefficients a_α and the bounded open subset $\Omega \subset \mathbb{R}^n$.

EXAMPLE 6.3. Let again $\Omega = (0,1)$ and consider
(6.27)
$$
\begin{cases}
\dfrac{\partial}{\partial t}\left(m(x)\dfrac{\partial u}{\partial t}\right) + (-1)^m \dfrac{\partial^{2m+1} u}{\partial x^{2m}\partial t} + A(x;D)u = f(x,t), \\
\qquad\qquad\qquad\qquad\qquad 0 < x < 1,\, 0 < t \le T, \\
\dfrac{\partial^{2i} u}{\partial x^{2i}}(x,t) = \dfrac{\partial^{2i+1} u}{\partial x^{2i}\partial t}(x,t) = 0, \quad 0 \le i \le m-1,\, x = 0,1,\, 0 < t \le T, \\
u(x,0) = u_0(x), \qquad\qquad 0 < x < 1, \\
m(x)\dfrac{\partial u}{\partial t}(x,0) = m(x)u_1(x), \qquad 0 < x < 1,
\end{cases}
$$

where $A(x;D)$, $m(x)$, $f(x,t)$ and u_0, u_1 are as in Example 6.2, $X = L^2(0,1)$, but this time $B = K^m$ with $K = -\frac{\partial^2}{\partial x^2}$, $\mathcal{D}(K) = H_0^1(0,1) \cap H^2(0,1)$. The boundary conditions $u^{(2i)}(0) = u^{(2i)}(1) = 0$ $(i = 0,1,\ldots,m-1)$ for all $u \in \mathcal{D}(B)$ imply that, if $(\lambda C + B)u = f \in X$, then the identity (6.26) holds equally and it is an easy matter to verify by means of analogous arguments that (6.4) and (6.5) are valid with $\alpha = 1$, $\beta = \frac{1}{2}$ and Theorems 6.1 and 6.2 translate into existence and regularity for the solutions to (6.27) which are differentiable even at $t = 0$ and satisfy the equation of (6.27). Again using the multiplier technique delivered in Example 3.1 we could treat the case where $X = L^p(0,1)$, $1 < p < \infty$.

EXAMPLE 6.4. Consider the problem
(6.28)

$$
\begin{cases}
\dfrac{\partial}{\partial t}\left(\left(1+\dfrac{\partial^2}{\partial x^2}\right)\dfrac{\partial u}{\partial t}\right)-\dfrac{\partial^3 u}{\partial x^2 \partial t}+A(x;D)u = f(x,t), \\
\qquad\qquad\qquad\qquad\qquad\qquad 0\le x\le\pi,\,0\le t\le T, \\
u(0,t)=u(\pi,t)=0,\qquad 0\le t\le T, \\
u(x,0)=u_0(x),\ \left(1+\dfrac{\partial^2}{\partial x^2}\right)\dfrac{\partial u}{\partial t}(x,0)=\left(1+\dfrac{\partial^2}{\partial x^2}\right)u_1(x),\,0\le x\le\pi.
\end{cases}
$$

Let us take $X=\{f\in\mathcal{C}([0,\pi]);\,f(0)=f(\pi)=0\}$,

$$
\begin{cases}
\mathcal{D}(C)=\{u\in\mathcal{C}^2([0,\pi]);\,u(0)=u(\pi)=u''(0)=u''(\pi)=0\}, \\
Cu=(1+\dfrac{\partial^2}{\partial x^2})u,
\end{cases}
$$

$B=-\frac{\partial^2}{\partial x^2}$, $\mathcal{D}(B)=\mathcal{D}(C)$, and let us take $A(x;D)=\sum_{k=0}^{2}a_k(x)\frac{\partial^k}{\partial x^k}$ a second order differential operator with coefficients $a_k(\cdot)$ continuous on $[0,\pi]$, $f\in\mathcal{C}^\theta([0,T];X)$, $0<\theta<1$. We have already seen (Example 3.10) that (6.4) and (6.5) hold with $\alpha=\beta=1$. Hence, if $u_0,\,u_1\in\mathcal{D}(C)$, and

$$
f(x,0)-A(x;D)u_0+u_1'' = w+w'',\quad 0\le x\le\pi,
$$

with $w\in\mathcal{D}(C)$, then (6.28) possesses a unique solution u with the regularity $\frac{\partial}{\partial t}((1+\frac{\partial^2}{\partial x^2})\frac{\partial u}{\partial t})$, $\frac{\partial^3 u}{\partial x^2 \partial t}\in\mathcal{C}^\theta([0,T];X)$, $0<\theta<1$.

EXAMPLE 6.5. This is an abstract version of the model considered in Example 6.1, where we consider the operators in a weak form from $H_0^1(\Omega)$ into $H^{-1}(\Omega)$.

Let V, H be two complex Hilbert spaces such that $V\subset H$ densely and continuously. Denoting by $\|\cdot\|$, $|\cdot|$ and $\|\cdot\|_*$ the norms in V, H and V' respectively, by $(\cdot,\cdot)$ the inner product in H and $\langle\cdot,\cdot\rangle$ the scalar product between V and V' such that $\langle u,v\rangle=(u,v)$ for all $u,v\in V$, we also assume that

(6.29) A, B are elements of $\mathcal{L}(V,V')$ and $\Re\langle Bu,u\rangle\ge a_0\|u\|^2$ for all $u\in V$, where $a_0>0$;

(6.30) $C\in\mathcal{L}(H)$ is a non negative operator.

For example, A, C could be the operators defined (in V') by two sesquilinear forms in V and H respectively, and B a positive (of parabolic type) operator in $\mathcal{L}(V,V')$.

If $\lambda Cu+Bu=f\in V'$, we have, taking the product with u,

$$
\lambda\langle Cu,u\rangle+\langle Bu,u\rangle=\langle f,u\rangle
$$

and hence, by our hypotheses,

$$\Re e(Cu, u) + \Re e\langle Bu, u\rangle = \Re e\langle f, u\rangle,$$

(6.29) and (6.30) yield $\Re e(Cu, u) + a_0\|u\|^2 \le \|f\|_*\|u\|$ and also

(6.31) $$\|u\| \le \tfrac{1}{a_0}\|(\lambda C + B)u\|_*.$$

Since $\langle B^*u, v\rangle = \langle u, Bv\rangle$ for all $u, v \in V$, an estimate like (6.31) is true for $\bar{\lambda}C + B^* = (\lambda C + B)^*$, permitting to conclude that $(\lambda C + B)^{-1} \in \mathcal{L}(V', V)$ for all λ, $\Re e\lambda > 0$, and

$$\|(\lambda C + B)^{-1}\|_{\mathcal{L}(V', V)} \le \tfrac{1}{a_0}.$$

On the other hand, $B \in \mathcal{L}(V, V')$ implies that

$$\|B(\lambda C + B)^{-1}\|_{\mathcal{L}(V')} \le \text{Const.}, \quad \Re e\lambda \ge 0.$$

It follows that (6.4) and (6.5) are verified with $\alpha = \beta = 1$ in a suitable sector $-\Sigma_1$.

Application to second order (in time) differential equations with differential operator coefficients in variational form is immediate (cf. J. L. Lions and E. Magenes [129]).

EXAMPLE 6.6. Let B be defined by means of $\mathcal{D}(B) = H^2(\mathbb{R}^n)$, $n \ge 1$, $Bu = -\Delta u + cu$, where $c(x) \ge c_0 > 0$ is a bounded continuous function on $\mathbb{R}^n$ and c_0 is sufficiently large. Hence B is selfadjoint in $L^2(\mathbb{R}^n)$ and positive. If C is multiplication by the potential $V(x)$, where $V(x) \ge 0$ is continuous and bounded on $\mathbb{R}^n$, we know (from Example 3.7) that (6.4) and (6.5) hold in $X = L^2(\mathbb{R}^n)$ with $\alpha = 1$, $\beta = \tfrac{1}{2}$.

If $A = \sum_{|\alpha|\le 2} a_\alpha(x)D^\alpha$, where the coefficients a_α are continuous and bounded functions from $\mathbb{R}^n$ to the complex plane, then Theorems 6.1 and 6.2 apply.

EXAMPLE 6.7. Let B be the $L^2(\mathbb{R}^n)$ realization of $I+(-1)^m\Delta^m$, $m \ge 2$, so that $\mathcal{D}(B) = H^{2m}(\mathbb{R}^n)$. Then it is well known that B is self adjoint and positive. Take $C = I - \Delta$ with $\mathcal{D}(C) = H^2(\mathbb{R}^n)$. Then, by taking the inner product in $L^2(\mathbb{R}^n)$ of $(\lambda C + B)u = f$ and u, we see that

$$(\lambda + 1)\|u\|_{L^2}^2 + \lambda\|\nabla u\|_{L^2}^2 + (-1)^m\langle \Delta^m u, u\rangle = \langle f, u\rangle.$$

Therefore, following the lines of proof in Example 3.3, one deduces that for all $\lambda \in -\Sigma_1$, $(|\lambda| + 1)\|u\|_{L^2} \le C\|f\|_{L^2}$, $(|\lambda| + 1)\|\nabla u\|_{L^2} \le C\|f\|_{L^2}$, $\|u\|_{H^m}^2 \le C\|f\|_{L^2}\|u\|_{L^2} \le C(|\lambda|+1)^{-1}\|f\|_{L^2}^2$. If we choose $m = 2$, we have the estimate

$$\|Cu\|_{L^2} \le k(|\lambda| + 1)^{-\frac{1}{2}}\|f\|_{L^2}.$$

Passage to the adjoint operator assures the existence of the inverse $(\lambda C + B)^{-1} \in \mathcal{L}(L^2(\mathbb{R}^n))$ and thus (6.4) and (6.5) hold with $\alpha = 1$, $\beta = \tfrac{1}{2}$.

In order to apply Theorems 6.1 and 6.2 it only needs to assume that A is the realization in $L^2(\mathbb{R}^n)$ of $A(x; D) = \sum_{|\alpha|\le 2m} a_\alpha(x)D^\alpha$, with coefficients continuous and bounded on $\mathbb{R}^n$.

EXAMPLE 6.8. Let Ω be a bounded domain in $\mathbb{R}^n$, $n \geq 1$, with a smooth boundary $\partial\Omega$. Let $1 < p < \infty$, $X = L^p(\Omega)$. Consider the differential operator with (complex) constant coefficients c_α

$$E(D)u = \sum_{|\alpha|=2m} c_\alpha D^\alpha u,$$

where $m \in \mathbb{N}$.

We are also given a normal system $\{B_j; \ j = 1, \ldots, m\}$ of boundary operators complemented with respect to $E(D)$, according to [172; pp. 333-334], so that, in particular

$$B_j u(x) = \sum_{|\alpha| \leq k_j} b_{j\alpha}(x) D^\alpha u(x), \quad b_{j\alpha} \in C^\infty(\partial\Omega),$$

$j = 1, \ldots, m$, $0 \leq k_1 < k_2 < \cdots < k_m < 2m$. It is to be mentioned that all these assumptions hold when

$$B_j u = \frac{\partial^{k+j-1} u}{\partial \nu^{k+j-1}},$$

where $k \in \{0, 1, \ldots, m\}$ is fixed. A remarkable result by R. Seeley [154] (see also [172; p. 321]) characterizes the complex interpolation spaces between X and

$$W^{2m}_{p,\{B_j\}} = \{u \in W^{2m}_p(\Omega); \ B_j u_{|\partial\Omega} = 0, \ j = 1, \ldots, m\}.$$

Precisely, if we are interested in the case where there does not exist a number k_j, $j = 1, \ldots, m$ such that $2\theta - \frac{1}{p} = k_j$, then

$$[X, W^{2m}_{p,\{B_j\}}]_\theta = W^{2m\theta}_{p,\{B_j\}}(\Omega)$$
$$= \{u \in W^{2m\theta}_p(\Omega); \ B_j u_{|\partial\Omega} = 0 \ for \ k_j < 2m\theta - \tfrac{1}{p}\}.$$

Moreover, the operator $-E$, where

$$\mathcal{D}(E) = W^{2m}_{p,\{B_j\}}(\Omega),$$
$$(Eu)(x) = E(D)u(x)$$

generates an analytic semigroup in X and the imaginary powers E^{it}, $t \in \mathbb{R}$ of E are bounded linear operators in X such that $\|E^{it}\|_{\mathcal{L}(X)} \leq Ce^{\gamma|t|}$, where $C > 0$, $\gamma \geq 0$ (see [172; p. 334]). For sake of simplicity (this can always be obtained by adding a suitable constant), we shall assume that E is positive, so that, in virtue of a famous result (see Section 0.3 and

[172; p. 103]), $\mathcal{D}(E^\sigma) = [X, \mathcal{D}(E^\alpha)]_{\frac{\sigma}{\alpha}}, 0 < \sigma < \alpha \, (\alpha > 0)$. It follows that if $2m\theta = h$ is a positive integer $< 2m$, then

$$[X, W^{2m}_{p,\{B_j\}}(\Omega)]_{\frac{h}{2m}} = \{u \in W^h_p(\Omega); B_j u_{|\partial\Omega} = 0 \text{ for } k_j < h - \tfrac{1}{p}\} = \mathcal{D}(E^{\frac{h}{2m}}).$$

These results indicate a procedure to be followed in order to reduce concrete partial differential equations to a form to which Theorems 6.3-6.6 apply with success.

Refering to the problem (6.2), it shall suffice to take an operator like E as B; concerning C, we have to assume that the pair (B, C) verifies (6.4) and (6.5) (as in Examples 6.1-6.6). At last, we take $A = UB$, where $-U$ generates an analytic semigroup in X with (6.8).

In what follows we give some examples to clarify the range of applications of Theorems 6.3-6.5 and Example 6.8.

EXAMPLE 6.9. Let Ω be the bounded domain in $\mathbb{R}^n$ as in Example 6.8, $X = L^p(\Omega)$, $1 < p < \infty$, and denote $W^2_p(\Omega) \cap \mathring{W}^1_p(\Omega)$ by Y.

If Δ is the Lapacian with $\mathcal{D}(\Delta) = Y$ and m is a positive integer, let $\mathcal{D}(\Delta^m) = \{u \in \mathcal{D}(\Delta^{m-1}); \Delta^{m-1}u \in Y\}$. Let m, k be two fixed positive integers such that $k < m$ and $m + 1 < 2k$. After introducing the operators A, B, C in the space X by

$$\mathcal{D}(A) = \mathcal{D}(\Delta^m), \; Au = (-1)^m \Delta^m u, \; u \in \mathcal{D}(A),$$
$$\mathcal{D}(B) = \mathcal{D}(\Delta^k), \; Au = (-1)^k \Delta^k u, \; u \in \mathcal{D}(B),$$
$$\mathcal{D}(C) = Y, \; Cu = (1 - \Delta)u, \; u \in Y,$$

it is well known that $-AB^{-1} = -U$ generates an analytic semigroup in X. Furthermore, from

$$(-1)^k \Delta^k (1 - \Delta)^{-1} = (-1)^{k-1}\Delta^{k-1} + (-1)^{k-1}[-1 + (1 - \Delta)^{-1}]\Delta^{k-2},$$

it follows that $-BC^{-1}$ generates another analytic semigroup in X, so that (6.4),(6.5) and (6.7) hold with $\alpha = \beta = \gamma = 1$.

In view of the above mentioned characterization by complex interpolation spaces (cf. [172; p. 103]), see Example 6.8, analyticity of the semigroup generated by $-BC^{-1}$ leads to

$$\|(BC^{-1} + \lambda)^{-1}\|_{\mathcal{L}(X, \mathcal{D}((-\Delta)^{(k-1)\theta}))} \leq C|\lambda|^{\theta - 1}, 0 < \theta < 1.$$

Since $\mathcal{D}(U) = \mathcal{D}(\Delta^{m-k})$, assumption (6.8) is verified with $\delta = (2k - (m + 1))(k - 1)^{-1}$. Notice that here k must necessarily be > 1.

EXAMPLE 6.10. Let us consider the critical boundary value problem
(6.32)

$$
\begin{cases}
\dfrac{\partial}{\partial t}\left(\mu(x)\dfrac{\partial u}{\partial t}\right) + (-1)^m \dfrac{\partial^{2m+1}u}{\partial x^{2m}\partial t} + (-1)^{m+\ell}\dfrac{\partial^{2(m+\ell)}u}{\partial x^{2(m+\ell)}} = f(x,t), \\[2mm]
\qquad\qquad 0 < x < 1,\ 0 \le t \le T, \\[2mm]
\dfrac{\partial^i u}{\partial x^i}(x,t) = \dfrac{\partial^{i+1}u}{\partial x^i \partial t}(x,t) = \dfrac{\partial^{2m+j}u}{\partial x^{2m+j}}(x,t) = 0, \\[2mm]
\qquad i = 0,1,\dots,m-1,\ j = 0,1,\dots,\ell-1,\ x \in \{0,1\},\ 0 \le t \le T, \\[2mm]
u(x,0) = u_0(x),\ \mu(x)\dfrac{\partial u}{\partial t}(x,0) = \mu(x)u_1(x), \quad 0 < x < 1,
\end{cases}
$$

where $\mu \in C^{2m}([0,1])$, $\mu(x) \ge 0$, on $[0,1]$, $f \in C([0,T]; L^2(0,1))$, $u_0, u_1 \in X = L^2(0,1)$. The operators B and C are defined as in Example 6.2, and therefore (6.4) and (6.5) are verified with $\alpha = 1$, $\beta = \frac{1}{2}$. The operator A is given by $A = UB$, where $U = (-1)^\ell \frac{\partial^{2\ell}}{\partial x^{2\ell}}$, $\mathcal{D}(U) = H_0^\ell(0,1) \cap H^{2\ell}(0,1)$. Thus (6.7) holds with $\gamma = 1$. In order to apply Theorem 6.5, in view of Remark 6.1, we require that $[X, \mathcal{D}(B)]_\tau \subset \mathcal{D}(U)$ for a suitable $\tau \in (0, \frac{1}{2})$ (and then $\delta = \frac{1}{2} - \tau$).

But from Example 6.8, $\tau = \frac{r}{2m}$, $r \in \{1,2,\dots,m-1\}$ implies

$$
[X, \mathcal{D}(B)]_{\frac{r}{2m}} = \{u \in H^r(0,1);\ u^{(j)}(0) = u^{(j)}(0) = 0,\ j = 0,1,\dots,r-1\}.
$$

Hence, if $m > 2\ell$, then (6.32) is solved according to Theorems 6.3-6.5, with $\alpha = \gamma = 1$, $\beta = \frac{1}{2}$.

More generally, if $U = (-1)^\ell \frac{\partial^{2\ell}}{\partial x^{2\ell}}$ with $\mathcal{D}(U) = \{u \in H^{2\ell}(0,1); u^{(i)}(0) = \cdots = u^{(i+\ell-1)}(0) = 0,\ u^{(i)}(1) = \cdots = u^{(i+\ell-1)}(1) = 0\}$, where $i \in \{0,1,\dots,\ell\}$, then the smallest space

$$
[X, \mathcal{D}(B)]_{\frac{m-1}{2m}}
$$
$$
= \{u \in H^{m-1}(0,1);\ u^{(j)}(0) = u^{(j)}(1) = 0,\ j = 0,1,\dots,m-2\}
$$

is embedded in $\mathcal{D}(U)$ if and only if $2\ell \le m-1$ and $i \le m-\ell-1$.

In particular, we have the following special case.

EXAMPLE 6.11. Let $\mu \in C^{2m}([0,1])$ be non negative, and let $f \in C([0,T]; L^2(0,1))$ and $u_0, u_1 \in L^2(0,1)$ be given respectively. Let B and C denote the same operators as in Example 6.2 and take $A = UB$, where $U = -\frac{d^2}{dx^2}$, $\mathcal{D}(U) = \{u \in H^2(0,1);\ u'(0) = u'(1) = 0\}$.

Then $-U$ generates an analytic semigroup in X. Moreover,

$$
[X, \mathcal{D}(B)]_{\frac{1}{m}} = \{u \in H^2(0,1);\ u^{(j)}(0) = u^{(j)}(1) = 0,\ j = 0,1\} \subset \mathcal{D}(U)
$$

and thus assumption $m > 2$ allows to conclude that in such a case the initial boundary value problem

$$
\begin{cases}
\dfrac{\partial}{\partial t}\left(\mu(x)\dfrac{\partial u}{\partial t}\right) + (-1)^m \dfrac{\partial^{2m+1}u}{\partial x^{2m}\partial t} + (-1)^{m+1}\dfrac{\partial^{2(m+1)}u}{\partial x^{2(m+1)}} = f(x,t), \\[2mm]
\qquad\qquad 0 < x < 1,\ \ 0 \le t \le T, \\[2mm]
\dfrac{\partial^i u}{\partial x^i}(x,t) = \dfrac{\partial^{i+1}u}{\partial x^i \partial t}(x,t) = \dfrac{\partial^{2m+1}u}{\partial x^{2m+1}}(x,t) = 0, \\[2mm]
\qquad i = 0,1,\dots,m-1,\ x \in \{0,1\},\ 0 \le t \le T, \\[2mm]
u(x,0) = u_0(x),\ \mu(x)\dfrac{\partial u}{\partial t}(x,0) = \mu(x)u_1(x), \quad 0 < x < 1,
\end{cases}
$$

possesses a unique strict solution u enjoying regularities

$$
\frac{\partial^{2(m+1)}u}{\partial x^{2(m+1)}},\ \frac{\partial^{2m+1}u}{\partial x^{2m}\partial t},\ \frac{\partial}{\partial t}\left(\mu\frac{\partial u}{\partial t}\right) \in C^{\theta-\frac12}([0,T];L^2(0,1)),
$$

provided that $\frac12 < \theta \le 1$, $f \in C^\theta([0,T];L^2(0,1))$, $u_0 \in H^{2(m+1)}(0,1) \cap H_0^m(0,1)$, $u_0^{(2m+1)}(0) = u_0^{(2m+1)}(1) = 0$, $u_1 \in H^{2m}(0,1) \cap H_0^m(0,1)$, $m > 2$, and

$$
f(x,0) + (-1)^m u_0^{2(m+1)} + (-1)^{m+1}u_1^{(2m)} = \mu(x)w,
$$
$$
w \in H^{2m}(0,1) \cap H_0^m(0,1).
$$

It only needs to observe that here we are in the situation described at the end of Example 6.10 with $i = \ell = 1$.

EXAMPLE 6.12. Let us consider the initial boundary value problem
(6.33)
$$
\begin{cases}
\dfrac{\partial}{\partial t}\left(\mu(x)\dfrac{\partial u}{\partial t}\right) + (-1)^m \dfrac{\partial^{2m+1}u}{\partial x^{2m}\partial t} + (-1)^{m+\ell}\dfrac{\partial^{2(m+\ell)}u}{\partial x^{2(m+\ell)}} = f(x,t), \\[2mm]
\qquad\qquad 0 < x < 1,\ \ 0 \le t \le T, \\[2mm]
\dfrac{\partial^{2i}u}{\partial x^{2i}}(x,t) = \dfrac{\partial^{2j+1}u}{\partial x^{2j}\partial t}(x,t) = 0, \\[2mm]
\qquad i = 0,1,\dots,m+\ell-1,\ j = 0,1,\dots,m-1,\ x \in \{0,1\},\ 0 \le t \le T, \\[2mm]
u(x,0) = u_0(x),\ \mu(x)\dfrac{\partial u}{\partial t}(x,0) = \mu(x)u_1(x), \quad 0 < x < 1,
\end{cases}
$$

where, as in Example 6.3, $A = K^{m+\ell}$, $B = K^m$, $Ku = -u'' = -\frac{d^2u}{dx^2}$, $\mathcal{D}(K) = H_0^1(0,1) \cap H^2(0,1)$, C is multiplication by μ in $L^2(0,1)$, $\mu \in C^{2m}([0,1])$, $\mu(x) \ge 0$ on $[0,1]$ and $\mu^{(2j+1)}(0) = \mu^{(2j+1)}(1) = 0$ for $j = 0,1,\dots,m-1$.

The assumptions on the function μ are introduced to ensure that $\mathcal{D}(B)$ is invariant under C and conditions (6.9) and (6.10) hold. Moreover,

$B = K^m$ yields that, in virtue of the already quoted properties of K (see [172; p. 434]), $B^{is} = K^{ism}$ is a bounded operator in $X = L^2(0,1)$ for all real numbers s. Since $A = B^{1+\frac{\ell}{m}}$, the condition (6.11) is clearly satisfied with $\tau = \frac{\ell}{m}$. Therefore, if $2\ell < m$, Theorem 6.5 yields (6.8).

We conclude that, if

$$f \in C^\theta([0,T]; L^2(0,1)), \frac{1}{2} < \theta < 1,$$

$$u_0 \in H^{2(\ell+m)}(0,1), u_0^{(2j)}(0) = u_0^{(2j)}(1) = 0, j = 0, 1, \ldots, \ell + m - 1,$$

$$u_1 \in H^{2m}(0,1), u_1^{(2i)}(0) = u_1^{(2i)}(1) = 0, i = 0, 1, \ldots, m - 1,$$

and

$$f(\cdot, 0) + (-1)^{m+\ell-1} u_0^{2(m+\ell)} + (-1)^{m+1} u_1^{(2m)} = \mu(\cdot)w,$$

where $w \in H^{2m}(0,1)$, $w^{(2i)}(0) = w^{(2i)}(1) = 0$, $i = 0, 1, \ldots, m - 1$, then problem (6.33) possesses a unique solution u with the same regularty as in Example 6.10.

We want to point out that in fact the differential operator B satisfies the hypotheses in Example 6.8, so that if h is a positive integer $< 2m$

$$[L^2(0,1), \mathcal{D}(B)]_{\frac{h}{2m}}$$
$$= \{u \in H^h(0,1); u^{(2j)}(0) = u^{(2j)}(1) = 0 \text{ for } 2j \leq h - 1\}.$$

Hence, if $h = m - 1$, then

$$[L^2(0,1), \mathcal{D}(B)]_{\frac{m-1}{2m}}$$
$$= \{u \in H^{m-1}(0,1); u^{(2j)}(0) = u^{(2j)}(1) = 0 \text{ for } 2j \leq m - 2\}.$$

But $\mathcal{D}(AB^{-1}) = \mathcal{D}(K^\ell) = \{u \in H^{2\ell}(0,1); u^{(2i)}(0) = u^{(2i)}(1) = 0, i = 0, 1, \ldots, \ell - 1\}$. It follows that $[L^2(0,1), \mathcal{D}(B)]_{\frac{m-1}{2m}} \subset \mathcal{D}(AB^{-1})$ if and only if $2\ell \leq m$.

Before continuing our concrete examples, we think it opportune to state first an observation of general character on the operator pencil $\sum_{j=0}^n \lambda^j A_j$.

Given A_n and A_{n-1}, let us assume that $A_{n-2} = B_1 A_{n-1}$, $A_{n-3} = B_2 B_1 A_{n-1}$, and $A_{n-j} = B_{j-1} B_{j-2} \cdots B_1 A_{n-1}$, $j = 2, \ldots, n$, where $-B_j$ generates an analytic semigroup in X. B_j is precisely U_{n-j} according to the notation we have used for $B_j = U_{n-j} = A_{n-j-1} A_{n-j}^{-1}$, $j = 1, \ldots, n$. Then assumptions (6.14) read $\mathcal{D}(B_j) \subset \mathcal{D}(B_{j+1})$ and in its turn (6.15) becomes

(6.34)
$$\|B_{j+1}(\lambda + B_j)^{-1}\|_{\mathcal{L}(X)} \leq k(1 + |\lambda|)^{-\delta}, \quad \delta > 0, |\lambda| \text{ large}, \lambda \in -\Sigma_\alpha,$$

(6.35)
$$\|B_1 A_n(\lambda A_n + A_{n-1})^{-1}\|_{\mathcal{L}(X)} \leq k(1 + |\lambda|)^{-\delta}, \quad \delta > 0, |\lambda| \text{ large}, \lambda \in -\Sigma_\alpha.$$

Now, (6.35) can be covered as in the preceding examples, (when $A_n = C$, $A_{n-1} = B$, $A_{n-2} = B_1 B = A$). Concerning (6.34), it shall suffice, for example, that $\mathcal{D}(B_{j+1}) \supset [X, \mathcal{D}(B_j)]_{\theta_j}$, $j = 2, \ldots, n-1$, $0 < \theta_j < 1$. In this case Theorem 6.7 is applicable with $1 - \beta_n < \theta < 1$.

EXAMPLE 6.13. Let a_j, $j = 0, 1, \ldots, n-1$ be a fixed positive munber and let $X = L^2(0,1)$. Given $f \in \mathcal{C}([0,T]; L^2(0,1))$, $m \in \mathcal{C}^\infty([0,1])$, $m(x) \geq 0$ on $[0,1]$, consider the equation

$$\frac{\partial}{\partial t}\left(m(x)\frac{\partial^{n-1}u}{\partial t^{n-1}}\right) + \sum_{j=0}^{n-1} A_j(D)\frac{\partial^j u}{\partial t^j} = f(x,t), \quad 0 < x < 1, \ 0 \leq t \leq T,$$

where $A_j(D) = (-1)^{r_1+r_2+\cdots+r_{n-j}} a_j \frac{\partial^{2(r_1+\cdots+r_{n-j})}}{\partial x^{2(r_1+\cdots+r_{n-j})}}$, $j = 0, 1, \ldots, n-1$, $r_1 > r_2 > \cdots > r_n \geq 1$. We take as A_n the multiplication operator by $m(x)$ in X, $\mathcal{D}(A_{n-1}) = H^{2r_1}(0,1) \cap H_0^{r_1}(0,1)$, $\mathcal{D}(B_1) = H^{2r_2}(0,1) \cap H_0^{r_2}(0,1)$, $A_{n-1}u = (-1)^{r_1}a_{n-1}u^{(2r_1)}$, $u \in \mathcal{D}(A_{n-1})$, $B_1 w = (-1)^{r_2}\frac{a_{n-2}}{a_{n-1}}w^{(2r_2)}$, $w \in \mathcal{D}(B_1)$.

According to the result in Example 6.10, if $r_1 > 2r_2$, then (6.35) holds. Let $B_2 u = (-1)^{r_3}\frac{a_{n-3}}{a_{n-2}}u^{(2r_3)}$, with $\mathcal{D}(B_2) = \{u \in H^{2r_3}(0,1); u^{(i)}(0) = \cdots = u^{(i+r_3-1)}(0) = 0, u^{(i)}(1) = \cdots = u^{(i+r_3-1)}(1) = 0\}$ for some $i \in \{0, 1, \ldots, r_3\}$. Let us choose $\theta_1 = \frac{2r_2-1}{2r_2}$, so that

$$[X, H^{2r_2}(0,1) \cap H_0^{r_2}(0,1)]_{\frac{2r_2-1}{2r_2}}$$
$$= \{u \in H^{2r_2-1}(0,1); u^{(j)}(0) = u^{(j)}(1) = 0, j \leq r_2 - 1, j < 2r_2 - 1\}$$
$$= \{u \in H^{2r_2-1}(0,1); u^{(j)}(0) = u^{(j)}(1) = 0, j = 0, 1, \ldots, r_2 - 1\}$$
$$\subset \mathcal{D}(B_2).$$

Hence we must have: $2r_2 - 1 \geq 2r_3$, $r_2 > r_3$, $i + r_3 - 1 \leq r_2 - 1$, that is, $i \leq r_2 - r_3$. This, for example, permits both Dirichlet and Neumann boundary conditions for B_2. The same procedure is applied to the successive operators $B_3, \ldots, B_n$, where $B_j u = (-1)^{r_{j+1}}\frac{a_{n-j-1}}{a_{n-j}}u^{(2r_{j+1})}$, for obtaining from B_k the admissible domains for B_{k+1}.

EXAMPLE 6.14. Let Ω be a bounded domain in $\mathbb{R}^m$, $m \geq 1$, with a $\mathcal{C}^\infty$ boundary $\partial\Omega$, take $X = L^p(\Omega)$, $1 < p < \infty$. Let Δ be the Laplacian in Ω with domain $Y = W_p^2(\Omega) \cap \mathring{W}_p^1(\Omega)$, as in Example 6.9. Given $n \in \mathbb{N}$, let $A_n = I - \Delta$, $\mathcal{D}(A_n) = Y$, $A_{n-1} = (-1)^{r_1}\Delta^{r_1}$, $\mathcal{D}(A_{n-1}) = \mathcal{D}(\Delta^{r_1}) = \{u \in \mathcal{D}(\Delta^{r_1-1}); \Delta^{r_1-1} \in Y\}$, $A_{n-2} = (-1)^{r_1+r_2}\Delta^{r_2}\Delta^{r_1} = (-1)^{r_1+r_2}\Delta^{r_1+r_2}$, $r_1 + r_2 + 1 < 2r_1$, that is, $r_1 > r_2 + 1$. Define $A_{n-j} = (-1)^{r_1+\cdots+r_j}\Delta^{r_1+\cdots+r_j}$, $j = 3, \ldots, m$, where $r_2 > r_3 > \cdots > r_n$. Then Theorem 6.7 works with $\alpha = \beta_k = \nu = \tau = 1$.

EXAMPLE 6.15. Let K be the operator introduced in Example 6.12, and thus $\mathcal{D}(K) = H^2(0,1) \cap H_0^1(0,1)$, $Ku = -u''$, $u \in \mathcal{D}(K)$ and $X = L^2(0,1)$. Let $m \in C^{2m_1}([0,1])$, $m(x) \geq 0$ on $[0,1]$, $m^{(2j+1)}(0) = m^{(2j+1)}(1) = 0$ for $j = 0, 1, \ldots, m_1 - 1$.

For $n \in \mathbb{N}$, $n \geq 2$, A_n denotes the operator of multiplication by $m(x)$, $A_{n-1} = K^{m_1}$, $A_{n-j} = K^{m_1 + \cdots + m_j}$, where $j = 2, \ldots, n$ and $m_1 > m_2 > \cdots > m_n$. Then $A_{n-2} = A_{n-1}^{1 + \frac{m_2}{m_1}}$, so that as in Example 6.12 we must require that in addition $2m_2 < m_1$. Therefore we arrive at the links $m_1 > 2m_2$, $m_2 > m_3 > \cdots > m_n$ and also $\alpha = 1$, $\tau = \frac{1}{2}$; hence, given $\theta \in (\frac{1}{2}, 1)$, Theorem 6.7 yields that, if $f \in C^\theta([0,T]; X)$, $u_0 \in H^{2(m_1 + \cdots + m_n)}(0,1)$, $u_0^{(2j)}(0) = u_0^{(2j)}(1) = 0$, $j = 0, 1, \ldots, m_1 + \cdots + m_n - 1$, $u_j \in H^{2(m_1 + \cdots + m_{n-j+1})}(0,1)$, $u_j^{(2i)}(0) = u_j^{(2i)}(1) = 0$, $i = 0, 1, \ldots, m_1 + \cdots + m_{n-j+1} - 1$, $j = 1, \ldots, n$, and u_n is related to the other data by means of

$$f(x,0) - \sum_{j=0}^{n-1} (-1)^{m_1 + \cdots + m_{n-j}} u_j^{2(m_1 + \cdots + m_{n-j})} = m(x) u_n$$

then the problem

$$
\begin{cases}
\dfrac{\partial}{\partial t}\left(m(x) \dfrac{\partial^{n-1} u}{\partial t^{n-1}} \right) + (-1)^{m_1} \dfrac{\partial^{2m_1 + n - 1} u}{\partial x^{2m_1} \partial t^{n-1}} \\
\qquad\qquad + (-1)^{m_1 + m_2} \dfrac{\partial^{2(m_1 + m_2) + n - 2} u}{\partial x^{2(m_1 + m_2)} \partial t^{n-2}} + \cdots \\
\qquad + (-1)^{m_1 + \cdots + m_n} \dfrac{\partial^{2(m_1 + \cdots + m_n)} u}{\partial x^{2(m_1 + \cdots + m_n)}} = f(x,t), \quad 0 < x < 1,\ 0 \leq t \leq T, \\[2mm]
\dfrac{\partial^{2i+j} u}{\partial x^{2i} \partial t^j}(x,t) = 0, \\
\qquad\qquad i = 0, 1, \ldots, m_1 + \cdots + m_{n-j} - 1,\ x \in \{0,1\},\ 0 \leq t \leq T, \\[2mm]
\dfrac{\partial^j u}{\partial t^j}(x,0) = u_j(x), \quad j = 0, \ldots, n-2,\ 0 < x < 1, \\[2mm]
m(x) \dfrac{\partial^{n-1} u}{\partial t^{n-1}}(x,0) = u_{n-1}(x), \quad 0 < x < 1,
\end{cases}
$$

possesses a unique solution u with the regularities

$$\frac{\partial^{2(m_1 + \cdots + m_j) + n - j} u}{\partial x^{2(m_1 + \cdots + m_j)} \partial t^{n-j}}, \quad \frac{\partial}{\partial t}\left(m(x) \frac{\partial^{n-1} u}{\partial t^{n-1}} \right) \in C^{\theta - \frac{1}{2}}([0,T]; X)$$

for $j = 1, \ldots, n$.

EXAMPLE 6.16. Let $m(x,t) \geq 0$ be a C^2 function on $[0,T] \times [0,1]$, such that

$$\left\|\tfrac{\partial m}{\partial x}(x,t)\right\| \leq km(x,t)^\rho, \quad \left\|\tfrac{\partial m}{\partial t}(x,t)\right\| \leq km(x,t)^\nu,$$

for some exponents $0 < \rho, \nu \leq 1$ and some constant $k > 0$. Denote by $C(t)$, $0 \leq t \leq T$, the operator of multiplication by $m(\cdot,t)$ in $X = L^2(0,1)$. For sake of simplicity, let us consider the time independent operator $B \equiv B(t)$ given by $\mathcal{D}(B) = H_0^1(0,1) \cap H^2(0,1)$, $Bu = -u''$, $u \in \mathcal{D}(B)$. Moreover, $\mathcal{D}(A(t)) = \mathcal{D}(B)$, where $(A(t)u)(x) = a(x,t)u''(x) + b(x,t)u'(x) + c(x,t)u(x)$ is a second order differential operator in the variable x, whose coefficients are assumed to be smooth.

Then it is seen in Example 3.3 that

$$\|C(t)(\lambda C(t) + B(t))^{-1}\|_{\mathcal{L}(X)} \leq K(1 + |\lambda|)^{-\frac{1}{2-\rho}}, \quad \lambda \in -\Sigma_1.$$

Hence Theorem 6.8 applies immediately with $\beta = \frac{1}{2-\rho}$, provided that $\nu(2 - \rho) + 2\rho > 2$ and $5\rho - 9 < (\nu - 4)(2 - \rho)^2$. It follows that in this case, if $\frac{1-\rho}{2-\rho} < \theta < \frac{\nu(2-\rho)+2(\rho-1)}{3-2\rho}$, for all $f \in C^\theta([0,T]; L^2(0,1))$, $u_0, u_1 \in H_0^1(0,1) \cap H^2(0,1)$ such that

$$f(x,0) - A(x,0;D)u_0 + u_1'' - \tfrac{\partial m}{\partial t}(x,0)u_1 = m(x,0)v$$

for a certain $v \in H_0^1(0,1) \cap H^2(0,1)$, then the problem

$$\begin{cases} \dfrac{\partial}{\partial t}\left(m(x,t)\dfrac{\partial u}{\partial t}\right) + A(x,t;D)u - \dfrac{\partial^3 u}{\partial x^2 \partial t} = f(x,t), \\[2mm] \qquad\qquad\qquad\qquad\qquad 0 < x < 1, \ 0 \leq t \leq T, \\[2mm] u(x,t) = \dfrac{\partial u}{\partial t}(x,t) = 0, \quad x \in \{0,1\}, \ 0 \leq t \leq T, \\[2mm] u(x,0) = u_0(x), \ m(x,0)\dfrac{\partial u}{\partial t}(x,0) = m(x,0)u_1(x), \quad 0 < x < 1, \end{cases}$$

possesses a unique solution $u = u(x,t)$ with the regularity

$$\tfrac{\partial}{\partial t}(m(x,t)\tfrac{\partial u}{\partial t}), \ \tfrac{\partial^3 u}{\partial x^2 \partial t}, \ A(x,t;D)u \in C^\omega([0,T]; L^2(0,1)),$$

where $\omega = \theta - \frac{1-\rho}{2-\rho}$.

EXAMPLE 6.17. Consider the initial boundary value problem (6.36)

$$\begin{cases} \dfrac{\partial}{\partial t}\left(m(t)\dfrac{\partial u}{\partial t}\right) - \dfrac{\partial^2 u}{\partial x^2} - \dfrac{\partial^3 u}{\partial x^2 \partial t} = f(x,t), \quad 0 < x < 1, \ 0 \leq t \leq T, \\[2mm] u(x,t) = \dfrac{\partial u}{\partial t}(x,t) = 0, \quad x \in \{0,1\}, \ 0 \leq t \leq T, \\[2mm] u(x,0) = u_0(x), \ m(0)\dfrac{\partial u}{\partial t}(x,0) = m(0)u_1(x), \quad 0 < x < 1, \end{cases}$$

where $m \in C^{1+\varphi}([0,T])$, $0 < \varphi \leq 1$, $|m'(t)| \leq km(t)^\nu$, $m(t) \geq 0$ on $[0,T]$, $f \in C([0,T]; L^2(0,1))$, $u_0, u_1 \in H_0^1(0,1) \cap H^2(0,1)$. Define B by $\mathcal{D}(B) = H_0^1(0,1) \cap H^2(0,1)$, $Bu = -u''$, $u \in \mathcal{D}(B)$. Then in view of the results in Chapter IV, all the assumptions in Theorem 6.9 are verified with $\alpha = \beta = 1$. Hence, for all $f \in C^\theta([0,T]; L^2(0,1))$, $\theta \in (0, \nu\varphi)$, and any $u_0, u_1 \in H_0^1(0,1) \cap H^2(0,1)$ satisfying

$$f(x,0) + u_0'' + u_1'' - m'(0)u_1 = m(0)w,$$

with $w \in \mathcal{D}(B) = H_0^1(0,1) \cap H^2(0,1)$, the problem (6.36) possesses a unique solution u with the regularity

$$\frac{\partial}{\partial t}\left(m(t)\frac{\partial u}{\partial t}\right), \ \frac{\partial^3 u}{\partial x^2 \partial t}, \ \frac{\partial^2 u}{\partial x^2} \in C^\theta([0,T]; L^2(0,1)).$$

Many other examples of more general character could be given, taking into account the results in Chapter IV, but we have only wanted to describe the types of applications that are allowed by our theory.

6.3 EQUATIONS OF HYPERBOLIC TYPE

In this section we shall confine ourselves to a particularly interesting class of second order implicit differential equations in Hilbert spaces, because we shall use in essential way the results obtained in Chapter II, precisely Theorem 2.10 and Corollary 2.11.

Moreover, on the ground of the concrete application to partial differential equations, we shall examine in more details the Cauchy problem in a Hilbert space H

$$(6.37) \qquad \begin{cases} C\dfrac{d^2u}{dt^2} + B\dfrac{du}{dt} + Au = f(t), \quad 0 \leq t \leq T, \\[2mm] u(0) = u_0, \ \dfrac{du}{dt}(0) = u_1, \end{cases}$$

rather than problem (6.2). To this end, we introduce the following definition of a solution to (6.37).

DEFINITION. Given $f \in C([0,T]; H)$ and $u_0 \in \mathcal{D}(A)$, $u_1 \in \mathcal{D}(B)$, a function $u \in C^2([0,T]; H)$ satisfying the equation and the initial conditions of (6.37) with the regularity $u \in C([0,T]; \mathcal{D}(A))$, $\frac{du}{dt} \in C([0,T]; \mathcal{D}(B))$, $\frac{d^2u}{dt^2} \in C([0,T]; \mathcal{D}(C))$ is called *a solution* to (6.37).

Henceforth we shall always assume that C is a bounded operator from H into itself. We are also given another Hilbert space V continuously and densely embedded in H, so that, by identifying H with its dual space H', we have: $V \subset H \subset V'$. The scalar product between V and V' is denoted by $\langle \cdot, \cdot \rangle$. Let us specify further our operators A, B and C by means of $A \in \mathcal{L}(V, V')$ with

(6.38) $\langle Au, v \rangle = \langle u, Av \rangle$ for all $u, v \in V$,

(6.39) $\langle Au, u \rangle \geq \omega \|u\|_V^2$ for all $u \in V$,

with some $\omega > 0$. Then it is well known (e. g. [167: Theorem 2.2.3]) that if A_H denotes the selfadjoint operator in H with domain

$$\mathcal{D}(A_H) = \{u \in V; \, Au \in H\}, \quad A_H u = Au \text{ for } u \in \mathcal{D}(A_H),$$

and $A^{\frac{1}{2}}$ is the square root of A_H, then $\mathcal{D}(A^{\frac{1}{2}}) = V$ and $\langle Au, v \rangle = (A^{\frac{1}{2}} u, A^{\frac{1}{2}} v)_H$ for all $u, v \in V$.

The hypotheses on B and C read:

(6.40) B is a closed linear operator in H with $\mathcal{D}(B) \supset \mathcal{D}(A^{\frac{1}{2}}) = V$;
(6.41) $\mathfrak{Re}\langle Bv, v \rangle \geq 0$ for all $v \in V$;
(6.42) $C = C^* \in \mathcal{L}(H)$, $C \geq 0$.

We begin with establishing the basic preliminary result.

LEMMA 6.10. *Let us assume (6.38)~(6.42). Then the problem*

(6.43) $$\begin{cases} C^{\frac{1}{2}} \dfrac{d}{dt}(C^{\frac{1}{2}} u') + B \dfrac{du}{dt} + A_H u = C^{\frac{1}{2}} f(t), \quad 0 \leq t \leq T, \\ u(0) = u_0, \quad C^{\frac{1}{2}} \dfrac{du}{dt}(0) = C^{\frac{1}{2}} u_1 \end{cases}$$

possesses a unique solution u provided that $f \in C^1(0, T]; H)$, $u_0 \in \mathcal{D}(A_H)$, $u_1 \in \mathcal{D}(A^{\frac{1}{2}})$ with the relation $A_H u_0 + B u_1 = C^{\frac{1}{2}} w$ for some $w \in H$.

PROOF. One easily checks that (6.43) can be equivalently rewritten in the form

$$\begin{pmatrix} I & 0 \\ 0 & C^{\frac{1}{2}} \end{pmatrix} \frac{d}{dt} \begin{pmatrix} I & 0 \\ 0 & C^{\frac{1}{2}} \end{pmatrix} \begin{pmatrix} u \\ u' \end{pmatrix} + \begin{pmatrix} 0 & -I \\ A_H & B \end{pmatrix} \begin{pmatrix} u \\ u' \end{pmatrix}$$
$$= \begin{pmatrix} 0 \\ C^{\frac{1}{2}} f(t) \end{pmatrix}, \quad 0 \leq t \leq T,$$

$$\begin{pmatrix} I & 0 \\ 0 & C^{\frac{1}{2}} \end{pmatrix} \begin{pmatrix} u(0) \\ u'(0) \end{pmatrix} = \begin{pmatrix} I & 0 \\ 0 & C^{\frac{1}{2}} \end{pmatrix} \begin{pmatrix} u_0 \\ u_1 \end{pmatrix}.$$

Take $X = \mathcal{D}(A^{\frac{1}{2}}) \times H$, where the inner product in X is given by

$$((x, y), (x_1, y_1))_X = (A^{\frac{1}{2}} x, A^{\frac{1}{2}} x_1)_H + (y, y_1)_H, \quad (x, y), (x_1, y_1) \in X,$$

and define operators M and L from X into itself by

$$\mathcal{D}(M) = X, \quad M(x,y) = (x, C^{\frac{1}{2}}y), \ (x,y) \in X,$$
$$\mathcal{D}(L) = \mathcal{D}(A_H) \times \mathcal{D}(A^{\frac{1}{2}}), \quad L(x,y) = (y, -A_H x - By), \ (x,y) \in \mathcal{D}(L).$$

In view of the hypotheses, L is a closed linear operator. In order to apply Theorem 2.10, we first estimate $\Re e(L(x,y),(x,y))_X$ for $(x,y) \in \mathcal{D}(L)$. We have:

$$\Re e(L(x,y),(x,y))_X = \Re e((y, -A_H x - By),(x,y))_X$$
$$= \Re e\{(A^{\frac{1}{2}}y, A^{\frac{1}{2}}x)_H - (A^{\frac{1}{2}}x, A^{\frac{1}{2}}y)_H - \langle By, y \rangle\} = -\Re e\langle By, y \rangle \le 0.$$

Hence, (2.10) holds with $\beta = 0$.

It remains to verify (2.11). Now, equation $(\lambda_0 M^* M - L)(x,y) = (f_1, f_2) \in X$ signifies

$$\lambda_0 x - y = f_1 \in \mathcal{D}(A^{\frac{1}{2}}), \ \lambda_0 Cy + A_H x + By = f_2 \in H.$$

But, for given $f_1 \in \mathcal{D}(A^{\frac{1}{2}})$, $f_2 \in H$ and sufficiently small positive λ_0, such a system has one solution, since in this situation all is reduced to the invertibility of the pencil $\lambda_0^2 C + \lambda_0 B + A_H$, and this is an obvious consequence of the hypotheses. It follows that for such a λ_0, $\mathcal{R}(\lambda_0 M^* M - L) = X$ and (2.11) holds true.

COROLLARY 6.11. *Under (6.38)~(6.42), the problem (6.43), where $u(0) = u_0$, $Cu'(0) = Cu_1$, possesses a solution u.*

Exploiting Lemma 6.10 and regularizing the solution of (6.43) by integration, we are able to establish a first existence result for (6.37) with $A = A_H$.

THEOREM 6.12. *Under assumptions (6.38)~(6.42), if $f(t) = C^{\frac{1}{2}}h(t)$, $0 \le t \le T$, $h \in \mathcal{C}^2([0,T];H)$ and $u_0, u_1 \in \mathcal{D}(A_H)$ satisfy*

$$C^{\frac{1}{2}}h(0) - Bu_1 - A_H u_0 = Cv_1, \ A_H u_1 + Bv_1 = C^{\frac{1}{2}}v_2,$$

for some $v_1 \in \mathcal{D}(A^{\frac{1}{2}})$, $v_2 \in H$, then problem (6.37) with $A = A_H$ possesses a solution.

PROOF. In view of Lemma 6.10 and Corollary 6.11, our assumptions yield that the problem

$$\begin{cases} \dfrac{d}{dt}(Cw') + Bw' + A_H w = C^{\frac{1}{2}}h'(t), & 0 \le t \le T, \\ w(0) = u_1, \ Cw'(0) = Cv_1, \end{cases}$$

has a solution w. Define

$$u(t) = \int_0^t w(s)ds + u_0, \quad 0 \le t \le T.$$

Then by integrating the equation above on $(0, t)$, we have:

$$Cw'(t) + Bw(t) + A_H \left(\int_0^t w(s)ds + u_0 - u_0 \right)$$
$$= C^{\frac{1}{2}}h(t) - C^{\frac{1}{2}}h(0) + Cv_1 + Bu_1, \ 0 \le t \le T$$

(notice that $\mathcal{D}(A^{\frac{1}{2}}) \subset \mathcal{D}(B)$ and the pair $(-w'(t), A_H w(t) + Bw'(t)) \in \mathcal{D}(A^{\frac{1}{2}}) \times H$). But $u'(t) = w(t)$, $u''(t) = w'(t)$, together with the given compatibility relations, imply that u is really the desired solution.

We now remove the assumption on the particular form $C^{\frac{1}{2}}h(\cdot)$ of the non homogeneous function f. This shall be achieved at the expense of higher time regularity for f. To this end we move in accordance with the procedure we have followed previously and first we establish the following lemma.

LEMMA 6.13. *Let us assume (6.38)~(6.42). For any $f \in C^2([0, T]; H)$, $u_0 \in \mathcal{D}(A_H)$, $u_1 \in \mathcal{D}(A^{\frac{1}{2}})$ satisfying the compatibility relation*

$$f(0) - A_H u_0 - Bu_1 = C^{\frac{1}{2}} u_2,$$

for a certain element u_2 of H, the problem

(6.44)
$$\begin{cases} C^{\frac{1}{2}} \dfrac{d}{dt}(C^{\frac{1}{2}} u') + B\dfrac{du}{dt} + A_H u = f(t), \quad 0 \le t \le T, \\ u(0) = u_0, \ C\dfrac{du}{dt}(0) = Cu_1, \end{cases}$$

possesses a solution.

PROOF. It is an immediate consequence of the arguments to be found in the proof of Lemma 6.10, applying Corollary 2.11. The compatibility assumption is an easy translation of the one in that result.

THEOREM 6.14. *Let A, B, C fulfil the assumptions (6.38) $\sim$ (6.42). Then for all $f \in C^3([0, T]; H)$, $u_0, u_1 \in \mathcal{D}(A_H)$ such that*

$$f(0) - A_H u_0 - Bu_1 = Cu_2, \ f'(0) - A_H u_1 - Bu_2 = C^{\frac{1}{2}} u_3$$

for some $u_2 \in \mathcal{D}(A^{\frac{1}{2}})$, $u_3 \in H$, problem (6.37) possesses a solution.

PROOF. Consider problem (6.44) with $f'(t)$ instead of $f(t)$ and u_1, u_2 instead of u_0, u_1 respectively. If $w(t)$ is the solution to this problem, guaranteed by Lemma 6.13, let us introduce

$$u(t) = \int_0^t w(s)ds + u_0.$$

Then one verifies that such a function is a solution of problem (6.37).

Our last result in this section before examples concernes solvability of the strongly degenerate first order equation

(6.45)
$$\begin{cases} B\dfrac{du}{dt} + A_H u = f(t), & 0 \le t \le T, \\ u(0) = u_0, \end{cases}$$

which can be viewed as a particular case of (6.37) with $C = 0$.

THEOREM 6.15. *Suppose that the operators A and B satisfy (6.38) $\sim$ (6.41). If $f \in C^3([0,T];H)$, $u_0 \in \mathcal{D}(A_H)$ and there are $u_1 \in \mathcal{D}(A_H)$, $u_2 \in \mathcal{D}(A^{\frac{1}{2}})$ such that*

$$f(0) - A_H u_0 - B u_1 = f'(0) - A_H u_1 - B u_2 = 0,$$

then problem (6.45) possesses a solution.

This affirmation may be compared with R. E. Showalter [155; Corollary 4.D], when A is supposed to be the Riesz map of Hilbert space V, B is a strictly monotone operator from $\mathcal{D}(B)(\subset V)$ into V' such that $A + B$ is a surjection from $\mathcal{D}(B)$ onto V'. But then only the weaker equation $(Bu)' + Au = f$ is solved instead of (6.45).

At last, we furnish a number of examples. The first one extends in some sense Example 2.3, that is a typical hyperbolic equation, to equations of hyperbolic-parabolic type.

EXAMPLE 6.18. Let Ω be a bounded open set of $\mathbb{R}^n$ with a smooth boundary Γ. If $T > 0$, denote $\Omega \times (0,T)$ by Q and $\Gamma \times (0,T)$ by Σ.

Let us consider the hyperbolic-parabolic problem

(6.46)
$$\begin{cases} m_1(x)\dfrac{\partial^2 u}{\partial t^2} + m_2(x)\dfrac{\partial u}{\partial t} - \Delta u = f(x,t), & (x,t) \in Q, \\ u(x,t) = 0, & (x,t) \in \Sigma, \\ u(x,0) = u_0(x), \ \dfrac{\partial u}{\partial t}(x,0) = u_1(x), & x \in \Omega, \end{cases}$$

where $m_1(x)$, $m_2(x)$ are two nonnegative continuous functions on $\overline{\Omega}$. Precisely, let us take $V = H_0^1(\Omega)$, $H = L^2(\Omega)$, $A = -\Delta$ in variational sense, so that $\mathcal{D}(A_H) = H^2(\Omega) \cap H_0^1(\Omega)$. Let C, B be the operators of multiplication by $m_1(x)$ and $m_2(x)$ in $L^2(\Omega)$ respectively. Suppose that $t \mapsto f(\cdot, t)$ defines a C^3 mapping from $[0,T]$ into H and that for given u_0, $u_1 \in \mathcal{D}(A_H)$ there exist $v_2 \in V$, $u_3 \in H$ satisfying

$$f(x,0) + \Delta u_0 - m_2(x)u_1 = m_1(x)u_2,$$

$$\tfrac{\partial f}{\partial t}(x,0) + \Delta u_1 - m_2(x)u_2 = m_1(x)^{\frac{1}{2}}u_3.$$

Then in virtue of Theorem 6.14 we deduce that (6.46) possesses a solution to be understood by translating the abstract model.

We also remark that the relationship between the initial data and the nonhomogeneous function $f(t)$ as described above are caused by the allowed large degeneration in the equation, that is hyperbolic where $m_1(x) > 0$, $m_2(x) = 0$, parabolic where $m_1(x) = 0$, $m_2(x) > 0$, and elliptic where $m_1(x) = m_2(x) = 0$.

We also notice that most of previous approaches to (6.46) guarantee only weak solutions and just in this higher regularity there is the main novelty of the results. We quote as references the following: A. Bensoussan, J. L. Lions and G. C. Papanicolau [27], O. A. de Lima [68], A. B. Maciel [132] and L. A. Medeiros [136].

EXAMPLE 6.19. Let Ω have the same meaning as in Example 6.18 and consider the two sesquilinear forms on $H_0^m(\Omega) \times H_0^m(\Omega)$ and $H_0^k(\Omega) \times H_0^k(\Omega)$ respectively, given by

$$a(u,v) = \sum_{|\alpha|,|\beta|=m} \int_\Omega a_{\alpha\beta}(x) D^\beta u \overline{D^\alpha v} dx + c(u,v)_{L^2}, \quad u, v \in H_0^m(\Omega),$$

$$b(u,v) = \sum_{|\gamma|,|\delta|=k} \int_\Omega b_{\gamma\delta}(x) D^\delta u \overline{D^\gamma v} dx, \quad u, v \in H_0^k(\Omega),$$

where the coefficients $a_{\alpha\beta}(x)$, $b_{\gamma\delta}(x)$ are continuous real valued functions on $\overline{\Omega}$ and $c > 0$ is a positive constant. We assume that the coercivity conditions

$$\Re \sum_{|\alpha|,|\beta|=m} a_{\alpha\beta}(x)\xi^{\alpha+\beta} \geq c_0|\xi|^{2m}, \quad \xi \in \mathbb{R}^n, \ x \in \overline{\Omega},$$

$$\Re \sum_{|\gamma|,|\delta|=k} b_{\gamma\delta}(x)\xi^{\gamma+\delta} \geq c_0|\xi|^{2k}, \quad \xi \in \mathbb{R}^n, \ x \in \overline{\Omega},$$

hold with some $c_0 > 0$ and that $a_{\alpha\beta}(x) = a_{\beta\alpha}(x)$, $b_{\gamma\delta}(x) = b_{\delta\gamma}(x)$. Then, it is well known that the differential operators A, $\tilde{B}$ defined by $a(\cdot,\cdot)$, $b(\cdot,\cdot)$ respectively, that is,

$$Au = \sum_{|\alpha|,|\beta|=m} (-1)^{|\alpha|} D^\alpha \{a_{\alpha\beta}(x) D^\beta u\},$$

$$\tilde{B}u = \sum_{|\gamma|,|\delta|=k} (-1)^{|\gamma|} D^\gamma \{b_{\gamma\delta}(x) D^\delta u\},$$

operate from $H_0^m(\Omega)$, $H_0^k(\Omega)$ into $H^{-m}(\Omega)$, $H^{-k}(\Omega)$ respectively. In its turn, A, $\tilde{B}$ arise their parts A_H and $B = \tilde{B}_H$ in the space $H = L^2(\Omega)$ with

$\mathcal{D}(A_H) = H^{2m}(\Omega) \cap H_0^m(\Omega)$, $\mathcal{D}(B) = H^{2k}(\Omega) \cap H_0^k(\Omega)$. Since $\mathcal{D}(A_H^{\frac{1}{2}}) = H_0^m(\Omega)$, the assumption $k \leq m$ guarantees (6.40). (6.38),(6.39) and (6.41) are also easily verified.

If $m(x)$ is a non negative continuous function on $\overline{\Omega}$ and C denotes the operator of multiplication by $m(x)$ in $X = L^2(\Omega)$, then Theorem 6.14 applies to the initial boundary value problem related to the equation

$$m(x)\frac{\partial^2 u}{\partial t^2} + B\frac{\partial u}{\partial t} + A_H u = f(x,t),$$

where $f \in \mathcal{C}^3([0,T]; L^2(\Omega))$.

EXAMPLE 6.20. Let $m(x)$ be a non negative continuous function on $[0,1]$ with C the non negative operator of multiplication in $L^2(0,1)$ associated to it. Take as operator B either

$$\mathcal{D}(B) = \{u \in H^1(0,1); \; u(0) = 0\}, \quad Bu = \tfrac{du}{dx},$$

or

$$\mathcal{D}(B) = \{u \in H^1(0,1); \; u(1) = 0\}, \quad Bu = -\tfrac{du}{dx}.$$

It is then easily seen that in both the cases $\Re e(Bu,u)_{L^2} \geq 0$. Let $H = L^2(0,1)$, $V = H_0^1(0,1)$. As in Example 6.18, $A = -\frac{d^2}{dx^2}$ so that $\mathcal{D}(A_H) = H^2(0,1) \cap H_0^1(\Omega)$. Therefore, all the assumptions (6.38)~(6.42) are verified and hence we deduce that the problems

$$\begin{cases} m(x)\dfrac{\partial^2 u}{\partial t^2} + \dfrac{\partial^2 u}{\partial x \partial t} - \dfrac{\partial^2 u}{\partial x^2} = f(x,t), \quad (x,t) \in (0,1) \times [0,T], \\[2mm] u(0,t) = u(1,t) = \dfrac{\partial u}{\partial t}(0,t) = 0, \quad t \in [0,T], \\[2mm] u(x,0) = u_0(x), \quad \dfrac{\partial u}{\partial t}(x,0) = u_1(x), \quad x \in (0,1), \end{cases}$$

and

$$\begin{cases} m(x)\dfrac{\partial^2 u}{\partial t^2} - \dfrac{\partial^2 u}{\partial x \partial t} - \dfrac{\partial^2 u}{\partial x^2} = f(x,t), \quad (x,t) \in (0,1) \times [0,T], \\[2mm] u(0,t) = u(1,t) = \dfrac{\partial u}{\partial t}(1,t) = 0, \quad t \in [0,T], \\[2mm] u(x,0) = u_0(x), \quad \dfrac{\partial u}{\partial t}(x,0) = u_1(x), \quad x \in (0,1), \end{cases}$$

admit solutions provided that $f \in \mathcal{C}^3([0,T]; L^2(0,1))$ and that the following compatibility conditions

$$f(x,0) + \frac{d^2 u_0}{dx^2} - \frac{du_1}{dx} = m(x)u_2, \quad \frac{\partial f}{\partial t}(x,0) + \frac{d^2 u_1}{dx^2} - \frac{du_2}{dx} = m(x)^{\frac{1}{2}}u_3,$$

and

$$f(x,0) + \frac{d^2u_0}{dx^2} + \frac{du_1}{dx} = m(x)u_2, \quad \frac{\partial f}{\partial t}(x,0) + \frac{d^2u_1}{dx^2} + \frac{du_2}{dx} = m(x)^{\frac{1}{2}}u_3,$$

respectively, are satisfied, where u_0, $u_1 \in H^2(0,1) \cap H_0^1(0,1)$, $u_2 \in H_0^1(0,1)$, $u_3 \in L^2(0,1)$.

Obviously, Theorem 6.15 ensures the existence of a solution to

$$\begin{cases} \dfrac{\partial^2 u}{\partial x \partial t} - \dfrac{\partial^2 u}{\partial x^2} = f(x,t), \quad (x,t) \in (0,1) \times [0,T], \\[2mm] u(0,t) = u(1,t) = \dfrac{\partial u}{\partial t}(0,t) = 0, \quad t \in [0,T], \\[2mm] u(x,0) = u_0(x), \qquad x \in (0,1), \end{cases}$$

under regularity and compatibility assumptions $f \in C^3([0,T]; L^2(0,1))$, $u_0 \in H^2(0,1) \cap H_0^1(0,1)$, and

$$f(x,0) + \frac{d^2u_0}{dx^2} - \frac{du_1}{dx} = 0, \quad \frac{\partial f}{\partial t}(x,0) + \frac{d^2u_1}{dx^2} - \frac{du_2}{dx} = 0,$$

with certain $u_1 \in H^2(0,1) \cap H_0^1(0,1)$, $u_2 \in H_0^1(0,1)$.

An analogous statement can be deduced relative to

$$\frac{\partial^2 u}{\partial x \partial t} + \frac{\partial^2 u}{\partial x^2} = f(x,t), \quad (x,t) \in (0,1) \times [0,T],$$

where this time u satisfies the boundary conditions

$$u(0,t) = u(1,t) = \frac{\partial u}{\partial t}(1,t) = 0, \quad t \in [0,T].$$

We observe that if one introduces the operator B by means of $\mathcal{D}(B) = H_0^1(0,1)$, $Bu = b\frac{du}{dx}$, when $u \in \mathcal{D}(B)$ and $b \in \mathbb{R} - \{0\}$, then (see [117; p. 145, p. 165]) it is closed and clearly $\Re e(Bu,u)_{L^2} = 0$ for all $u \in \mathcal{D}(B)$. Therefore, we are allowed to conclude with results analogous to the preceding ones, relative to this new differential operator. Notice that B has no bounded inverse (Example 3.14 of T. Kato [117; p. 153]).

EXAMPLE 6.21. Let $p \in C^\infty((0,1))$ be a non negative infinitely differentiable function defined in $(0,1)$ such that

$$\lim_{x \to 0+} \frac{p(x)}{x} = C_0 > 0, \quad \lim_{x \to 1-} \frac{p(x)}{1-x} = C_1 > 0.$$

Define

$$B_0 u = -\frac{d}{dx}\left(p(x)\frac{du}{dx}\right), \quad u \in \mathcal{D}(B_0) = C^\infty((0,1)).$$

B_0 is a Legendre differential operator, according to [172; pp. 430-431]. It is well known (see [172; p. 441]) that B_0 is essentially selfadjoint in $X = L^2(0,1)$ and its closure B is an operator with pure point spectrum, its eigenvalues are non negative and the domain $\mathcal{D}(B)$ coincides with the Sobolev space with weight $W_2^2((0,1), p^2)$; that is, $\mathcal{D}(B) = \{u \in \mathcal{D}'((0,1)); \sum_{|\alpha| \leq 2}(\int_0^1 p(x)^2|u^{(\alpha)}(x)|^2 dx)^{\frac{1}{2}} < \infty\}$ (see [172; p. 445]). Hence, B coincides with the Friedriches' extension of the minimal operator B° given by

$$B^\circ u = -\frac{d}{dx}\left(p(x)\frac{du}{dx}\right), \quad u \in \mathcal{C}_0^\infty((0,1)) = \mathcal{D}(B^\circ).$$

Moreover, since B_0 is symmetric and clearly

$$(B_0 u, u)_{L^2} = \int_0^1 p(x)|u'(x)|^2 dx \geq 0, \quad u \in \mathcal{D}(B_0),$$

it follows that B is a selfadjoint non negative operator in X. Therefore, Theorems 6.14 and 6.15 permit to solve the problem

$$\begin{cases} m(x)\dfrac{\partial^2 u}{\partial t^2} - \dfrac{\partial}{\partial x}\left(p(x)\dfrac{\partial^2 u}{\partial x \partial t}\right) + \dfrac{\partial^4 u}{\partial x^4} = f(x,t), \quad (x,t) \in (0,1) \times [0,T], \\ u(0,t) = \dfrac{\partial u}{\partial x}(0,t) = u(1,t) = \dfrac{\partial u}{\partial t}(1,t) = 0, \quad 0 \leq t \leq T, \\ u(x,0) = u_0(x), \dfrac{\partial u}{\partial t}(x,0) = u_1(x), \quad 0 < x < 1, \end{cases}$$

and

$$\begin{cases} \dfrac{\partial}{\partial x}\left(p(x)\dfrac{\partial^2 u}{\partial x \partial t}\right) = \dfrac{\partial^4 u}{\partial x^4} + f(x,t), \quad (x,t) \in (0,1) \times [0,T], \\ u(0,t) = \dfrac{\partial u}{\partial x}(0,t) = u(1,t) = \dfrac{\partial u}{\partial x}(1,t) = 0, \quad 0 \leq t \leq T, \\ u(x,0) = u_0(x), \quad 0 < x < 1, \end{cases}$$

as well, with $m \in \mathcal{C}([0,1])$, $m(x) \geq 0$.

It suffices to observe that in both the cases $\mathcal{D}(A_H) = H^4(0,1) \cap H_0^2(0,1)$ so that $\mathcal{D}(A^{\frac{1}{2}}) \subset \mathcal{D}(B)$. More generally, in virtue of some results due to M. A. Naimark [139], if $q \in \mathcal{C}^\infty((0,1))$, $q(x) > 0$, satisfies

$$\lim_{x \to 0+} \frac{q(x)}{x^{k_0}} > 0, \quad \lim_{x \to 1-} \frac{q(x)}{(1-x)^{k_1}} > 0,$$

where k_0, k_1 are two positive integers, then the operator

$$\tilde{B}_0 u = -\frac{d}{dx}\left(q(x)\frac{du}{dx}\right), \quad u \in \mathcal{D}(\tilde{B}_0) = \mathcal{C}^\infty((0,1)),$$

is essentially selfadjoint in $L^2(0,1)$ so that its closure $B = \tilde{B}_0^{**}$ is selfadjoint. Hence, our general theory applies to

$$A_H = (-1)^k \frac{\partial^{2k}}{\partial x^{2k}},$$
$$\mathcal{D}(A_H) = H^{2k}(0,1) \cap H_0^k(0,1),\ k \geq 2,$$

C multiplication by a continuous non negative function $m(x)$ on $[0,1]$.

EXAMPLE 6.22. In many interesting cases the operator B entering in (6.37) has the expression $B = D^*ED$, where D, E are two closed linear operators from the complex Hilbert space H into itself, $E \geq 0$ is everywhere defined (so that it is bounded) and $\mathcal{D}(D)$ is dense in H. Then, $B = D^*E^{\frac{1}{2}}E^{\frac{1}{2}}D$.

Now $(E^{\frac{1}{2}}D)^* = D^*E^{\frac{1}{2}}$ and hence if $E^{\frac{1}{2}}D$ is closed (and this happens if, for example, E has a bounded inverse), then $B = (E^{\frac{1}{2}}D)^*(E^{\frac{1}{2}}D)$ is selfadjoint ≥ 0 (see T. Kato [117; p. 275]).

Let us describe two concrete examples. Let $a(x)$ be a real valued C^1 function on $[0,1]$, $a(x) \geq a_0 > 0$ for all $x \in [0,1]$, and let E be the operator of multiplication by $a(x)$ in $L^2(0,1)$. Take D the maximal operator in $L^2(0,1)$ associated to $\frac{d}{dx}$, that is, $\mathcal{D}(D) = H^1(0,1)$, $Du = u' = \frac{du}{dx}$, $u \in H^1(0,1)$. Then D is a closed densely defined operator in $L^2(0,1)$ and its adjoint D^* coincides with $D^*u = -u'$, and $\mathcal{D}(D^*) = H_0^1(0,1)$, (see [117; p. 145, p. 169]). Hence, $\mathcal{D}(B) = \{u \in H^2(0,1); u'(0) = u'(1) = 0\}$, $Bu = -\frac{d}{dx}(a(x)\frac{du}{dx})$, $u \in \mathcal{D}(B)$. On the other hand, maintaining the same meaning to E, since $D^{**} = D$, also another possible choice for B is given by the operator $\mathcal{D}(B) = H^2(0,1) \cap H_0^1(0,1)$, $Bu = -\frac{d}{dx}(a(x)\frac{du}{dx})$, $u \in \mathcal{D}(B)$. If $A_H = (-1)^m \frac{d^{2m}}{dx^{2m}}$, $\mathcal{D}(A_H) = H^{2m}(0,1) \cap H_0^m(0,1)$, $m \geq 1$, since $\mathcal{D}(A^{\frac{1}{2}}) = \{u \in H^m(0,1); u^{(j)}(0) = u^{(j)}(1) = 0, j = 0,1,\dots,m-1\}$ (see Example 6.8), $m \geq 2$ guarantees that all our hypotheses on A and B hold.

Obviously, different choices of A are allowed, like the ones described in Example 6.8, provided that $\mathcal{D}(A^{\frac{1}{2}})$, completely characterized in that example, be embedded into $\mathcal{D}(B)$.

What happens when B itself degenerates, in the sense that $E^{\frac{1}{2}}D$ is not closed ($E^{\frac{1}{2}}$ can have no bounded inverse)? Then we take

$$B = (E^{\frac{1}{2}}D)^*\overline{(E^{\frac{1}{2}}D)} = (E^{\frac{1}{2}}D)^*(E^{\frac{1}{2}}D),$$

provided that $E^{\frac{1}{2}}D$ is closable and this holds if, for example, $\mathcal{D}(D^*E^{\frac{1}{2}})$ is dense in H. In fact, to see this it suffices to show that if $\mathcal{B} \in \mathcal{L}(H)$ and $\mathcal{A}$ is a closed densely defined operator from H into H with $\overline{\mathcal{D}(\mathcal{A}\mathcal{B})} = H$, then $\mathcal{B}^*\mathcal{A}^* = (\mathcal{A}\mathcal{B})^*$. But $graph(\mathcal{B}^*\mathcal{A}^*) \subset graph((\mathcal{A}\mathcal{B})^*)$ and thus we prove that inclusion is in fact identity; otherwise, there exists $0 \neq x_0 \in \mathcal{D}((\mathcal{A}\mathcal{B})^*)$

such that $\langle x, x_0 \rangle + \langle \overline{\mathcal{B}^* \mathcal{A}^*} x, (\mathcal{AB})^* x_0 \rangle = 0$ for all $x \in \mathcal{D}(\overline{\mathcal{B}^* \mathcal{A}^*})$. Hence, $x \mapsto \langle (\mathcal{B}^* \mathcal{A}^*) x, (\mathcal{AB})^* x_0 \rangle$ has a bounded extension to H. It follows that $(\mathcal{AB})^* x_0 \in \mathcal{D}((\mathcal{B}^* \mathcal{A}^*)^*) = \mathcal{D}(\overline{\mathcal{AB}})$ and $\langle x, x_0 \rangle + \langle x, (\mathcal{AB})(\mathcal{AB})^* x_0 \rangle = 0$ for all $x \in H$. Taking $x = x_0$, it follows that $x_0 = 0$, thus establishing the affirmation.

Of course, it remains to characterize the closure of $E^{\frac{1}{2}} D$. We exploit the closedness of the operator $DE^{\frac{1}{2}}$, assuming that the commutator $[E^{\frac{1}{2}}, D] = E^{\frac{1}{2}} D - DE^{\frac{1}{2}}$ (defined on the maximal domain) has a bounded extension from H into itself. Then, $\overline{E^{\frac{1}{2}} D} x = y$ if and only if there exists a sequence $x_n \in \mathcal{D}(D)$, $x_n \to x$ in H, $E^{\frac{1}{2}} D x_n \to y$ in H. But, $E^{\frac{1}{2}} D x_n = [E^{\frac{1}{2}}, D] x_n + DE^{\frac{1}{2}} x_n$ implies together with the preceding assumptions that $DE^{\frac{1}{2}} x_n \to y - [E^{\frac{1}{2}}, D] x$ and $y = [E^{\frac{1}{2}}, D] x + DE^{\frac{1}{2}} x$.

Let us apply this remark. Assume $a(x)$ a real valued $\mathcal{C}^1$ function on $[0, 1]$ such that $a(x) > 0$ for $0 < x < 1$, $a(0) = a(1) = 0$, $a(\cdot)^{\frac{1}{2}} \in \mathcal{C}^1([0, 1])$, and let E be multiplication by $a(x)$ in $X = L^2(0, 1)$. If D is the operator of derivation in $L^2(0, 1)$ with domain $H^1(0, 1)$, then $[E^{\frac{1}{2}}, D] u = -(a(x)^{\frac{1}{2}})' u$ and thus $\overline{E^{\frac{1}{2}} D} u = f$ if $a^{\frac{1}{2}} u \in H^1(0, 1)$ and $f = (a^{\frac{1}{2}} u)' - (a^{\frac{1}{2}})' u$. Call $a^{\frac{1}{2}} u = v \in H^1(0, 1)$. Then $f = v' - \frac{(a^{\frac{1}{2}})'}{a^{\frac{1}{2}}} v = \left(\frac{v}{a^{\frac{1}{2}}} \right)' a^{\frac{1}{2}} = a^{\frac{1}{2}} u' \ (\in L^2(0, 1))$. Therefore, $B = (\overline{E^{\frac{1}{2}} D})^* \overline{E^{\frac{1}{2}} D} = D^* E^{\frac{1}{2}} E^{\frac{1}{2}} D$ signifies $Bu = -\frac{d}{dx}(a(x) \frac{du}{dx})$ with $u \in L^2(0, 1)$ and

$$\int_0^1 a(x) |u'(x)|^2 dx < \infty, \quad \int_0^1 |\frac{d}{dx}(a(x) \frac{du}{dx})|^2 dx < \infty,$$

$$\lim_{x \to 0+} a(x) u'(x) = \lim_{x \to 1-} a(x) u'(x) = 0.$$

It may be verified, see M. Campiti, G. Metafune and D. Pallara [46], that $u \in \mathcal{D}(B)$ reads equivalently: $u \in L^2(0, 1)$, u is locally absolutely continuous in $(0, 1)$ and $au' \in H_0^1(0, 1)$.

If we take $D = -\frac{d}{dx}$ in $H = L^2(0, 1)$ with domain $H_0^1(0, 1)$, then analogous reasonings yield that $B_1 = (\overline{E^{\frac{1}{2}} D})^* \overline{E^{\frac{1}{2}} D}$ means $B_1 u = -\frac{d}{dx}(a(x) \frac{du}{dx})$ with $u \in L^2(0, 1)$

$$\int_0^1 a(x) |u'(x)|^2 dx < \infty, \quad \int_0^1 |\frac{d}{dx}(a(x) \frac{du}{dx})|^2 dx < \infty,$$

$$\lim_{x \to 0+} a(x)^{\frac{1}{2}} u(x) = \lim_{x \to 1-} a(x)^{\frac{1}{2}} u(x) = 0.$$

On the other hand, for $u \in \mathcal{D}(B_1)$ we have $au' \in H^1(0, 1)$ so that $a(x) u'(x)$ tends to a limit as $x \to 0, 1$. Let $\lambda = \lim_{x \to 0+} a(x) u'(x)$. If $\lambda \neq 0$, then $u'(x) \sim \frac{\lambda}{a(x)}$ in a neighbourhood of 0. Hence $a(x) |u'(x)|^2 \sim \frac{\lambda^2}{a(x)}$ can not be integrable on $(0, \delta)$, contradicting $u \in \mathcal{D}(B_1)$. Therefore, since a corresponding argument applies to $x = 1$, we conclude that $\mathcal{D}(B_1) \subset \mathcal{D}(B)$ and B is a closed extension of B_1. But then B and B_1 in fact coincide.

Notice that if $|a'(x)| \leq Ca(x)^{\frac{1}{2}}$ is assumed, then we could substitute the condition $\int_0^1 |\frac{d}{dx}(a(x)\frac{du}{dx})|^2 dx < \infty$ with $\int_0^1 a(x)^2 |\frac{d^2u}{dx^2}|^2 dx < \infty$.

Degenerate parabolic operators shall be studied in more details in the next chapter.

These results say what types of operators A_H are permitted in order to apply Theorems 6.14 and 6.15 in this situation. We want to point out that the preceding observations allow to solve doubly degenerate equations of parabolic type in $L^2(0,1)$ like

$$\frac{\partial}{\partial t}(m(x)u) = \frac{\partial}{\partial x}\left(a(x)\frac{\partial u}{\partial x}\right) - ku + f(x,t), \quad (x,t) \in (0,1) \times [0,T],$$

where $m(x) \geq 0$ on $[0,1]$ is a continuous function, k is suitably large. Arguing as in Example 3.3, we deduce that the involved operators satisfy (3.7) with $\alpha = 1$, $\beta = \frac{1}{2}$ ($\gamma = 0$).

6.4 EQUATIONS OF HYPERBOLIC TYPE (CONTINUED)

In this section our aim is to extend the treatment of problem (6.37) to the case where the operator C is unbounded as acting in the complex Hilbert space H. Since in this situation the arguments in the paragraph above do not work, we shall give some conditions on the operators A, B, C (that could be generalized) implying that, on the contrary, the general method used for in Chapter V applies.

We begin with listing the assumptions for our first result:

(6.47) $A = A^* > 0$ is a positive self adjoint operator in H;

(6.48) B is a densely defined closed linear operator in H such that

$$\Re e(Bv, v)_H \geq 0, \quad \Re e(B^*f, f)_H \geq 0 \quad \text{for all } v \in \mathcal{D}(B), \ f \in \mathcal{D}(B^*);$$

(6.49) $C = C^* \geq$ is a non negative self adjoint operator in H;

(6.50) $\mathcal{D}(A^{\frac{1}{2}}) \subset \mathcal{D}(B) \cap \mathcal{D}(C)$;

(6.51) B^* is A-bounded with A-bound equal to zero.

Notice that under (6.47) $\|A^{\frac{1}{2}}v\|_H \geq c_0\|v\|_H$ for all $v \in \mathcal{D}(A^{\frac{1}{2}})$, where $c_0 > 0$. In order to obtain the needed resolvent estimates for all $u \in \mathcal{D}(A) = \mathcal{D}(P(\lambda))$, $P(\lambda) = \lambda^2 C + \lambda B + A$, $\Re e\lambda > 0$, let $P(\lambda)u = f \in H$ and call $\lambda u = v$, so that $\lambda Cv + Bv + \lambda^{-1}Av = f$. Multiplying this equation by v we obtain that

$$\lambda(Cv, v)_H + (Bv, v)_H + \frac{\bar{\lambda}}{|\lambda|^2}(Av, v)_H = (f, v)_H$$

and hence, taking into account $(6.47)\sim(6.49)$, we have:

$$\Re\lambda\|C^{\frac{1}{2}}v\|_H^2 + \Re(Bv,v)_H + \tfrac{\Re\lambda}{|\lambda|^2}\|A^{\frac{1}{2}}v\|_H^2 = \Re(f,v)_H.$$

It follows that there exist two positive constants C_1, C_2 such that

$$C_1\tfrac{\Re\lambda}{|\lambda|^2}\|A^{\frac{1}{2}}v\|_H \le \|f\|_H,$$

(6.52)
$$C_2\tfrac{\Re\lambda}{|\lambda|^2}\|v\|_H \le \|f\|_H.$$

Therefore, (6.52) yields that

$$C_2\tfrac{\Re\lambda}{|\lambda|}\|u\|_H \le \|f\|_H.$$

In virtue of (6.50), there exists $C_3 > 0$ such that

(6.53) $$C_3\tfrac{\Re\lambda}{|\lambda|^2}\|Cv\|_H = C_3\tfrac{\Re\lambda}{|\lambda|}\|Cu\|_H \le \|f\|_H$$

or else

(6.54) $$C_3\|Cu\|_H \le \tfrac{|\lambda|}{\Re\lambda}\|f\|_H.$$

(6.50) again guarantees that

$$C_4\|Bu\|_H \le \tfrac{|\lambda|}{\Re\lambda}\|f\|_H.$$

On the other hand, we know (see Example 5.2 and J. Weidmann [175; p. 109]) that $(\lambda B + A)^* = \bar{\lambda}B^* + A$ (in view of (6.50) and (6.51)). Moreover $\mathcal{D}(A + \lambda B) = \mathcal{D}(A)$, so that in its turn

$$(\lambda^2 C + (\lambda B + A))^* = \bar{\lambda}^2 C + (\lambda B + A)^* = \bar{\lambda}^2 C + \bar{\lambda}B^* + A, \quad \Re\lambda > 0.$$

Hence, we are allowed to repeat the same argument as above to deduce that $P(\lambda)^*$ is one-to-one. This implies that $P(\lambda)$ has a bounded inverse for $\Re\lambda > 0$.

Let $\Re\lambda \ge a_0 > 0$. Then (6.53) and (6.54) say us that

$$\|AP(\lambda)^{-1}\|_{\mathcal{L}(H)} \le k(1 + |\lambda|)^3, \quad \|BP(\lambda)^{-1}\|_{\mathcal{L}(H)} \le k(1 + |\lambda|),$$
$$\|CP(\lambda)^{-1}\|_{\mathcal{L}(H)} \le k(1 + |\lambda|), \quad \Re\lambda \ge a_0.$$

According to Section 5.7 and Theorem 5.10, we can translate problem (6.37) in the product space $X = (\mathcal{D}(B) \cap \mathcal{D}(C)) \times H$, where we take $\mathcal{D}(M) = (\mathcal{D}(B)\cap\mathcal{D}(C)) \times \mathcal{D}(C)$, $\mathcal{D}(L) = \mathcal{D}(A) \times (\mathcal{D}(B)\cap\mathcal{D}(C))$. By using the same notation as in Theorem 5.10, we have $\nu = 3$ and $m = 5$. Therefore the smallest integer h given in Theorem 5.10 is $h = 8$. Let

$$\mathcal{A} = \begin{pmatrix} -A^{-1}B & -A^{-1} \\ C & 0 \end{pmatrix}$$

be the operator matrix from X into itself associated to ML^{-1}. We have:

THEOREM 6.16. *Let us assume (6.47)~(6.51). If $f \in C^8([0,T];H)$ and $u_0 \in \mathcal{D}(A)$, $u_1 \in \mathcal{D}(B) \cap \mathcal{D}(C)$ satisfy the compatibility relation*

$$(6.55) \qquad (u_1, -Au_0 - Bu_1) + \sum_{j=0}^{7} \mathcal{A}^j(0, f^{(j)}(0)) \in \mathcal{R}(\mathcal{A}^8),$$

then problem (6.37) has a unique strict solution u, that is, Au, Bu', $Cu'' \in C([0,T];H)$ and the above resolvent estimates.

PROOF. It is an easy consequence of Theorem 5.10 and the resolvent estimates above.

Notice that the very strong relation (6.55) means that the Cauchy problem for highly degenerate equations is not well posed in the sense that (6.37) has one solution for a given function which is sufficiently smooth in time, but the initial values $u(0)$ and $u'(0)$ cannot be arbitrarily fixed since they depend on $f^{(j)}(0)$, $j = 0, 1, \ldots, 7$.

If we content ourselves of integrated solutions of (6.37), all compatibility relations can in fact be disregarded, according to the following theorem.

THEOREM 6.17. *Let us assume (6.47)~(6.51). For all $f \in C([0,T];H)$ and any $u_0 \in \mathcal{D}(B) \cap \mathcal{D}(C)$, $u_1 \in \mathcal{D}(C)$, the problem (6.37) has a unique 7-integrated solution.*

PROOF. It suffices to use Theorem 5.28 with $m = 5$, $\beta = 0$.

The result in Theorem 6.16 can be a bit improved if we require something more on B, precisely

$$(6.56) \qquad B = B^* > 0 \text{ and } \mathcal{D}(B^{\frac{1}{2}}) \subset \mathcal{D}(C), \quad \mathcal{D}(A^{\frac{1}{2}}) \subset \mathcal{D}(B).$$

THEOREM 6.18. *Let us assume (6.47)~(6.49) and (6.56). Then*

$$\|CP(\lambda)^{-1}\|_{\mathcal{L}(X)} \le k(1 + |\lambda|)^{-1}, \quad \Re\lambda \ge a_0,$$
$$\|BP(\lambda)^{-1}\|_{\mathcal{L}(X)} \le k(1 + |\lambda|), \quad \Re\lambda \ge a_0 > 0,$$

and (6.37) has a unique solution for all $f \in C^7([0,T];H)$ and all $u_0 \in \mathcal{D}(A)$, $u_1 \in \mathcal{D}(B)$ satisfying

$$(u_1, -Au_0 - Bu_1) + \sum_{j=0}^{6} \mathcal{A}^j(0, f^{(j)}(0)) \in \mathcal{R}(\mathcal{A}^7),$$

where $\mathcal{A}(x,y) = (-A^{-1}Bx - A^{-1}y, Cx)$, $(x,y) \in \mathcal{D}(B) \times H$.

PROOF. If $P(\lambda)u = f$, so that $\lambda Cv + Bv + \lambda^{-1}Av = f$, $\lambda u = v$, where $\Re\lambda > 0$ and $f \in H$, $u \in \mathcal{D}(A)$, then $(Bv, v)_H = \|B^{\frac{1}{2}}v\|_H^2 \le$

$k\|f\|_H\|v\|_H \le k'\|f\|_H\|B^{\frac{1}{2}}v\|_H$, and hence $|\lambda|\|B^{\frac{1}{2}}u\|_H \le C_1\|f\|_H$. In view of (6.56), we also deduce that $|\lambda|\|Cu\|_H \le C_2\|f\|_H$. Moreover the estimate $\frac{\Re e\lambda}{|\lambda|}\|A^{\frac{1}{2}}u\|_H \le C_3\|f\|_H$ says, again in virtue of (6.56), that

$$\|Bu\|_H \le C_4 \frac{|\lambda|}{\Re e\lambda}\|f\|_H$$

On the other hand, well known properties of fractional powers of positive operators imply that $P(\lambda)^* = P(\overline{\lambda})$, $\Re e\lambda > 0$. Combining these facts, we conclude that $P(\lambda)$ has the properties described in the statement.

Last part of the theorem is an immediate consequence of Theorem 5.10 and Section 5.7 with $m = 4$, $b = 0$ by an obvious change-of-variable argument as in Theorem 6.2.

EXAMPLE 6.23. Let Ω be a smooth bounded domain in $\mathbb{R}^n$, $n \ge 1$, and $r(x)$ be a smooth positive function on Ω such that for all x near the boundary $\partial\Omega$, $r(x)$ coincides with the distance from x to $\partial\Omega$. If $m(x) = \frac{1}{r(x)}$, $x \in \Omega$, let C be multiplication by $m(x)$ in $H = L^2(\Omega)$, so that $\mathcal{D}(C) = L^2_{m^2}(\Omega) = \{u \in L^2(\Omega); \int_\Omega m(x)^2|u(x)|^2dx < \infty\}$.

We also observe (see R. W. Carroll and R. E. Showalter [48; p. 204]) that

$$\|Cu\|_H \le k\|u\|_V, \quad \text{for all } u \in V = H^1_0(\Omega).$$

If $A = (-1)^m\Delta^m$, $\mathcal{D}(A) = H^{2m}(\Omega) \cap H^m_0(\Omega)$, $m \ge 1$, then A is self adjoint positive in H and $\mathcal{D}(A^{\frac{1}{2}}) \subset \mathcal{D}(C)$. Let B be multiplication by $\pi(x)$ which is a continuous non negative function on $\overline{\Omega}$ or more generally π is continuous ≥ 0 on Ω but $\int_\Omega \pi(x)^2|u(x)|^2dx \le k\|u\|^2_V$ for all $u \in V$. Hence all assumptions (6.47)$\sim$(6.51) are verified and equations

$$m(x)\frac{\partial^2 u}{\partial t^2} + \pi(x)\frac{\partial u}{\partial t} + (-1)^m\Delta^m u = f(x,t), \quad (x,t) \in \Omega \times [0,T],$$

$$\frac{\partial^2 v}{\partial t^2} + \beta(x)\frac{\partial u}{\partial t} + (-1)^m r(x)\Delta^m v = g(x,t), \quad (x,t) \in \Omega \times [0,T],$$

where $g \in \mathcal{C}([0,T]; L^2_{m^2}(\Omega))$ and either $\frac{\beta(x)}{r(x)}$ admits a continuous extension to $\overline{\Omega}$ or $\int_\Omega \frac{\beta(x)^2}{r(x)^2}|u(x)|^2dx \le k\|u\|^2_V$ for all $u \in V$, can be handled with the aid of Theorems 6.16 and 6.17. This result is obviously weaker than the one described in Example 6.18, where $m(x)$ is assumed continuous on $\overline{\Omega}$.

We again remark that the solutions treated in R. W. Carroll and R. E. Showalter [48; pp. 204-205] would take their values in $H^{-m}(\Omega)$.

EXAMPLE 6.24. Let $\Omega = (0,1)$ and $m(x) = \frac{1}{x(1-x)}$. Again C is multiplication by $m(x)$ in $X = L^2(0,1)$, and $A = -\frac{d^2}{dx^2}$ with $\mathcal{D}(A) = H^2(0,1) \cap H^1_0(0,1)$.

If B is defined by $\mathcal{D}(B) = \{u \in H^1(0,1); \ u(0) = 0\}$, $Bu = u'$, then it is easily seen that $B^* = -B_1$, where $\mathcal{D}(B_1) = \{u \in H^1(0,1); \ u(1) = 0\}$, $B_1 u = u'$. Furthermore, $\Re e(Bu, u)_H \geq 0$ and $\Re e(B^* f, f)_H \geq 0$ on $\mathcal{D}(B)$ and $\mathcal{D}(B^*)$ respectively.

Since $\mathcal{D}(A^{\frac{1}{2}}) = H_0^1(0,1) \subset \mathcal{D}(B) \cap \mathcal{D}(C)$ and since B^* has A-bound equal to zero, we conclude that Theorems 6.16 and 6.17 are applicable to the equation

$$\frac{\partial^2 u}{\partial t^2} + x(1-x)\frac{\partial^2 u}{\partial x \partial t} - x(1-x)\frac{\partial^2 u}{\partial x^2} = g(x,t), \quad (x,t) \in (0,1) \times [0,T],$$

where $g \in C([0,T]; L^2_{1/x^2(1-x)^2}(0,1))$. Notice that by Example 6.23 we are already able to treat an equation like $\frac{\partial^2 u}{\partial t^2} = x(1-x)\frac{\partial^2 u}{\partial x^2} + g(x,t)$ with $g \in C([0,T]; L^2_{1/x^2(1-x)^2}(0,1))$.

EXAMPLE 6.25. Let B be either the operator (of first derivative) introduced in Example 6.24 or $B = \frac{d^2}{dx^2}$, $\mathcal{D}(B) = H^2(0,1) \cap H_0^1(0,1)$. Let C be the operator in $L^2(0,1)$ given by $\mathcal{D}(C) = H^2(0,1) \cap H_0^1(0,1)$, $Cu = u - u''$. Therefore, (6.47),(6.48) and (6.49) are verified.

Let $\mathcal{D}(A) = H^4(0,1) \cap H_0^2(0,1)$, $Au = u^{(4)}$. Since $\mathcal{D}(A^{\frac{1}{2}}) \subset \{u \in H^2(0,1); \ u(0) = u(1) = 0\}$ and B is A-bounded with A-bound equal to zero, we can solve in the ambient space $L^2(0,1)$ some equations like

$$\left(1 - \frac{\partial^2}{\partial x^2}\right)\frac{\partial^2 u}{\partial t^2} + \frac{\partial^2 u}{\partial x \partial t} + \frac{\partial^4 u}{\partial x^4} = f(x,t),$$

$$\left(1 - \frac{\partial^2}{\partial x^2}\right)\frac{\partial^2 u}{\partial t^2} - \frac{\partial^3 u}{\partial x^2 \partial t} + \frac{\partial^4 u}{\partial x^4} = f(x,t).$$

6.5 ABSTRACT ELLIPTIC EQUATIONS

In this section we show that interesting degenerate elliptic partial differential equations can be handled by the operational method introduced in Section 3.5, too.

Let L and M be two closed linear operators in a Banach space X satisfying

$$(6.57) \qquad \|M(\lambda M - L)^{-1}\|_{\mathcal{L}(X)} \leq \frac{C}{(1+|\lambda|)^\beta}, \quad \lambda \in \Sigma,$$

where $\Sigma = \{\lambda \in \mathbb{C}; \ \Re e \lambda \leq c(1+|\Im m\lambda|)^\alpha\}$, $c > 0$, $0 < \beta \leq \alpha \leq 1$. Consider the problem

$$(6.58) \qquad \begin{cases} -\dfrac{d^2}{dt^2}(Mu) + Lu = f(t), & 0 \leq t \leq 1, \\ Mu(0) = Mu(1) = 0, \end{cases}$$

in X, with $f \in \mathcal{C}([0,1]; X) = E$.

To begin with, we introduce the operator $B_1 u = u'' = \frac{d^2 u}{dt^2}$, where

$$\mathcal{D}(B_1) = \{u \in E;\ u', u'' \in E,\ u(0) = u(1) = 0\}.$$

Then it is easily seen that

$$\|(\lambda - B_1)^{-1}\|_{\mathcal{L}(E)} \le \frac{1}{|\lambda| \cos \frac{\theta}{2}} = \frac{1}{|\sqrt{\lambda}| \Re e \sqrt{\lambda}},$$

where $\theta = \arg \lambda$ and $\lambda \notin (-\infty, 0]$. Fix $d > 0$ and define $B = -B_1 + d$. Then $\lambda - B$ has a bounded inverse for all $\lambda \notin [d, \infty)$. Further, if $\Re e \lambda \le \frac{d}{2}$, then the estimate

$$\|(\lambda - B)^{-1}\|_{\mathcal{L}(E)} \le \frac{1}{\cos \frac{\varphi}{2} |\lambda - d|},$$

where $\varphi = \arg(d - \lambda)$, holds and this implies that, in particular, B has the property

$$\|(\lambda - B)^{-1}\|_{\mathcal{L}(E)} \le \frac{C}{1 + |\Re e \lambda|}, \qquad \Re e \lambda \le a_0,$$

with some $a_0 > 0$.

Problem (6.58) can be equivalently formulated as an operational equation

$$BMu + (L - dM)u = f$$

if we identify L, M with the operators induced by them in E in the natural way. Now

$$\|M(\lambda M - L + dM)^{-1}\|_{\mathcal{L}(E)} = \|M((\lambda + d)M - L)^{-1}\|_{\mathcal{L}(E)} \le \frac{C}{(1 + |\lambda|)^\beta},$$

provided that $\Re e \lambda \le c(1 + |\Im m \lambda|)^\alpha - d$. It suffices to take d sufficiently small to have desired estimate in a (possibly different) region Σ': $\Re e \lambda \le c'(1 + |\Im m \lambda|)^\alpha$, $0 < c' < c$.

This then means that the results for operational equations obtained in Section 3.5 are applicable and hence we need a characterization of $(E, \mathcal{D}(B))_{\theta,\infty}$ in order to exploit Theorem 3.24. But this is furnished by Theorem 3.1.29 of A. Lunardi [131; p. 107], where scalar valued functions are concerned, but extension to X valued functions is straightforward. Indeed,

$$(E, \mathcal{D}(B))_{\theta,\infty} = \begin{cases} \{u \in \mathcal{C}^{2\theta}([0,1]; X);\ u(0) = u(1) = 0\}, & \theta \ne \frac{1}{2}, \\ \{u \in \mathfrak{C}^1([0,1]; X);\ u(0) = u(1) = 0\}, & \theta = \frac{1}{2}. \end{cases}$$

Here, $\mathfrak{C}^1([0,1]; X)$ denotes the Zygmund space of order 1, that is,

$$\mathfrak{C}^1([0,1]; X) = \{f \in \mathcal{C}([0,1]; X); \sup_{0 \le s \ne t \le 1} \frac{\left\|f(s)+f(t)-2f(\frac{s+t}{2})\right\|_X}{|s-t|} < \infty\}.$$

We conclude from Theorem 3.24 that if $2\alpha+\beta > 2$, $\frac{2-\alpha-\beta}{\alpha} < \theta < 1$, then for all $f \in \mathcal{C}^{2\theta}([0,1]; X)$ if $\theta \ne \frac{1}{2}$, and $f \in \mathfrak{C}^1([0,1]; X)$ if $\theta = \frac{1}{2}$, $f(0) = f(1) = 0$, problem (6.58) has a unique solution u satisfying $\frac{d^2}{dt^2}(Mu) \in (E, \mathcal{D}(B))_{\omega,\infty}$, $\omega = \alpha\theta + \alpha + \beta - 2$.

We describe a simple example related to Tricomi equation of elliptic type.

EXAMPLE 6.26. Let $X = L^2(0,1)$. Let $\mathcal{D}(L) = H^2(0,1) \cap H_0^1(0,1)$, $Lu = -u''$ and $M \in \mathcal{L}(X)$ be multiplication by a given function $m \in \mathcal{C}([0,1])$ such that $m(x) \ge 0$ on $[0,1]$. Then problem (6.58) reads

$$\begin{cases} \dfrac{\partial^2 u}{\partial x^2} + \dfrac{\partial^2}{\partial y^2}(m(x)u) = f(x,y), & 0 \le x, y \le 1, \\ m(x)u(x,0) = m(x)u(x,1) = 0, & 0 \le x \le 1, \\ u(0,y) = u(1,y) = 0, & 0 \le y \le 1. \end{cases}$$

Notice that in fact in this case $\alpha = 1$, $\beta = \frac{1}{2}$.

One could ask for a corresponding problem in the space $X = \mathcal{C}([0,1])$ also. In view of properties of B and Example 5.10, we are allowed to apply Theorem 5.1 with $m = 1$, $p = 0$, $k = 3$, and hence for

$$f \in \mathcal{C}^6([0,1]^2), \quad \frac{\partial^{2i}}{\partial y^{2i}}f(x,0), \quad \frac{\partial^{2i}}{\partial y^{2i}}f(x,1) = 0, \quad i = 0,1,2,$$

the problem above has a strict solution in $\mathcal{C}([0,1]^2)$.

6.6 ABSTRACT ELLIPTIC-HYPERBOLIC EQUATIONS

Until now the indicated methods did not work for time depending operator coefficient equations of the type $\frac{d^2 u}{dt^2} = A(t)u$, $t \ge 0$. In this section we prove that generators of C_0 groups of linear operators can be applied with success to solve some particular, but very interesting, degenerate problems and problems of mixed type, as well.

In particular, we shall indicate application to the famous Tricomi equation

(6.59)
$$\begin{cases} \dfrac{\partial^2 u}{\partial t^2} = t^m \dfrac{\partial^2 u}{\partial x^2}, & -\infty < x, t < \infty, \\ u(x,0) = u_0(x), & -\infty < x < \infty, \\ \dfrac{\partial u}{\partial t}(x,0) = u_1(x), & -\infty < x < \infty, \end{cases}$$

where m is an odd positive integer.

Consider the abstract equation

(6.60)
$$\begin{cases} \dfrac{d^2u}{dt^2} = t^m A^2 u, \quad 0 \le t < \infty, \\[2mm] u(0) = u_0, \quad \dfrac{du}{dt}(0) = u_1, \end{cases}$$

in a Banach space X, where A is a closed linear operator in X, $m > 0$ is a positive number, $u_0 \in \mathcal{D}(A^2)$, $u_1 \in X$.

DEFINITION. A function

$$u \in \mathcal{C}^2([0,\infty); X) \cap \mathcal{C}([0,\infty); \mathcal{D}(A)) \cap \mathcal{C}((0,\infty); \mathcal{D}(A^2))$$

satisfying the equation and the initial conditions in (6.60) is called a *solution* to (6.60).

Our first result comes from A. Favini [79] and it is based upon some properties of the Airy function. It has some analogy with the treatment of Euler-Poisson-Darboux equation developed by J. A. Donaldson [69]; see R. W. Carroll and R. E. Showalter [48; pp. 66-80]. Properties of the functions Z_ν suggest, in fact, to seek a solution to (6.60) as a linear combination of the two functions

$$v(t) = t \int_{-1}^{1} e^{-\frac{2}{m+2} t^{(\frac{m}{2}+1)} As} (1 - s^2)^{-\frac{m}{2(m+2)}} v_1 \, ds$$

$$w(t) = \int_{-1}^{1} e^{-\frac{2}{m+2} t^{(\frac{m}{2}+1)} As} (1 - s^2)^{-\frac{m+4}{2(m+2)}} v_0 \, ds,$$

for certain $v_0, v_1 \in \mathcal{D}(A^2)$. We then have the theorem.

THEOREM 6.26. *Assume that A generates a C_0-group on X. Then, for all $u_0, u_1 \in \mathcal{D}(A^2)$, (6.60) possesses a solution.*

PROOF. Since $\|e^{tA}\|_{\mathcal{L}(X)} \le M e^{\omega|t|}$, $t \in \mathbb{R}$, summability of functions entering under the sign of integral in $v(t)$ and $w(t)$ depends only on the coefficients $(1 - s^2)^{-\frac{m}{2(m+2)}}$ and $(1 - s^2)^{-\frac{m+4}{2(m+2)}}$ on $(-1, 1)$. Therefore, the first function is summable on $(-1, 1)$ if and only if $m \in (-\infty, -4) \cup (-2, \infty)$ and the second one is summable on $(-1, 1)$ if and only if $m \in (-\infty, -2) \cup (0, \infty)$. Hence, $v(t)$ and $w(t)$ are well defined and continuous on $(0, \infty)$ for all $m \in (-\infty, -4) \cup (0, \infty)$.

Let us suppose $m > 0$ and $v_0, v_1 \in \mathcal{D}(A^2)$. Then we shall prove that $v(t)$ and $w(t)$ are twice strongly differentiable. First, we begin with a formal calculation, which gives

$$v'(t) = \int_{-1}^{1} e^{-\frac{2}{m+2} t^{(\frac{m}{2}+1)} As} (1 - s^2)^{-\frac{m}{2(m+2)}} v_1 \, ds$$

$$- t^{\frac{m}{2}+1} \int_{-1}^{1} e^{-\frac{2}{m+2} t^{(\frac{m}{2}+1)} As} s (1 - s^2)^{-\frac{m}{2(m+2)}} A v_1 \, ds.$$

So that,

$$v''(t) = -t^{\frac{m}{2}} \int_{-1}^{1} e^{-\frac{2}{m+2}t^{(\frac{m}{2}+1)}As} s(1-s^2)^{-\frac{m}{2(m+2)}} Av_1 ds$$

$$- \frac{m+2}{2} t^{\frac{m}{2}} \int_{-1}^{1} e^{-\frac{2}{m+2}t^{(\frac{m}{2}+1)}As} s(1-s^2)^{-\frac{m}{2(m+2)}} Av_1 ds$$

$$+ t^{m+1} \int_{-1}^{1} e^{-\frac{2}{m+2}t^{(\frac{m}{2}+1)}As} s^2(1-s^2)^{-\frac{m}{2(m+2)}} A^2 v_1 ds$$

$$= t^{\frac{m}{2}} \int_{-1}^{1} e^{-\frac{2}{m+2}t^{(\frac{m}{2}+1)}As} \frac{d}{ds}\left(\frac{m+2}{2}(1-s^2)^{-\frac{m}{2(m+2)}+1}\right) Av_1 ds$$

$$+ t^{m+1} \int_{-1}^{1} e^{-\frac{2}{m+2}t^{(\frac{m}{2}+1)}As} s^2(1-s^2)^{-\frac{m}{2(m+2)}} A^2 v_1 ds.$$

Integration by parts in the first integral yields

$$v''(t) = t^{m+1} \int_{-1}^{1} e^{-\frac{2}{m+2}t^{(\frac{m}{2}+1)}As}(1-s^2)^{-\frac{m}{2(m+2)}+1} A^2 v_1 ds$$

$$+ t^{m+1} \int_{-1}^{1} e^{-\frac{2}{m+2}t^{(\frac{m}{2}+1)}As}(1-s^2)^{-\frac{m}{2(m+2)}} A^2 v_1 ds$$

$$- t^{m+1} \int_{-1}^{1} e^{-\frac{2}{m+2}t^{(\frac{m}{2}+1)}As}(1-s^2)^{-\frac{m}{2(m+2)}+1} A^2 v_1 ds = t^m A^2 v(t),$$

because of $v_1 \in \mathcal{D}(A^2)$. On the other hand, it is easily seen that $v_1 \in \mathcal{D}(A^2)$ assures that all passages are correct and thus $v(t)$ satisfies the equation in (6.60) and has continuous second derivative on $[0, \infty)$, with $v(0) = 0$; moreover,

$$v'(0) = \int_{-1}^{1}(1-s^2)^{-\frac{m}{2(m+2)}} v_1 ds = C_1 v_1.$$

Let us turn to study $w(t)$. We have:

$$w'(t) = -t^{\frac{m}{2}} \int_{-1}^{1} e^{-\frac{2}{m+2}t^{(\frac{m}{2}+1)}As} s(1-s^2)^{-\frac{m+4}{2(m+2)}} Av_0 ds.$$

If $m \geq 2$, then clearly $w'(t)$ is differentiable in $t = 0$, too, but more delicate analysis is needed for studying this problem if $0 < m < 2$. To this end, we observe that $w'(0) = 0$, so that

$$t^{-1}w'(t) = -t^{\frac{m}{2}-1} \int_{-1}^{1} e^{-\frac{2}{m+2}t^{(\frac{m}{2}+1)}As} s(1-s^2)^{-\frac{m+4}{2(m+2)}} Av_0 ds$$

$$= \frac{2}{m} t^m \int_{-1}^{1} e^{-\frac{2}{m+2}t^{(\frac{m}{2}+1)}As}(1-s^2)^{\frac{m}{2(m+2)}} A^2 v_0 ds \quad \to 0,$$

as $t \to 0+$. Therefore, there exists $w''(0) = 0$. On the other hand,

$$w(0) = \left(\int_{-1}^{1} (1 - s^2)^{-\frac{m+4}{2(m+2)}} ds \right) v_0 = C_0 v_0.$$

It follows that $w''(t) = t^m A^2 w(t)$ for $t = 0$. Let us consider the case $t > 0$.
Then,

$$w''(t) = -\frac{m}{2} t^{\frac{m}{2}-1} \int_{-1}^{1} e^{-\frac{2}{m+2} t^{(\frac{m}{2}+1)} As} s(1 - s^2)^{-\frac{m+4}{2(m+2)}} A v_0 ds$$

$$+ t^m A^2 \int_{-1}^{1} e^{-\frac{2}{m+2} t^{(\frac{m}{2}+1)} As} s^2 (1 - s^2)^{-\frac{m+4}{2(m+2)}} v_0 ds$$

$$= t^m A^2 \int_{-1}^{1} e^{-\frac{2}{m+2} t^{(\frac{m}{2}+1)} As} (s^2 - 1 + 1)(1 - s^2)^{-\frac{m+4}{2(m+2)}} v_0 ds$$

$$- \frac{m}{2} t^{\frac{m}{2}-1} \int_{-1}^{1} e^{-\frac{2}{m+2} t^{(\frac{m}{2}+1)} As} s(1 - s^2)^{-\frac{m+4}{2(m+2)}} A v_0 ds,$$

so that, by integration by parts, it is obtained that

$$w''(t) = t^m A^2 w(t), \quad t \geq 0.$$

Thus the function $u(t)$ given by

$$u(t) = C_0^{-1} \int_{-1}^{1} e^{-\frac{2}{m+2} t^{(\frac{m}{2}+1)} As} (1 - s^2)^{-\frac{m+4}{2(m+2)}} u_0 ds$$

$$+ C_1^{-1} t \int_{-1}^{1} e^{-\frac{2}{m+2} t^{(\frac{m}{2}+1)} As} (1 - s^2)^{-\frac{m}{2(m+2)}} u_1 ds$$

is a solution to (6.60), as affirmed.

EXAMPLE 6.27. Let $m > 0$ and let X be an arbitrary Banach space be-
tween $L^p(\mathbb{R})$, $1 < p < \infty$, $UCB(\mathbb{R}) = \{f : \mathbb{R} \to \mathbb{C};\ f$ is uniformly continuous
and bounded on $\mathbb{R}\}$ with $\|f\|_\infty = \sup_x |f(x)|$, $\mathcal{C}(\overline{\mathbb{R}}) = \{f \in \mathcal{C}(\mathbb{R});\ f(x)$ has
a finite limits as $|x| \to \infty\}$ with norm $\|f\|_\infty$.
   If $\mathcal{D}(A) = \{f \in X;\ f \in AC_{\ell oc}(\mathbb{R}),\ f' \in X\}$, $Af = f'$, where $AC_{\ell oc}(\mathbb{R})$
denotes the space of all functions $f : \mathbb{R} \to \mathbb{C}$ that are locally absolutely
continuous on $\mathbb{R}$, then it is well known that A generates the $\mathcal{C}_0$ group of
translations $(T(t)f)(x) = f(x + t)$, $f \in X$, and

$$\begin{cases} \mathcal{D}(A^r) = \{f \in X;\ f', f'', \dots, f^{(r-1)} \in AC_{\ell oc}(\mathbb{R}) \cap X,\ f^{(r)} \in X\} \\ A^r f = f^{(r)}. \end{cases}$$

See P. L. Butzer and H. Berens [39; p. 52]. Therefore Theorem 6.26 allows to affirm that if u_0, $u_1 \in \mathcal{D}(A^2)$, then problem (6.60) possesses a solution u written in the form

$$u(x,t) = C_0 \int_{-1}^{1} (1-s^2)^{-\frac{m+4}{2(m+2)}} u_0(x - \tfrac{2}{m+2}t^{\frac{m}{2}+1}s)ds$$

$$+ C_1 t \int_{-1}^{1} (1-s^2)^{-\frac{m}{2(m+2)}} u_1(x - \tfrac{2}{m+2}t^{\frac{m}{2}+1}s)ds.$$

If $s = 2\tau - 1$, we have the expression

$$u(x,t) = C_0' \int_{-1}^{1} \{\tau(1-\tau)\}^{-\frac{m+4}{2(m+2)}} u_0(x + \tfrac{2}{m+2}t^{\frac{m}{2}+1}(1-2\tau))d\tau$$

$$+ C_1' t \int_{-1}^{1} \{\tau(1-\tau)\}^{-\frac{m}{2(m+2)}} u_1(x + \tfrac{2}{m+2}t^{\frac{m}{2}+1}(1-2\tau))d\tau,$$

that is the Bitsadze formula for generalized Tricomi equations in the hyperbolic case (see A. V. Bitsadze [32]).

In particular, if $m = 1$, we obtain the properly said Tricomi equation and

$$u(x,t) = C_0 \int_{-1}^{1} (1-s^2)^{-\frac{5}{6}} u_0(x+\tfrac{2}{3}t^{\frac{3}{2}}s)ds + C_1 t \int_{-1}^{1} (1-s^2)^{-\frac{1}{6}} u_1(x+\tfrac{2}{3}t^{\frac{3}{2}}s)ds,$$

becomes, after the change of variable $\xi = x + \tfrac{2}{3}t^{\frac{3}{2}}s$,

$$u(x,t) = C_0'(x + \tfrac{2}{3}t^{\frac{3}{2}} - x + \tfrac{2}{3}t^{\frac{3}{2}})$$

$$\times \int_{x-\frac{2}{3}t^{\frac{3}{2}}}^{x+\frac{2}{3}t^{\frac{3}{2}}} \{(x + \tfrac{2}{3}t^{\frac{3}{2}} - \xi)(\xi - (x - \tfrac{2}{3}t^{\frac{3}{2}}))\}^{-\frac{5}{6}} u_0(\xi)d\xi$$

$$+ C_1' \int_{x-\frac{2}{3}t^{\frac{3}{2}}}^{x+\frac{2}{3}t^{\frac{3}{2}}} \{(x + \tfrac{2}{3}t^{\frac{3}{2}} - \xi)(\xi - (x - \tfrac{2}{3}t^{\frac{3}{2}}))\}^{-\frac{1}{6}} u_1(\xi)d\xi,$$

that is precisely formula (6) in L. Schwartz [153; p. 349].

Before studying the "elliptic" equation

$$u''(t) = t^m Bu(t), \quad t \in [0,T], \, m > 0,$$

we recall that if B is a densely defined closed linear operator in the complex Banach space X such that $\|(\xi + B)^{-1}\|_{\mathcal{L}(X)} \leq \frac{M}{\xi}$, $\xi > 0$, then B is of type (ω, M) for some $\omega \in (0, \pi)$, i.e. $\{\lambda \in \mathbb{C}; \, |\arg \lambda| < \pi - \omega\} \subset \rho(-B)$ and $\|(\lambda + B)^{-1}\|_{\mathcal{L}(X)} \leq \frac{M_\varepsilon}{|\lambda|}$, $|\arg \lambda| \leq \pi - \omega - \varepsilon$. Moreover, also in this case, a bit more general than the one outlined in Section 0.3, $-B^{\frac{1}{2}}$ is well defined and generates an analytic semigroup in X. See S. G. Krein [121; p. 125].

LEMMA 6.27. *Let $T > 0$ fixed and suppose B an operator of type (ω, M). Then the problem*

(6.61)
$$\begin{cases} \dfrac{d^2u}{dt^2} = Bu, \quad [-T, T], \\[2mm] u(0) = e^{-TB^{\frac{1}{2}}} v_0, \quad \dfrac{du}{dt}(0) = 0, \end{cases}$$

where $v_0 \in \mathcal{D}(B)$, possesses a twice continuously differentiable solution u such that $u \in \mathcal{C}([-T, T]; \mathcal{D}(B))$.

PROOF. It is enough to introduce the function

$$u(t) = \tfrac{1}{2}(e^{-(T-t)B^{\frac{1}{2}}} + e^{-(T+t)B^{\frac{1}{2}}})v_0, \quad -T \le t \le T,$$

to check that all conditions in (6.61) are satisfied.

The lemma above shall allow us to solve a second order Cauchy problem for an abstract elliptic equation, provided that the initial data are sufficiently regular. The result reads as follows.

THEOREM 6.28. *Let B be an operator of type (ω, M) and let $m > 0$, $T_0, T_1 > \frac{2}{m+2}$. If $u_i = e^{-T_i B^{\frac{1}{2}}} v_i$, $v_i \in \mathcal{D}(B)$, $i = 0, 1$, then the problem*

(6.62)
$$\begin{cases} \dfrac{d^2v}{dt^2} = t^m Bv, \quad 0 \le t \le T, \\[2mm] v(0) = u_0, \quad \dfrac{dv}{dt}(0) = u_1, \end{cases}$$

possesses a solution v such that $v \in \mathcal{C}([0, T]; \mathcal{D}(B)) \cap \mathcal{C}^2([0, T]; X)$.

PROOF. First of all, we observe that the position $\frac{t}{T} = \tau$ transforms (6.62) into

$$\begin{cases} \dfrac{d^2w}{d\tau^2}(\tau) = \tau^m T^{m+2} Bw(\tau), \quad 0 \le \tau \le 1, \\[2mm] w(0) = u_0, \quad \dfrac{dw}{d\tau}(0) = Tu_1, \end{cases}$$

and the operator $T^{m+2}B$ is obviously of type (ω, M). Therefore, we can assume $T = 1$ without loss of generality.

Let $T_0 > \frac{2}{m+2}$ and define

$$f(s, t) = -\frac{2}{m+2} st^{\frac{m}{2}+1}, \quad -1 \le s \le 1, \ 0 \le t \le 1.$$

Then $f([-1, 1] \times [0, 1]) = [-\frac{2}{m+2}, \frac{2}{m+2}]$ and $T_0 + \frac{2}{m+2} st^{\frac{m}{2}+1} \ge T_0 - \frac{2}{m+2} > 0$, $-1 \le s \le 1$, $0 \le t \le 1$. Hence, the function

$$u(-\tfrac{2}{m+2} st^{\frac{m}{2}+1}) = \tfrac{1}{2}(e^{-(T_0 + \frac{2}{m+2} st^{\frac{m}{2}+1})B^{\frac{1}{2}}} + e^{-(T_0 - \frac{2}{m+2} st^{\frac{m}{2}+1})B^{\frac{1}{2}}})v_0$$

is well defined in $(s, t) \in [-1, 1] \times [0, 1]$. Let us introduce

$$\varphi(t) = \int_{-1}^{1} u(-\tfrac{2}{m+2} st^{\frac{m}{2}+1})(1 - s^2)^{-\frac{m+4}{2(m+2)}} ds, \quad 0 \le t \le 1,$$

(see the proof of Theorem 6.26). We have:

$$\varphi'(t) = -t^{\frac{m}{2}} \int_{-1}^{1} u'(-\tfrac{2}{m+2} st^{\frac{m}{2}+1}) s(1 - s^2)^{-\frac{m+4}{2(m+2)}} ds, \quad 0 \le t \le 1,$$

and, in particular, $\varphi'(0) = 0$. It is easily seen that $\varphi''(0) = 0$ and thus $\varphi(t)$ satisfies the equation in (6.62) at $t = 0$. Moreover, if $0 < t \le 1$, then

$$\varphi''(t) = -\tfrac{m}{2} t^{\frac{m}{2}-1} \int_{-1}^{1} u'(-\tfrac{2}{m+2} st^{\frac{m}{2}+1}) s(1 - s^2)^{-\frac{m+4}{2(m+2)}} ds$$

$$+ t^m \int_{-1}^{1} u''(-\tfrac{2}{m+2} st^{\frac{m}{2}+1}) s^2 (1 - s^2)^{-\frac{m+4}{2(m+2)}} ds$$

$$= t^m \int_{-1}^{1} u''(-\tfrac{2}{m+2} st^{\frac{m}{2}+1})(s^2 - 1 + 1)(1 - s^2)^{-\frac{m+4}{2(m+2)}} ds$$

$$+ \tfrac{m+2}{2} t^{\frac{m}{2}-1} \int_{-1}^{1} u'(-\tfrac{2}{m+2} st^{\frac{m}{2}+1}) \tfrac{\partial}{\partial s}(1 - s^2)^{-\frac{m+4}{2(m+2)}+1} ds.$$

Hence integration by parts yields

$$\varphi''(t) = t^m \int_{-1}^{1} u''(-\tfrac{2}{m+2} st^{\frac{m}{2}+1})(1 - s^2)^{-\frac{m+4}{2(m+2)}} ds$$

$$- t^m \int_{-1}^{1} u''(-\tfrac{2}{m+2} st^{\frac{m}{2}+1})(1 - s^2)^{-\frac{m+4}{2(m+2)}+1} ds$$

$$+ t^m \int_{-1}^{1} u''(-\tfrac{2}{m+2} st^{\frac{m}{2}+1})(1 - s^2)^{-\frac{m+4}{2(m+2)}+1} ds = t^m B\varphi(t).$$

But, since $u(0) = e^{-T_0 B^{\frac{1}{2}}} v_0 = u_0$, $u'(0) = 0$, it follows that

$$\varphi(0) = \int_{-1}^{1} (1 - t^2)^{-\frac{m+4}{2(m+2)}} dt\, u_0 = C_0 u_0, \quad \varphi'(0) = 0.$$

Let

$$u_1(\tau) = \tfrac{1}{2}(e^{-(T_1 - \tau)B^{\frac{1}{2}}} + e^{-(T_1 + \tau)B^{\frac{1}{2}}}) v_1, \quad -T_1 \le \tau \le T_1.$$

Then

$$\psi(t) = t \int_{-1}^{1} u_1(-\tfrac{2}{m+2} st^{\frac{m}{2}+1})(1 - s^2)^{-\frac{m}{2(m+2)}} ds, \quad 0 \le t \le 1,$$

is well defined and continuous on $[0, 1]$. Moreover, $\psi \in C^2([0, 1]; X)$ with

$$\psi'(t) = \int_{-1}^{1} u_1(-\tfrac{2}{m+2} st^{\frac{m}{2}+1})(1 - s^2)^{-\frac{m}{2(m+2)}} ds$$

$$- t^{\frac{m}{2}+1} \int_{-1}^{1} u_1'(-\tfrac{2}{m+2} st^{\frac{m}{2}+1}) s (1 - s^2)^{-\frac{m}{2(m+2)}} ds,$$

$$\psi''(t) = -t^{m+1} \int_{-1}^{1} u_1''(-\tfrac{2}{m+2} st^{\frac{m}{2}+1})(1 - s^2)^{-\frac{m}{2(m+2)}+1} ds$$

$$+ t^{m+1} \int_{-1}^{1} u_1''(-\tfrac{2}{m+2} st^{\frac{m}{2}+1})(1 - s^2)^{-\frac{m}{2(m+2)}} ds$$

$$- (\tfrac{m}{2} + 2) t^{\frac{m}{2}} \int_{-1}^{1} u_1'(-\tfrac{2}{m+2} st^{\frac{m}{2}+1}) s (1 - s^2)^{-\frac{m}{2(m+2)}} ds.$$

Integration by parts then yields that

$$\psi''(t) = t^m B\psi(t) - t^{m+1} \int_{-1}^{1} u_1''(-\tfrac{2}{m+2} st^{\frac{m}{2}+1})(1 - s^2)^{-\frac{m}{2(m+2)}+1} ds$$

$$+ t^{\frac{m}{2}} \int_{-1}^{1} u_1''(-\tfrac{2}{m+2} st^{\frac{m}{2}+1})(1 - s^2)^{-\frac{m}{2(m+2)}+1} t^{\frac{m}{2}+1} ds = t^m B\psi(t).$$

In addition, $\psi(0) = 0$, $\psi'(0) = (\int_{-1}^{1}(1 - s^2)^{-\frac{m}{2(m+2)}} ds)u_1 = C_1 v_1$. It thus follows that
$$v(t) = C_0^{-1} \varphi(t) + C_1^{-1} \psi(t), \quad 0 \le t \le 1,$$

is a solution to (6.62) and this completes the proof of the theorem.

If m is an even positive integer, the solution $u = u(t)$ to (6.60) on $[0, \infty)$ becomes in fact a solution of the problem on all of $(-\infty, \infty)$. Analogously, in this case, the solution $v = v(t)$ ensured by Theorem 6.28 satisfies the equation of (6.62) on $[-T, T]$ when $B = -A^2$, for we know, from Theorem 1.15 of [138; A-II, p. 36], that, if A generates a C_0 group, then its square A^2 generates an analytic semigroup of angle $\frac{\pi}{2}$, so that $B = -A^2$ is an operator of type (ω, M), for certain ω, $M \in \mathbb{R}^+$.

The situation changes if m is odd. We shall say that $u = u(t)$ is a solution to the problem

(6.63)
$$\begin{cases} \dfrac{d^2 u}{dt^2} = t^m A^2 u, & -T \le t \le T, \\[2mm] u(0) = u_0, \quad \dfrac{du}{dt}(0) = u_1, \end{cases}$$

if $u \in C^2([-T, T]; X) \cap C([-T, T]; \mathcal{D}(A^2))$ satisfies all the conditions in (6.63).

Here, A is assumed to generate a C_0 group of linear operators in X and $u_0 \in \mathcal{D}(A^2)$, $u_1 \in X$ are given; m is a positive odd integer. Then $-B = A^2$ generates an analytic semigroup.

Consider (6.63) on the interval $[-T, 0]$. The change of variable $t = -s$, $s \in [0, T]$ transforms the problem into

(6.64)
$$\begin{cases} \dfrac{d^2v}{ds^2}(s) = -s^m A^2 v(s), & 0 \le s \le T, \\[2mm] v(0) = u_0, \quad \dfrac{dv}{ds}(0) = -u_1. \end{cases}$$

Since $-A^2$ is an operator of type (ω, M), Theorem 6.28 holds and hence a solution to (6.64) exists provided that

$$u_0 = e^{-T_0(-A^2)^{\frac{1}{2}}} v_0, \ u_1 = e^{-T_1(-A^2)^{\frac{1}{2}}} v_1$$

with $v_i \in \mathcal{D}(A^2)$, $i = 0, 1$, $T_0, T_1 > \frac{2}{m+2}$. Thus the solution u to (6.63) on $[-T, 0]$ exists under these assumptions.

On the other hand, since the semigroup generated by $-(-A^2)^{\frac{1}{2}}$ is analytic, and u_0, $u_1 \in \mathcal{D}(((-A^2)^{\frac{1}{2}})^2) = \mathcal{D}(A^2)$, Theorem 6.26 shows that the problem (6.63) on $[0, T]$ possesses a solution $\bar{u}(t)$. If we define

$$w(t) = \begin{cases} u(t), & \text{if } -T \le t \le 0, \\ \bar{u}(t), & \text{if } 0 \le t \le T, \end{cases}$$

we have proved that $w(t)$ satisfies all the conditions of (6.63).

Summaring on, we establish the following statement refining A. Favini [79; pp. 238 -239].

THEOREM 6.29. *Suppose that A generates a C_0 group in X, and let m be an odd positive integer. If $T > 0$ and $u_i = e^{-T_i(-A^2)^{\frac{1}{2}}} v_i$, $v_i \in \mathcal{D}(A^2)$, $i = 0, 1$, with $T_i > \frac{2}{m+2}$, $i = 0, 1$, then the problem (6.63) possesses a solution.*

Notice that in the case of Example 6.27 it is seen that

$$(e^{-tB^{\frac{1}{2}}} u)(x) = (e^{-t(-A^2)^{\frac{1}{2}}} u)(x) = \frac{\pi}{t} \int_{-\infty}^{\infty} \frac{u(y)}{t^2 + (x - y)^2} dy.$$

Hence, if v_i in Theorem 6.29 belongs to $H_p^2(\mathbb{R})$, $1 < p < \infty$, then $u_i = u_i(x)$ is the value in (x, T_i) of a harmonic function whose trace on the real axis is an element $v_i = v_i(x) \in H_p^2(\mathbb{R})$.

In this manner we have given conditions on $u_i(x)$ sufficient to ensure that the Tricomi equation (6.59) have solutions. Here the space X is chosen according to Example 6.27.

DEGENERATE PARABOLIC EQUATIONS:
SOME ALTERNATIVE APPROACHES

The degenerate parabolic operators treated in Chapter III, Example 3.10, have had in last decades a relevant place in the mathematical research for their applications, both in physical, chemical, economical sciences and in probability theory. Our preceding results were obtained on reducing the resolvent $(\lambda + BA)^{-1}$ to the form $(\lambda B^{-1} + A)^{-1} B^{-1}$ and this forced us to assumptions on the multiplication operator B that do not hold in some important concrete cases. This is due to the fact that the natural domain of BA is too small. Hence in this chapter we shall describe more direct approaches to degenerate parabolic operators that allow us to handle those situations and to establish in some cases the analyticity of the semigroups generated by them in spaces L^2 with weight, in spaces L^p, in spaces $W^{1,p}$, and above all, in spaces of continuous functions.

In particular, we shall consider the one-dimensional degenerate operator $\alpha(x)u'' + \beta(x)u'$, $\alpha(0) = \alpha(1) = 0$, $\alpha(x) > 0$ on $(0, 1)$, α, $\beta \in \mathcal{C}([0, 1])$, on the closed unit interval $[0, 1]$, with different boundary conditions, whose importance in approximation theory and in probability theory is well known.

7.1 GENERATION THEOREMS FOR $\alpha u''$

We begin our discussion by proving the following result to be found in V. Barbu, A. Favini and S. Romanelli [22]. To this end, we first need some notations.

If Ω is an open bounded domain of $\mathbb{R}^n$, $n \geq 1$, with smooth boundary $\partial\Omega$ and

(7.1) $\alpha \in \mathcal{C}(\overline{\Omega})$, $\alpha(x) > 0$ for any $x \in \Omega$, $\alpha(x) = 0$ for $x \in \partial\Omega$, then two Hilbert spaces $L^2_{1/\alpha}(\Omega)$ and $L^2_\alpha(\Omega)$ are defined by

$$L^2_{1/\alpha}(\Omega) = \{u \in L^2(\Omega);\ \alpha^{-\frac{1}{2}}u \in L^2(\Omega)\},$$

$$\langle u, v \rangle_{1/\alpha} = \int_\Omega \alpha(x)^{-1}u(x)\overline{v(x)}dx, \quad u,\, v \in L^2_{1/\alpha}(\Omega),$$

$$L^2_\alpha(\Omega) = \{u \in L^2_{loc}(\Omega);\ \sqrt{\alpha}u \in L^2(\Omega)\},$$

$$\langle u, v \rangle_\alpha = \int_\Omega \alpha(x)u(x)\overline{v(x)}dx, \quad u,\, v \in L^2_\alpha(\Omega),$$

respectively. In addition, we shall denote by $H^1_\alpha(\Omega)$ the completion of $\mathcal{C}^\infty_0(\Omega)$ with respect to the norm

$$\|u\|_{1,\alpha} = (\|u\|^2_{L^2_{1/\alpha}} + \|\nabla u\|^2_{L^2})^{\frac{1}{2}},$$

and by $H^2_\alpha(\Omega)$ the completion of $\mathcal{C}^\infty_0(\Omega)$ with respect to

$$\|u\|_{2,\alpha} = (\|u\|^2_{1,\alpha} + \|\Delta u\|^2_{L^2_\alpha})^{\frac{1}{2}}.$$

Then the announced result reads as follows.

THEOREM 7.1. *If α satisfies (7.1), then the operator A given by*

$$Au = \alpha\Delta u, \quad u \in \mathcal{D}(A) = H^2_\alpha(\Omega),$$

is m-dissipative and self adjoint in the space $L^2_{1/\alpha}(\Omega)$.

PROOF. We establish the Green formula

(7.2) $$\int_\Omega \Delta u\bar\phi dx = -\int_\Omega \nabla u \cdot \nabla \bar\phi dx, \quad \phi \in H^1_\alpha(\Omega).$$

Indeed, approximate ϕ by a sequence $\{\phi_n\} \subset \mathcal{C}^\infty_0(\Omega)$ in the norm $\|\cdot\|_{1,\alpha}$ and observe that

$$\int_\Omega \alpha^{\frac{1}{2}}\Delta u(\alpha^{-\frac{1}{2}}\bar\phi_n)dx = -\int_\Omega \nabla u \cdot \nabla \bar\phi_n dx$$

for each $n \in \mathbb{N}$. Then letting $n \to \infty$ yields the desired formula. By (7.2) we immediately recognize that A is dissipative in $L^2_{1/\alpha}(\Omega)$, or more precisely

$$\langle Au, u \rangle_\alpha = \int_\Omega \Delta u\bar u dx = -\int_\Omega |\nabla u|^2 dx$$

for all $u \in \mathcal{D}(A)$. It remains to prove that the range $\mathcal{R}(1 - A)$ of $1 - A$ is all of $L^2_{1/\alpha}(\Omega)$. To accomplish this, we notice that $1 - A$ is continuous and coercive from $H_\alpha^1(\Omega)$ to its dual space $H_\alpha^1(\Omega)^*$; in fact, $1 - A$ has a continuous extension having this property. Since (7.2) holds, we have:

$$|\langle Au, v\rangle_\alpha| \leq \|\nabla u\|_{L^2}\|\nabla v\|_{L^2} \leq \|u\|_{1,\alpha}\|v\|_{1,\alpha}$$

and

$$\langle(1-A)u, u\rangle_\alpha = \|u\|_{L^2}^2 + \|\nabla u\|_{L^2}^2 = \|u\|_{1/\alpha}^2.$$

Hence, by the Lax-Milgram Theorem, we conclude that

$$\mathcal{R}(1-A) = (H_\alpha^1(\Omega))^* \supset L_{1/\alpha}^2(\Omega),$$

as claimed.

Since A is symmetric and m-dissipative, it is self adjoint in $L_{1/\alpha}^2(\Omega)$.

Of course, Theorem 7.1 implies that A generates an analytic semigroup on $L_{1/\alpha}^2(\Omega)$.

A more precise description of $\mathcal{D}(A)$ can be obtained if $\alpha(x)$ has a weak singularity of the type we have considered in Example 3.8.

THEOREM 7.2. Let $\alpha(x)$ satisfy (7.1) and

$$(7.3) \qquad \int_\Omega \alpha(x)^{-q}dx < \infty, \quad q = \begin{cases} \frac{n}{2}, & \text{if } n > 2, \\ \text{arbitrary in } (1,\infty), & \text{if } n = 2, \\ 1, & \text{if } n = 1. \end{cases}$$

Then,

$$\mathcal{D}(A) = \{u \in H_\alpha^1(\Omega); \sqrt{\alpha}\Delta u \in L^2(\Omega)\}$$
$$\subset \{u \in H_0^1(\Omega); \Delta u \in L^1(\Omega)\} \subset W_0^{1,p^*}(\Omega),$$

with $p^* \in (1, \frac{n}{n-1})$.

PROOF. To begin with, we observe that $H_0^1(\Omega) \subset L_{1/\alpha}^2(\Omega)$. Indeed, we obviously have

$$\int_\Omega \alpha^{-1}|u|^2dx \leq \left(\int_\Omega |u|^p dx\right)^{\frac{2}{p}}\left(\int_\Omega \alpha^{-q}dx\right)^{\frac{1}{q}}, \quad u \in L^p(\Omega),$$

when $2 < p < \infty$, $q = \frac{p}{p-2}$.

If p varies in $(2,\infty)$, then q describes all of $(1,\infty)$. Moreover, given $q \in (1,\infty)$, then $p = \frac{2q}{q-1}$. Therefore, if $u \in L^{\frac{2q}{q-1}}(\Omega)$, $q \in (1,\infty)$ and (7.3) holds, then

$$\int_\Omega \alpha(x)^{-1}|u(x)|^2dx \leq \left(\int_\Omega |u(x)|^{\frac{2q}{q-1}}dx\right)^{\frac{q-1}{q}}\left(\int_\Omega \alpha(x)^{-q}dx\right)^{\frac{1}{q}}.$$

Since we have assumed that Ω is regular, it has the cone property and hence the Sobolev imbedding theorem says, in particular, that

(i) $\qquad\qquad H^1(\Omega) \subset L^p(\Omega), \ 2 \le p \le \dfrac{2n}{n-2}, \quad \text{if } n > 2,$

(ii) $\qquad\qquad H^1(\Omega) \subset L^p(\Omega), \ 2 < p < \infty, \quad \text{if } n = 2,$

(iii) $\qquad\qquad H^1(\Omega) \subset \mathcal{C}(\overline{\Omega}), \quad \text{if } n = 1.$

In the case (i), $H^1(\Omega) \subset L^{\frac{2q}{q-1}}(\Omega)$ provided that $2 \le \frac{2q}{q-1} \le \frac{2n}{n-2}$, and this means $q \ge \frac{n}{2}$. In the case (ii), $H^1(\Omega) \subset L^{\frac{2q}{q-1}}(\Omega)$ provided that $\frac{2q}{q-1} > 2$, that is, q is arbitrary in $(1,\infty)$. Finally in the case (iii), it is however more suitable, instead of using (iii), to observe that

$$\int_\Omega \alpha(x)^{-1}|u(x)|^2 dx \le (\sup_\Omega |u(x)|)^2 \int_\Omega \alpha^{-1} dx < \infty.$$

To conclude the proof, we only need to notice that $\mathcal{D}(A) = \{u \in H^1_\alpha(\Omega);\ \Delta u \in L^2_\alpha(\Omega)\}$ and

$$\|\Delta u\|_{L^1} \le \|\sqrt{\alpha}\Delta u\|_{L^2}\|\alpha^{-1}\|_{L^1}^{\frac{1}{2}}, \quad u \in \mathcal{D}(A).$$

This implies (see H. Brezis and W. Strauss [37]) that $\mathcal{D}(A) \subset W_0^{1,p^*}(\Omega)$, where $p^* < \frac{n}{n-1}$.

The next two theorems exploit Theorem 7.2 to get that operators corresponding to A acting in some spaces of continuous functions on $[0,1]$ do generate infinitely differentiable semigroups. Of course, since we apply Theorem 7.2, this forces us to assume that $\alpha(x)$ is only weakly degenerate, according to (7.3) in the case $n = 1$. Really, we shall assume a bit more than (7.3), precisely

(7.4) $\quad \alpha(x)$ satisfies (7.1) with $\Omega = (0,1)$ and $\int_0^1 \alpha(x)^{-s} dx < \infty$ with some $s > 1$.

Condition (7.4) is equivalent to assume that $\alpha(x)$ satisfies

(7.5) $\qquad\qquad \displaystyle\int_0^1 \alpha(x)^{-\frac{p}{2-p}} dx < \infty,$

where p $(= \frac{2s}{s+1}) \in (1,2)$. Here and henceforth $\mathcal{C}_0([0,1])$ shall denote the Banach space

$$\mathcal{C}_0([0,1]) = \{u \in \mathcal{C}([0,1]);\ u(0) = u(1) = 0\},$$

endowed with the supremum norm.

Let us introduce the operator $Bu = \alpha u''$ with domain $\mathcal{D}(B)$ given by

$$\mathcal{D}(B) = \{u \in C^1([0,1]) \cap C^2((0,1)); \ u(0) = u(1) = 0,$$
$$\alpha(x)u''(x) \to 0 \text{ as } x \to 0+ \text{ and } 1-\}.$$

Assumption (7.4), implying summability of $\alpha(x)^{-1}$ on $(0,1)$, yields that $\mathcal{D}(B) \subset \mathcal{D}(A)$, where $\mathcal{D}(A)$ is described in Theorem 7.2 with $n = 1$. Indeed,

$$\int_0^1 \alpha(x)|u''(x)|^2 dx \le \int_0^1 \alpha(x)^{-1} dx \|\alpha u''\|_C^2.$$

On the other hand, by (7.4) or (7.5) if we take $q = \frac{2}{2-p}$, it is seen that for all $u \in \mathcal{D}(A)$

$$\int_0^1 |u''(x)|^p dx = \int_0^1 \alpha(x)^{-\frac{p}{2}} (\alpha(x)|u''(x)|^2)^{\frac{p}{2}} dx$$
$$\le \left(\int_0^1 \alpha(x)^{-\frac{pq}{2}} dx \right)^{\frac{1}{q}} \left(\int_0^1 \alpha(x)|u''(x)|^2 dx \right)^{\frac{p}{2}}$$
$$\le C \left(\int_0^1 \alpha(x)|u''(x)|^2 dx \right)^{\frac{p}{2}},$$

and therefore

$$\|u''\|_{L^p} \le C \left(\int_0^1 \alpha(x)|u''(x)|^2 dx \right)^{\frac{1}{2}}$$

with a suitable constant C.

Recalling Theorem 7.2, this estimate says that for all $f \in L^2_{1/\alpha}(0,1)$ and any $\lambda \in \mathbb{C}$ in the sector Σ': $\Re \lambda + |\Im \lambda| \ge \varepsilon_0 > 0$, the solution u to $(\lambda - A)u = f$ has in fact the regularity $u \in W^{2,p}(0,1)$, when $p = \frac{2s}{s+1} \in (1,2)$.

The Sobolev imbedding theorem (notice that here again we use the particular dimension $n = 1$) implies that $u \in C^1([0,1])$. Now we multiply both the members of $(\lambda - A)u = f$ by $-\overline{u}''(x)$ and integrate the obtained identity on $(0,1)$, so that

$$\lambda \int_0^1 |u'(x)|^2 dx + \int_0^1 \alpha(x)|u''(x)|^2 dx = - \int_0^1 f(x)\overline{u}''(x) dx.$$

Taking the real part and the imaginary part, we deduce

$$\Re \lambda \|u'\|_{L^2}^2 + \|\sqrt{\alpha} u''\|_{L^2}^2 \le \|f\|_{L^2_{1/\alpha}} \|\sqrt{\alpha} u''\|_{L^2},$$
$$|\Im \lambda| \int_0^1 |u'(x)|^2 dx = \left| \Im \int_0^1 f(x)\overline{u}''(x) dx \right|,$$

and hence

$$(\Re\lambda + |\Im\lambda|)\|u'\|_{L^2}^2 + \|\sqrt{\alpha}u''\|_{L^2}^2 \le 2\|f\|_{L^2_{1/\alpha}}\|\sqrt{\alpha}u''\|_{L^2}.$$

Therefore, since $\lambda \in \Sigma'$,

$$(|\Re\lambda| + |\Im\lambda|)\|u'\|_{L^2}^2 + \|\sqrt{\alpha}u''\|_{L^2}^2 \le C\|f\|_{L^2_{1/\alpha}}^2.$$

It follows that there exists $K > 0$ such that

$$K(1 + |\lambda|)\|u'\|_{L^2}^2 \le \|f\|_{L^2_{1/\alpha}}^2.$$

The Poincaré inequality combined with the Sobolev imbedding theorem then allows to conclude that

(7.6) $$C(1 + |\lambda|)\|u\|_{\mathcal{C}}^2 \le \|f\|_{L^2_{1/\alpha}}^2, \quad f \in L^2_{1/\alpha}(0,1).$$

Since, according to (7.4), $\int_0^1 \alpha(x)^{-1}|f(x)|^2 dx \le m\|f\|_{\mathcal{C}}^2$, (7.6) implies that for all $f \in \mathcal{C}_0([0,1])$ we have:

$$C_1(1 + |\lambda|)\|u\|_{\mathcal{C}_0}^2 \le \|f\|_{\mathcal{C}_0}^2.$$

The final step is to show that the solution u to $(\lambda - A)u = f$, guaranteed by Theorem 7.2 for all $f \in L^2_{1/\alpha}(0,1)$, coincides with the solution to $(\lambda - B)u = f$ when $f \in \mathcal{C}_0([0,1])$. But this is obvious, since we have already seen that the solution u to the former equation belongs to $\mathcal{C}_0([0,1])$ and all what remains to do is to observe that $u \in \mathcal{C}^1([0,1])$, and this has been previously obtained, too.

We notice however that the last property $u \in \mathcal{C}^1([0,1])$ could be deduced in a direct way observing that, if $u \in \mathcal{D}(A)$ and $0 < x_1 < x_2 < 1$, then $u'(x_2) - u'(x_1) = \int_{x_1}^{x_2} \alpha(x)^{-\frac{1}{2}}\alpha(x)^{\frac{1}{2}}u''(x)dx$ yields

$$|u'(x_2) - u'(x_1)| \le C \left(\int_{x_1}^{x_2} \alpha(x)^{-s}dx\right)^{\frac{1}{2}} \left(\int_0^1 \alpha(x)|u''(x)|^2 dx\right)^{\frac{1}{2}}.$$

Assumption (7.4) allows to conclude as desired.

Summarizing, we have proven that the operator B introduced previously satisfies the resolvent estimates

$$\|(\lambda - B)^{-1}f\|_{\mathcal{C}_0} \le \frac{K\|f\|_{\mathcal{C}_0}}{(1 + |\lambda|)^{\frac{1}{2}}}, \quad f \in \mathcal{C}_0([0,1]), \; \lambda \in \Sigma', \; |\lambda| \text{ large}.$$

Taking into account Theorem 3.1, we establish the following statement.

THEOREM 7.3. *Let $\alpha(x)$ satisfy (7.1) and (7.4). Then, the operator B generates an infinitely differentiable semigroup e^{tB}, $t > 0$, in the space $X = \mathcal{C}_0([0,1])$ and $\|e^{tB}\|_{\mathcal{L}(X)} \leq Kt^{-\frac{1}{2}}e^{\delta t}$, $t > 0$, where K, δ are suitable positive numbers.*

Next goal is to show an analogous generation result for the so called Ventcel's boundary conditions, see Ph. Clément and C. A. Timmermans [53], precisely for the operator $V = \alpha(x)u''$ with domain $\mathcal{D}(V)$ described by

$$\mathcal{D}(V) = \{u \in \mathcal{C}([0,1]) \cap \mathcal{C}^2(0,1); \; \alpha(x)u''(x) \to 0 \text{ as } x \to 0+ \text{ and } 1-\}.$$

THEOREM 7.4. *Let us assume (7.1) and (7.4). Then V generates an infinitely differentiable semigroup in $X = \mathcal{C}([0,1])$ with*

$$\|(\lambda - V)^{-1}\|_{\mathcal{L}(X)} \leq \frac{K}{(1 + |\lambda|)^{\frac{1}{2}}}, \quad \lambda \in \Sigma', \; |\lambda| \text{ large.}$$

PROOF. Given $g \in \mathcal{C}([0,1])$, let us introduce

$$h(x) = g(x) - (1 - x)g(0) - xg(1), \quad x \in [0,1],$$

so that $h \in \mathcal{C}_0([0,1])$. Hence, in virtue of Theorem 7.3, for all $\lambda \in \Sigma'$, $|\lambda|$ large, the equation $\lambda v - \alpha v'' = h$ possesses a unique solution $v \in \mathcal{D}(B)$ with

$$\|v\|_{\mathcal{C}_0} \leq K(1 + |\lambda|)^{-\frac{1}{2}}\|h\|_{\mathcal{C}_0}.$$

It then follows that

$$\lambda(v + \tfrac{(1-x)g(0)+xg(1)}{\lambda}) - \alpha(v + \tfrac{(1-x)g(0)+xg(1)}{\lambda})'' = g(x).$$

Therefore, if $u = v + \lambda^{-1}((1 - x)g(0) + xg(1))$, then clearly $u \in \mathcal{D}(V)$ and

$$\|u\|_X \leq C(1 + |\lambda|)^{-\frac{1}{2}}\|h\|_{\mathcal{C}_0} + C|\lambda|^{-1}\|g\|_X \leq C(1 + |\lambda|)^{-\frac{1}{2}}\|g\|_X.$$

The proof is complete, since it is known (by Ph. Clément and C. A. Timmermans [53]) that V is densely defined.

EXAMPLE 7.1. The function $\alpha(x) = x^m(1-x)^s$, $x \in [0,1]$, satisfies (7.4) if and only if $0 < m, s < 1$.

Notice that such a function is not differentiable on $[0,1]$. Moreover, Theorem 7.4 does not apply to $\alpha(x) = x(1-x)$, that is the most interesting case in applications to probability theory. As we recalled in the introduction it is known by Ph. Clément and C. A. Timmermans [53] that if $\alpha, \beta \in C([0,1])$ are real valued and (7.1) holds, then the operator $(W, \mathcal{D}(W))$ with

$$
\begin{cases}
\mathcal{D}(W) = \{u \in C([0,1]) \cap C^2(0,1); \ \alpha u'' + \beta u \to 0 \\
\qquad\qquad\qquad\qquad\qquad\qquad \text{as } x \to 0+ \text{ and } x \to 1-\}, \\
Wu = \alpha(x)u'' + \beta(x)u'
\end{cases}
$$

generates a C_0-semigroup in $C([0,1])$ if and only if $w(x) = \exp\{-\int_{\frac{1}{2}}^{x} \frac{\beta(s)}{\alpha(s)} ds\}$ satisfies $w \in L^1(0,\frac{1}{2})$ or $\int_0^{\frac{1}{2}} w(x) \int_0^{x} \alpha(t)^{-1} w(t)^{-1} dt dx = \infty$ or both and $w \in L^1(\frac{1}{2}, 1)$ or $\int_{\frac{1}{2}}^{1} w(x) \int_x^1 \alpha(t)^{-1} w(t)^{-1} dt dx = \infty$ or both.

Notice that in fact the operator V in Theorem 7.4 does generate a C_0-semigroup in $C([0,1])$. In what follows we shall investigate the regularity of the semigroup generated by W both the cases where $\alpha(x)$ has simple zeros at $x = 0$ and $x = 1$ or $\alpha(x)$ has higher order zeros.

Let us begin our discussion with the more important (and difficult) situation of simple zeros. To this end we shall use a method inspired by G. Fichera [103] and to avoid non essential complications, here we shall take

$$\alpha(x) \equiv x(1-x), \qquad \beta(x) \equiv 0.$$

In that paper the operator $x(1-x)u''(x)$ with Ventcel's boundary conditions is studied in various function spaces, including $C([0,1])$, reducing these boundary conditions into Dirichlet boundary conditions. More precisely, Fichera observes that if $v = v(x,t)$ solves the Cauchy-Dirichlet problem

(7.7)
$$
\begin{cases}
\dfrac{\partial v}{\partial t} = x(1-x)\dfrac{\partial^2 v}{\partial x^2}, & (x,t) \in (0,1) \times (0,\infty), \\
v(0,t) = v(1,t) = 0, & t \in (0,\infty), \\
v(x,0) = f(x) - (1-x)f(0) - xf(1) = h(x), & x \in (0,1),
\end{cases}
$$

with $f \in C([0,1])$, then $u(x,t) = v(x,t) + (1-x)f(0) + xf(1)$ solves just

$$
\begin{cases}
\dfrac{\partial u}{\partial t} = x(1-x)\dfrac{\partial^2 u}{\partial x^2}, & (x,t) \in (0,1) \times (0,\infty), \\
x(1-x)\dfrac{\partial^2 u}{\partial x^2} \to 0, & \text{as } x \to 0+ \text{ and } x \to 1-, \\
u(x,0) = f(x), & x \in (0,1).
\end{cases}
$$

On the other hand, he shows that the solution to (7.7) is necessarily given by

$$v(x,t) = \sum_{k=1}^{\infty} \varphi_k(x) e^{-\lambda_k t} \int_0^1 \frac{h(s)\varphi_k(s)}{s(1-s)} ds,$$

where $\varphi_k(x) = xF(k+1, -k, 2; x)$, F being the hypergeometric function, and $\lambda_k = k(k+1)$, so that λ_k and $\varphi_k(x)$ are the only eigenvalues and the only eigenfunctions of the problem

$$x(1-x)w''(x) + \lambda w(x) = 0, \quad x \in (0,1),$$
$$w(0) = w(1) = 0.$$

Solvability of (7.7) both in $H_0^1(0,1)$ and in $C_0([0,1])$ is proven in a direct way. The choice of $H_0^1(0,1)$ is motivated from demanding a value of $x(1-x)u''$ on the boundary. Notice that if one considers the non degenerate operator $Cu = u''$ with domain $H^2(0,1) \cap H_0^1(0,1)$ in the space $L^2(0,1)$, then $\mathcal{D}((-C)^{\frac{1}{2}}) = H_0^1(0,1)$ and the part C_1 of C in $H_0^1(0,1)$, i.e.

$$\begin{cases} \mathcal{D}(C_1) = \{u \in H^2(0,1) \cap H_0^1(0,1); \ u'' \in H_0^1(0,1)\}, \\ C_1 u = Cu, \end{cases}$$

generates an analytic semigroup in $H_0^1(0,1)(= \mathcal{D}((-C)^{\frac{1}{2}}))$. Hence, $\mathcal{D}(C_1) = \{u \in H^3(0,1) \cap H_0^1(0,1); \ u''(0) = u''(1) = 0\}$. But then also the operator C_2 defined by

$$\begin{cases} \mathcal{D}(C_2) = \{u \in H^3(0,1); \ u''(0) = u''(1) = 0\}, \\ C_2 u = Cu = u'', \end{cases}$$

generates an analytic semigroup in $H^1(0,1)$. Indeed,

$$\lambda u - u'' = f \in H^1(0,1), \quad u \in \mathcal{D}(C_2), \quad \lambda \in \Sigma',$$

is written $\lambda v - v'' = g \in H_0^1(0,1)$, where $v = u - \frac{(1-x)f(0)+xf(1)}{\lambda} \in \mathcal{D}(C_2)$ and $g(x) = f(x) - (1-x)f(0) - xf(1)$. Therefore, the analiticity property follows easily from the corresponding property of C_1.

Of course, in the degenerate case, care must be taken to define the operator analogous to C_1.

To accomplish this we introduce the operator K in the space $H_0^1(0,1)$ by means of

$$\begin{cases} \mathcal{D}(K) = \{u \in H_0^1(0,1); \ u'' \text{ exists (in the sense of distributuions)} \\ \qquad\qquad\qquad \text{with } x(1-x)u'' \in H_0^1(0,1)\}, \\ Ku = x(1-x)u''. \end{cases}$$

We have

THEOREM 7.5. $(K, \mathcal{D}(K))$ *(or shortly, K) generates an analytic semi-group in $H_0^1(0,1)$.*

PROOF. First of all, we consider $H_0^1(0,1)$ endowed with the inner product

$$\langle u, v \rangle = \int_0^1 u'(x)\overline{v'(x)}dx, \quad u, v \in H_0^1(0,1),$$

that is equivalent to the usual one $\langle u, v \rangle + \int_0^1 u(x)\overline{v(x)}dx$.

Let $\lambda u - \alpha(x)u'' = f \in H_0^1(0,1)$, where $\alpha(x) = x(1-x)$, $\lambda \in \mathbb{C}$, $\Re\lambda > 0$, $u \in \mathcal{D}(K)$. We observe that, if $u \in \mathcal{D}(K)$, then

$$\int_0^1 |u''(x)\overline{u(x)}|dx = \int_0^1 \sqrt{x(1-x)}|u''(x)|\frac{|u(x)|}{\sqrt{x(1-x)}}dx$$

implies that $u''\overline{u}$ is summmmable on $(0,1)$, since Schwarz inequality gives

$$\int_0^1 \frac{|u(x)|^2}{x(1-x)}dx = \int_0^{\frac{1}{2}} \frac{1}{x(1-x)}\left|\int_0^x u'(y)dy\right|^2 dx$$
$$+ \int_{\frac{1}{2}}^1 \frac{1}{x(1-x)}\left|\int_x^1 u'(y)dy\right|^2 dx$$
$$\leq 2\int_0^{\frac{1}{2}}\int_0^x |u'(y)|^2 dy dx + 2\int_{\frac{1}{2}}^1\int_x^1 |u'(y)|^2 dy dx$$
$$\leq \int_0^{\frac{1}{2}} |u'(y)|^2 dy + \int_{\frac{1}{2}}^1 |u'(y)|^2 dy = \|u'\|_{L^2}^2 \leq \|u\|_{H_0^1}^2,$$

and

$$\int_0^1 |\sqrt{x(1-x)}u''(x)|^2 dx = \int_0^1 x(1-x)|u''(x)|^2 dx$$
$$= \int_0^1 \frac{|x(1-x)u''(x)|^2}{x(1-x)}dx \leq C\|x(1-x)u''\|_{H_0^1}^2.$$

Then, by letting $a \to 0+$, $b \to 1-$ into the identity

$$\int_a^b u''(x)\overline{u(x)}dx = [u'(x)\overline{u(x)}]_{x=a}^{x=b} - \int_a^b |u'(x)|^2 dx$$

we recognize that there exist the limits

$$\lim_{x\to0+} u'(x)\overline{u(x)} \quad \text{and} \quad \lim_{x\to1-} u'(x)\overline{u(x)}$$

for all $u \in \mathcal{D}(K)$ and they vanish. Indeed, since $x(1-x)u'' = g \in H_0^1(0,1)$, $g(x) = \int_0^x g'(t)dt$; so that

$$|g(x)| \leq \sqrt{x}\|g\|_{H_0^1}.$$

Analogously, since $g(x) = -\int_x^1 g'(t)dt$, we deduce that

$$|g(x)| \leq \sqrt{1-x}\|g\|_{H_0^1}.$$

Hence, from $u'(x_1) - u'(x_2) = \int_{x_1}^{x_2} u''(t)dt = \int_{x_1}^{x_2} \frac{g(t)}{t(1-t)}dt$, we infer that, for $0 < x_2 < x_1 \leq \frac{1}{2}$,

$$|u'(x_1) - u'(x_2)| \leq 2\int_{x_2}^{x_1} t^{-1}|g(t)|dt \leq 4\|g\|_{H_0^1}|\sqrt{x_1} - \sqrt{x_2}|,$$

and, for $\frac{1}{2} \leq x_2 < x_1 \leq 1$,

$$|u'(x_1) - u'(x_2)| \leq 4\|g\|_{H_0^1}|\sqrt{1-x_1} - \sqrt{1-x_2}|.$$

Therefore, the function u' admits the limits

$$\lim_{x \to 0+} u'(x) \quad \text{and} \quad \lim_{x \to 1-} u'(x).$$

It follows that $[u'\bar{u}]_0^1 = 0$. Taking real and imaginary parts in $\int_0^1\{\frac{\lambda u \bar{u}}{\alpha} - u''\bar{u}\}dx = \int_0^1 \frac{f\bar{u}}{\alpha}dx$, we deduce that

$$\Re\mathrm{e}\lambda \int_0^1 \frac{|u(x)|^2}{x(1-x)}dx + \|u'\|_{L^2}^2 = \Re\mathrm{e}\int_0^1 \frac{f(x)\overline{u(x)}}{x(1-x)}dx,$$

$$|\Im\mathrm{m}\lambda|\int_0^1 \frac{|u(x)|^2}{x(1-x)}dx = \left|\Im\mathrm{m}\int_0^1 \frac{f(x)\overline{u(x)}}{x(1-x)}dx\right|,$$

and this yields

$$(\Re\mathrm{e}\lambda + |\Im\mathrm{m}\lambda|)\int_0^1 \frac{|u(x)|^2}{x(1-x)}dx + \|u'\|_{L^2}^2 \leq C\|f\|_{H_0^1}\left(\int_0^1 \frac{|u(x)|^2}{x(1-x)}dx\right)^{\frac{1}{2}}.$$

Therefore, if $\Re\mathrm{e}\lambda + |\Im\mathrm{m}\lambda| \geq \varepsilon_0 > 0$, the estimate

$$\|u\|_{H_0^1} \leq C\|f\|_{H_0^1}$$

is obtained for some constant C. On the other hand, multiplying $\lambda u - \alpha u'' = f$ by $-u''$ and integrating on $(0,1)$, we obtain again after the trick to take real parts and imaginary parts of the corresponding equality

$$(\Re\lambda + |\Im\lambda|) \int_0^1 |u'(x)|^2 dx + \int_0^1 \alpha(x)|u''(x)|^2 dx$$

$$= -\Re \int_0^1 f(x)\overline{u}''(x)dx + \left| \Im \int_0^1 f(x)\overline{u}''(x)dx \right|.$$

We observe that the integral $\int_0^1 f(x)\overline{u}''(x)dx$ is well defined, because

$$(7.8) \quad \int_0^1 f(x)\overline{u}''(x)dx \leq \left(\int_0^1 \frac{|f(x)|^2}{x(1-x)} dx \right)^{\frac{1}{2}} \left(\int_0^1 x(1-x)|u''(x)|^2 dx \right)^{\frac{1}{2}}$$

$$\leq C\|f\|_{H_0^1} \|x(1-x)u''\|_{H_0^1}.$$

Moreover $\int_0^1 u(x)\overline{u}''(x)dx = -\int_0^1 |u'(x)|^2 dx$ for all $u \in \mathcal{D}(K)$, since in view of the preceding remark, for every $f \in H_0^1(0,1)$ and all $u \in \mathcal{D}(K)$,

$$\int_0^1 f\overline{u}'' dx = [f\overline{u}']_{x=0}^{x=1} - \int_0^1 f'\overline{u}' dx = - \int_0^1 f'\overline{u}' dx.$$

Hence, if $\lambda \in \Sigma'$: $\Re\lambda + |\Im\lambda| \geq \varepsilon_0 > 0$, the estimate

$$(7.9) \qquad \left(\int_0^1 x(1-x)|u''(x)|^2 dx \right)^{\frac{1}{2}} \leq C\|f\|_{H_0^1}$$

is proved.

Consequently (7.8) and (7.9) yield

$$|\lambda| \|u\|_{H_0^1}^2 \leq C\|f\|_{H_0^1}^2,$$

that is $\|u\|_{H_0^1} \leq C(1 + |\lambda|)^{-\frac{1}{2}} \|(\lambda - K)u\|_{H_0^1}$, $\lambda \in \Sigma$. In fact the stronger estimate

$$\|u\|_{H_0^1} \leq C(1 + |\lambda|)^{-1} \|(\lambda - K)u\|_{H_0^1}$$

holds, in view of the fact that for all $u \in \mathcal{D}(K)$ and $f \in H_0^1(0,1)$ one has $\int_0^1 f(x)\overline{u}''(x)dx = -\int_0^1 f'(x)\overline{u}'(x)dx$. Hence, by (7.8) and (7.9),

$$|\lambda| \|u\|_{H_0^1} \leq C\|f\|_{H_0^1}$$

as desired.

We continue the proof observing that K is symmetric, for

$$\langle Ku, v\rangle_{H_0^1} = \int_0^1 \frac{d}{dx}(\alpha(x)u''(x))\overline{v}'(x)dx = -\int_0^1 \alpha(x)u''(x)\overline{v}''(x)dx$$

$$= -[u'(x)\alpha(x)\overline{v''(x)}]_{x=0}^{x=1} + \int_0^1 u'(x)\frac{d}{dx}(\alpha(x)\overline{v''(x)})dx = \langle u, Kv\rangle_{H_0^1}$$

always in view of properties of $u \in \mathcal{D}(K)$. Notice that any element $u \in \mathcal{D}(K)$ necessarily satisfies $\int_0^1 \alpha(x)|u''(x)|^2 dx < \infty$. Furthermore, if Y denotes the completion of $C_0^\infty(0,1)$ with respect to

$$\|u\|_Y = \left\{ \|u\|_{H^1}^2 + \int_0^1 x(1-x)|u''(x)|^2 dx \right\}^{\frac{1}{2}},$$

then, for all $u \in Y$,

$$\langle (1-K)u, u\rangle_{H_0^1} = \int_0^1 |u'(x)|^2 dx - \int_0^1 \frac{d}{dx}(x(1-x)u''(x))\overline{u}'(x)dx = \|u\|_Y^2,$$

and hence $\mathcal{R}(1-K) \supset H_0^1(0,1)$ because $\|u\|_Y^2 = a(u,u)$, where

$$a(u,v) = \int_0^1 u'(x)\overline{v'(x)}dx + \int_0^1 x(1-x)u''(x)\overline{v''(x)}dx$$

is a sesquilinear form which is continuous on $Y \times Y$ and coercive.

It only remains to recall that the operator $\tilde{B}$ associated to $a(\cdot, \cdot)$ (see [167: Theorems 2.22 and 2.23]) is an isomorphism from Y to its dual Y' and the part B of $\tilde{B}$ in $H_0^1(0,1)$ is positive definite and selfadjoint. But $\mathcal{D}(B) = \{u \in Y; \tilde{B}u \in H_0^1(0,1)\}$, so that B is precisely $I - K$ and it is onto $H_0^1(0,1)$.

This concludes the proof.

In order to describe the result corresponding to Theorem 7.4 for Ventcel's boundary conditions, we introduce the operator K_1 by

$$\begin{cases} \mathcal{D}(K_1) = \{u \in H^1(0,1); \ u'' \text{ exists in the sense of distributions and} \\ \qquad\qquad\qquad x(1-x)u'' \in H_0^1(0,1)\} \\ K_1 u = x(1-x)u''. \end{cases}$$

Then we prove the following generation theorem.

THEOREM 7.6. *The operator K_1 generates an analytic semigroup in $H^1(0,1)$.*

PROOF. Let $f \in H^1(0,1)$; so that, if $f(x) - (1-x)f(0) - xf(1) = h(x)$ then $h \in H_0^1(0,1)$. Theorem 7.5 implies that, for all $\lambda \in \Sigma'$ there exists a unique $u \in \mathcal{D}(K)$ such that

(7.10) $$\lambda u - x(1-x)u'' = h.$$

Therefore, $u \in H_0^1(0,1)$, $x(1-x)u'' \in H_0^1(0,1)$. Now (7.10) is written

$$\lambda(u + \tfrac{1-x}{\lambda}f(0) + \tfrac{x}{\lambda}f(1)) - x(1-x)(u + \tfrac{1-x}{\lambda}f(0) + \tfrac{x}{\lambda}f(1))'' = f(x)$$

and thus $v(x) = u(x) + \tfrac{1-x}{\lambda}f(0) + \tfrac{x}{\lambda}f(1) \in H^1(0,1)$ and solves $\lambda v - x(1-x)v'' = f$ with Ventcel's boundary conditions. Moreover,

$$\|v\|_{H^1} \leq \|u\|_{H^1} + \|\tfrac{1-x}{\lambda}f(0) + \tfrac{x}{\lambda}f(1)\|_{H^1}$$
$$\leq C(1+|\lambda|)^{-1}\|f\|_{H^1} + |\lambda|^{-1}\|f\|_{H^1} \leq C(1+|\lambda|)^{-1}\|f\|_{H^1}.$$

This completes the proof.

As a by-product, we derive the following regularity result for a related degenerate differential operator with Neumann boundary conditions; see the remark after Example 6.22. Take again $\alpha(x) = x(1-x)$ and consider the operator $(K_2, \mathcal{D}(K_2))$ given by

$$\begin{cases} \mathcal{D}(K_2) = \{u \in L^2(0,1); \ u \text{ is locally absolutely continuous in } (0,1) \\ \qquad\qquad \text{and } \alpha u' \in H_0^1(0,1)\}, \\ K_2 u = \dfrac{d}{dx}\left(\alpha(x)\dfrac{du}{dx}\right). \end{cases}$$

Then K_2 is a closed densely defined operator on $L^2(0,1)$. We prove

THEOREM 7.7. *The operator K_2 generates an analytic semigroup in $L^2(0,1)$.*

PROOF. First of all, if $\lambda v - (\alpha v')' = 0$, where $v \in \mathcal{D}(K_2)$, then $w = \alpha v' \in H_0^1(0,1)$ satisfies $\lambda w - \alpha w'' = 0$, so that Theorem 7.5 applies, implying that $w = \alpha v' = v = 0$ provided that $\Re\lambda$ is sufficiently large. It follows that if the solution v to
$$\lambda v - K_2 v = g \in L^2(0,1)$$
exists, it is unique. Let $f(x) = \int_0^x g(t)dt$, so that $f \in H^1(0,1)$ and $f+c = h$, c being an arbitrary constant, furnishes all the solutions $h \in H^1(0,1)$ to $h' = g$.

From Theorem 7.6 we deduce that the solution u to

$$\lambda u(x) - \alpha(x)u''(x) = h(x), \quad 0 < x < 1, \, \Re\lambda > k,$$

with $\alpha u'' \in H_0^1(0,1)$, coincides with $u = u_1 + \frac{c}{\lambda}$, where u_1 satisfies

$$\lambda u_1(x) - \alpha(x)u_1''(x) = f(x), \quad 0 < x < 1,$$

and $\alpha u_1'' \in H_0^1(0,1)$. One also has

$$\|u_1\|_{H^1} \leq C|\lambda|^{-1}\|f\|_{H^1} \leq C_1|\lambda|^{-1}\|g\|_{L^2}.$$

It follows that $v = u' = u_1' \in L^2(0,1)$ solves

$$\lambda v(x) - (\alpha(x)v'(x))' = g(x),$$

$\alpha v' \in H_0^1(0,1)$ and

$$\|v\|_{L^2} = \|u_1'\|_{L^2} = \|u'\|_{L^2} \leq \frac{C}{|\lambda|}\|g\|_{L^2},$$

thus concluding the proof.

The technique of passing from Ventcel's boundary conditions for $\alpha u''$ in $H^1(0,1)$ to generalized Neumann boundary conditions for $(\alpha u')'$ in $L^2(0,1)$ could be extended to $L^p(0,1)$, $1 < p < \infty$. In the sequel we shall show that the inverse process works too.

Results analogous to ours for Neumann boundary conditions were recently established in Theorem 2.9 of M. Campiti, G. Metafune and D. Pallara [46]; see also our Remark following Example 6.22.

Let $1 < p < \infty$. Then the operator A_p is defined by

$$\begin{cases} \mathcal{D}(A_p) = \{u \in W_{loc}^{1,p}(0,1) \cap L^p(0,1); \, \alpha u' \in W_0^{1,p}(0,1)\}, \\ A_p u = (\alpha u')'. \end{cases}$$

We have

THEOREM 7.8. *If $\alpha \in C^1([0,1])$, $\alpha(0) = \alpha(1) = 0$, $\alpha(x) > 0$ for $0 < x < 1$, then A_p generates a C_0-analytic semigroup on $L^p(0,1)$, $1 < p < \infty$.*

PROOF. Let $\Re\lambda > 0$ and consider $\lambda u - A_p u = f \in L^p(0,1)$. Then for all $0 < a < b < 1$ one has

$$(7.10) \quad \lambda \int_a^b |u(x)|^p dx - [\alpha(x)u'(x)\overline{u(x)}|u(x)|^{p-2}]_{x=a}^{x=b}$$

$$+ \int_a^b \alpha(x)|u'(x)|^2|u(x)|^{p-2}dx$$

$$+ (p-2)\int_a^b \alpha(x)u'(x)\overline{u(x)}|u(x)|^{p-4}\Re(u'(x)\overline{u(x)})dx$$

$$= \int_a^b f(x)\overline{u(x)}|u(x)|^{p-2}dx.$$

Taking real parts and imaginary parts in (7.10) we easily obtain that

$$(7.11) \quad \Re\lambda\|u\|_p^p - \Re[\alpha(x)u'(x)\overline{u(x)}|u(x)|^{p-2}]_{x=a}^{x=b}$$

$$+ \int_a^b \alpha(x)|u'(x)|^2|u(x)|^{p-2}dx$$

$$+ (p-2)\int_a^b \alpha(x)\{\Re(u'(x)\overline{u(x)})\}^2|u(x)|^{p-4}dx$$

$$= \Re\int_a^b f(x)\overline{u(x)}|u(x)|^{p-2}dx,$$

$$(7.12) \quad \Im\lambda\|u\|^p - \Im[\alpha(x)u'(x)\overline{u(x)}|u(x)|^{p-2}]_{x=a}^{x=b}$$

$$+ (p-2)\int_a^b \alpha(x)\Im(u'(x)\overline{u(x)})\Re(u'(x)\overline{u(x)})|u(x)|^{p-4}dx$$

$$= \Im\int_a^b f(x)\overline{u(x)}|u(x)|^{p-2}dx.$$

From (7.11) it follows that there exist the limits

$$\lim_{x\to 0+} \Re\alpha(x)u'(x)\overline{u(x)}|u(x)|^{p-2}, \quad \lim_{x\to 1-} \Re\alpha(x)u'(x)\overline{u(x)}|u(x)|^{p-2}.$$

Consider the first one and call it a_0. Suppose $a_0 \neq 0$. For all $u \in \mathcal{D}(A_p)$, we have

$$\alpha(x)u'(x) = \int_0^x \frac{d}{dy}(\alpha(y)u'(y))dy$$

so that

$$|\alpha(x)u'(x)\overline{u(x)}| \leq |u(x)|\int_0^x \left|\frac{d}{dy}(\alpha(y)u'(y))\right|dy \leq Cx^{\frac{1}{p'}}|u(x)|,$$

when $\frac{1}{p} + \frac{1}{p'} = 1$. Therefore,

$$|\alpha(x)\Re(u'(x)\overline{u(x)})| \leq Cx^{\frac{1}{p'}}|u(x)|.$$

Now $\alpha(x)|\Re(u'(x)\overline{u(x)})||u(x)|^{p-2} \sim |a_0|$ near 0, so that

$$|u(x)|^{p-2} \sim \frac{|a_0|}{\alpha(x)|\Re(u'(x)\overline{u(x)})|} \geq \frac{c'}{x^{\frac{1}{p'}}|u(x)|};$$

this implies that

$$|u(x)|^{p-1} \geq \frac{C'}{x^{\frac{1}{p'}}}$$

in a neighbourhood $(0, \delta)$ of 0. Hence, we must have

$$\int_0^\delta |u(x)|^p dx \geq C' \int_0^\delta \frac{dx}{x} dx = +\infty,$$

a contradiction to $u \in \mathcal{D}(A_p)$. Analogously,

$$\lim_{b \to 1-} \Re e \alpha(x) u'(x) \overline{u(x)} |u(x)|^{p-2} = 0.$$

Letting $a \to 0+$ and $b \to 1-$ in (7.11) we deduce that

$$(7.13) \quad \Re\lambda\|u\|_p^p + \int_0^1 \alpha(x)|u'(x)|^2|u(x)|^{p-2}dx$$

$$+ (p-2)\int_0^1 \alpha(x)(\Re e(u'(x)\overline{u(x)}))^2|u(x)|^{p-2}dx = \Re e \int_0^1 f(x)|u(x)|^{p-2}dx.$$

In particular,

$$(7.14) \qquad\qquad \Re\lambda\|u\|_p \leq \|(\lambda - A_p)u\|_p, \qquad \Re\lambda > 0,$$

so that A_p is dissipative for all $p \in (1, \infty)$. Notice that (7.13) is equivalently written

$$\Re\lambda\|u\|_p^p + (p-1)\int_0^1 \alpha(x)(\Re e(u'(x)\overline{u(x)}))^2|u(x)|^{p-4}dx$$

$$+ \int_0^1 \alpha(x)(\Im m(u'(x)\overline{u(x)}))^2|u(x)|^{p-4}dx = \Re e \int_0^1 f(x)|u(x)|^{p-2}dx,$$

and necessarily $\int_0^1 \alpha(x)|u'(x)|^2|u(x)|^{p-2}dx < \infty$ for all $u \in \mathcal{D}(A_p)$. From (7.12) we deduce that $\Im m[\alpha(x)u'(x)\overline{u(x)}|u(x)|^{p-2}]_{x=a}^{x=b} \to 0$ as $a \to 0+$ and $b \to 1-$ and

$$|\Im m\lambda|\|u\|_p^p \leq |p-2| \left(\int_0^1 \alpha(x)(\Re e(u'(x)\overline{u(x)}))^2|u(x)|^{p-4}dx \right)^{\frac{1}{2}}$$

$$\times \left(\int_0^1 \alpha(x)(\Im m(u'(x)\overline{u(x)}))^2|u(x)|^{p-4}dx \right)^{\frac{1}{2}}$$

$$+ \|f\|_p\|u\|_p^{p-1} \leq \left(\frac{|p-2|}{\sqrt{p-1}} + 1 \right) \|f\|_p\|u\|_p^{p-1}.$$

Hence there exists a positive constant C_p such that
(7.15)
$$C_p(1 + |\lambda|)\|u\|_p \leq \|(\lambda - A_p)u\|_p, \quad u \in \mathcal{D}(A_p), \Re\lambda > 0, 1 < p < \infty.$$

In view of (7.14) it remains to prove that $I - A_p$ has a dense range in $L^p(0,1)$. To this end, we need the following result on the operator $(A_\infty, \mathcal{D}(A_\infty))$, where

$$\begin{cases} \mathcal{D}(A_\infty) = \{u \in \mathcal{C}([0,1]) \cap \mathcal{C}^1((0,1)); \ \alpha u' \in \mathcal{C}^1([0,1])\}, \\ A_\infty u = (\alpha u')'. \end{cases}$$

See M. Campiti, G. Metafune and D. Pallara [46]. Sometimes we shall use A instead of A_∞.

LEMMA. $(A, \mathcal{D}(A_\infty))$ *generates a* $\mathcal{C}_0$-*contraction semigroup on* $\mathcal{C}([0,1])$.

PROOF. Each $u \in \mathcal{D}(A_\infty)$ has the property $\lim_{x\to 0,1} \alpha(x)u'(x) = 0$. In fact, for $0 < x_1 < x_2 < 1$,

$$\alpha(x_2)u'(x_2) - \alpha(x_1)u'(x_1) = \int_{x_1}^{x_2} \frac{d}{dy}(\alpha(y)u'(y))dy,$$

so that $\lim_{x\to 0,1} \alpha(x)u'(x)$ exist. If $\alpha(x)u'(x) \to \mu \neq 0$ as $x \to 0$, then $u'(x) \sim \frac{\mu}{\alpha(x)}$, so that $u(x) - u(\varepsilon) = \int_\varepsilon^x \frac{\mu}{\alpha(y)}(1 + \frac{o(1)}{\mu})dy$ contradicts $u \in \mathcal{C}([0,1])(0 < \varepsilon < x < 1)$, since $\frac{1}{\alpha(x)}$ is not integrable on $(0,\delta)$. Hence $\mathcal{D}(A_\infty) \subset \mathcal{D}(A_p)$, $1 < p < \infty$.

Moreover, letting $p \to \infty$ in (7.14), we deduce that

$$\Re e\lambda \|u\|_\mathcal{C} \leq \|(\lambda - A)u\|_\mathcal{C}, \quad \Re e\lambda > 0.$$

$I - A$ is onto $\mathcal{C}([0,1])$ in view of a result due to Theorem 3 of C. A. Timmermans [171; p. 411], according which $I - A$ is surjective if and only if

$$-\int_0^{\frac{1}{2}} R(x)dx = \int_{\frac{1}{2}}^1 R(x)dx = +\infty,$$

where $R(x) = (2x - 1)/2\alpha(x)$, $x \in (0,1)$.

This completes the proof of the lemma.

End of proof to Theorem 7.8.

Since $(I - A)(D_\infty(A)) = \mathcal{C}([0,1])$ is dense in $L^p(0,1)$,

$$(I - A_p)(D_p(A)) \supset (I - A)(D_\infty(A))$$

is everywhere dense in $L^p(0,1)$. Since it is easily seen that A_p is a closed operator, we deduce that A_p generates a $\mathcal{C}_0$-contraction semigroup in $L^p(0,1)$ that is analytic in virtue of (7.15).

If $1 < p < \infty$, let $B_p u = \alpha u''$, where $\alpha \in \mathcal{C}^1([0,1])$, $\alpha(0) = \alpha(1) = 0$, $\alpha(x) > 0$ on $(0,1)$ and

$$\mathcal{D}(B_p) = \{u \in W^{1,p}(0,1) \cap W^{2,p}_{loc}(0,1); B_p u \in W^{1,p}(0,1)$$
$$\text{and } \lim_{x\to 0,1} \alpha(x)u''(x) = 0\}.$$

Then we have

THEOREM 7.9. *Let $1 < p < \infty$, $\alpha \in \mathcal{C}^1([0,1])$, $\alpha > 0$ on $(0,1)$, $\alpha(0) = \alpha(1) = 0$. Then B_p generates a $\mathcal{C}_0$-analytic semigroup on $W^{1,p}(0,1)$.*

PROOF. Let $F \in W^{1,p}(0,1)$ and take $f = F' \in L^p(0,1)$. From Theorem 7.8, for every λ with $\Re\lambda$ sufficiently large there is $v \in \mathcal{D}(A_p)$ such that

(7.16) $$\lambda v - (\alpha v')' = f = F'$$

with $\|v\|_p \le C(1+|\lambda|)^{-1}\|f\|_p \le C(1+|\lambda|)^{-1}\|F\|_{W^{1,p}}$, and

$$\alpha(x)v'(x) \to 0 \quad \text{as} \quad x \to 0+, \ x \to 1-.$$

Let $u(x) = \int_0^x v(y)dy + \frac{F(0)}{\lambda}$. Then $u' = v$, so that $u \in \mathcal{D}(B_p)$ and

$$\lambda u - \alpha u'' = F, \quad 0 < x < 1.$$

Moreover,

$$\|u'\|_p = \|v\|_p \le C(1+|\lambda|)^{-1}\|F\|_{W^{1,p}},$$

and

$$\|u\|_p \le \left(\int_0^1 \left|\int_0^x v(y)dy\right|^p dx\right)^{\frac{1}{p}} + \frac{|F(0)|}{|\lambda|} \le \|v\|_p + \frac{|F(0)|}{|\lambda|}$$
$$\le C(1+|\lambda|)^{-1}\|F\|_{W^{1,p}} + |\lambda|^{-1}\|F\|_{W^{1,p}} \le C'(1+|\lambda|)^{-1}\|F\|_{W^{1,p}},$$

by the Sobolev embedding theorem.

This proves that $(\lambda - B_p)u = F$ has a (necessarily unique) solution $u \in \mathcal{D}(B_p)$ for all $F \in W^{1,p}(0,1)$ and for any λ with $\Re\lambda$ sufficiently large. The last estimate also shows that B_p generates an analytic semigroup in $W^{1,p}(0,1)$.

THEOREM 7.10. *Let B_ν be the operator in $\mathcal{C}^1([0,1])$ given by*

$$\begin{cases} \mathcal{D}(B_\nu) = \{u \in \mathcal{C}^1([0,1]) \cap \mathcal{C}^2(0,1); \ x(1-x)u'' \in \mathcal{C}^1([0,1])\}, \\ B_\nu u = x(1-x)u'' \end{cases}$$

Then B_ν generates a $\mathcal{C}_0$ (infinitely) differentiable semigroup in $\mathcal{C}^1([0,1])$.

PROOF. To begin with, we observe that in view of M. Campiti, G. Metafune and D. Pallara [46; Theorem 3.3], the operator $(A, \mathcal{D}(A_\infty))$, $Au = (x(1-x)u')'$ in the Lemma, generates a $\mathcal{C}_0$-differentiable contraction semigroup in $\mathcal{C}([0,1])$. The argument that follows is then similar to the one for Theorem 7.9.

Let $F \in \mathcal{C}^1([0,1])$, $\lambda > 0$, and consider $\lambda u - Au = F' \in \mathcal{C}([0,1])$. Then it has a unique solution $u \in \mathcal{D}(A_\infty)$ with

$$\|u\|_{\mathcal{C}([0,1])} \le \lambda^{-1}\|F'\|_{\mathcal{C}([0,1])}.$$

Integrating the preceding equation on $(0, t)$, we deduce

$$\lambda \int_0^x u(y)dy - x(1-x)u'(x) = F(x) - F(0),$$

that is

$$\lambda \left[\int_0^x u(y)dy + \frac{F(0)}{\lambda} \right] - x(1-x)\frac{d^2}{dx^2}\left[\int_0^x u(y)dy + \frac{F(0)}{\lambda} \right] = F(x).$$

If $w(x) = \int_0^x u(y)dy + \frac{F(0)}{\lambda}$, then $w \in \mathcal{C}^1([0,1])$, $x(1-x)w''(x) = x(1-x)u'(x) \to 0$ as $x \to 0, 1$ and $x(1-x)w'' \in \mathcal{C}^1([0,1])$, so that $w \in \mathcal{D}(B_\nu)$.

On the other hand, if

$$\lambda w - x(1-x)w'' = 0, \quad w \in \mathcal{D}(B_\nu),$$

then

$$\lambda w' - \frac{d}{dx}(x(1-x)w'') = 0, \quad w' \in \mathcal{C}([0,1])$$

implies that $w' \equiv 0$. Hence, $w'' \equiv 0$ and w itself $\equiv 0$. This yields uniqueness of solution. Now we need the estimate of $\|w\|_{\mathcal{C}^1([0,1])}$. To this aim we observe that the norm

$$\|w\|_1 = \max\{|w(0)|, \|w'\|_{\mathcal{C}([0,1])}\}$$

is equivalent to the usual norm

$$\|w\|_{\mathcal{C}^1} = \max\{\|w\|_{\mathcal{C}([0,1])}, \|w\|_{\mathcal{C}([0,1])}\},$$

because $w(x) = w(0) + \int_0^x w'(t)dt$ implies

$$|w(x)| \leq |w(0)| + \int_0^x |w'(t)|dt \leq |w(0)| + \|w'\|_{\mathcal{C}([0,1])} \leq 2\|w\|_1.$$

Since $w(x) = \frac{F(0)}{\lambda} + \int_0^x u(y)dy$, we have

$$\|w\|_1 = \max\{|w(0)|, \|w^{-1}\|_{\mathcal{C}([0,1])}\} = \max\{\frac{|F(0)|}{\lambda}, \|u\|_{\mathcal{C}([0,1])}\}$$

$$\leq \max\{\frac{|F(0)|}{\lambda}, \frac{\|F'\|_{\mathcal{C}([0,1])}}{\lambda}\} = \frac{1}{\lambda}\max\{|F(0)|, \|F'\|_{\mathcal{C}([0,1])}\} = \frac{1}{\lambda}\|F\|_1.$$

Then B_ν generates a $\mathcal{C}_0$-contraction semigroup in $\mathcal{C}^1([0,1])$. Since $(A, \mathcal{D}(A_\infty))$ generates a differential semigroup, we know that for all λ with

$\Re\lambda \leq 0$ and $\Re\lambda \geq a - b\log|\Im\mathrm{m}\lambda|$, where $b > 0$, (see A. Pazy [145; p. 54]), one has

$$\|u\|_{\mathcal{C}([0,1])} \leq C(1 + |\Im\mathrm{m}\lambda|)\|F'\|_{\mathcal{C}([0,1])}$$

with a suitable constant C. Since for these λ's away from 0

$$|w(0)| = \frac{1}{|\lambda|}|F(0)| \leq C(1 + |\Im\mathrm{m}\lambda|)\|F\|_1,$$

we have

$$\max\{|w(0)|, \|w'\|_{\mathcal{C}([0,1])}\} \leq C(1 + |\Im\mathrm{m}\lambda|)\|F\|_1,$$

so that the semigroup generated by $(B_\nu, \mathcal{D}(B_\nu))$ is in fact differentiable in $\mathcal{C}^1([0,1])$, according to the above mentioned criterion by A. Pazy [145].

7.2 GENERATION THEOREMS FOR $\alpha u'' + \beta u'$: A FIRST APPROACH

In this paragraph we study the operator $Au = \alpha u'' + \beta u'$ with $\beta \not\equiv 0$. It must be observed that the term $\beta u'$ is not a small perturbation of $\alpha u''$ and therefore the arguments in 7.1 cannot be applied.

In order to avoid cumbersome new notations for the operators we shall sometimes use the same capital letters as in 7.1 to denote the new operators.

Let us recall that $\mathcal{C}_0([0,1])$ denotes the Banach space of the complex valued continuous functions on $[0,1]$ vanishing at 0 and 1.

Let α, β be two given real valued functions in $\mathcal{C}([0,1])$ such that

(7.17) $\alpha(x) > 0$ for $x \in (0,1)$, $\alpha(0) = \alpha(1) = 0 = \beta(0) = \beta(1)$

and consider in the space $\mathcal{C}_0([0,1])$ the operator

(7.18) $\quad \begin{cases} A_0 u = \alpha u'' + \beta u', & u \in \mathcal{D}(A_0), \\ \mathcal{D}(A_0) = \{u \in \mathcal{C}_0([0,1]) \cap \mathcal{C}^2(0,1); \ A_0 u \in \mathcal{C}_0([0,1])\} \end{cases}$

Our first aim in what follows is to give some conditions on α and β guaranteeing that A_0 generates either an analytic $\mathcal{C}_0$-semigroup or a differentiable semigroup that is real analytic in $\mathcal{C}_0([0,1])$.

Besides accomplishing this goal, we shall also prove some generation results extending Theorems 7.1, 7.5 and 7.6 to $\beta(x) \not\equiv 0$.

In addition to (7.17) we shall assume that

(7.19) $\dfrac{\beta}{\alpha} \in L^1(0,1), \qquad \dfrac{x(1-x)}{\alpha} \in L^1(0,1).$

We define
(7.20)

$$w(x) = \alpha(x)\exp\left(-\int_0^x \frac{\beta(s)}{\alpha(s)}ds\right), \quad \eta(x) = \exp\left(\int_0^x \frac{\beta(s)}{\alpha(s)}ds\right), \quad 0 < x < 1.$$

For sake of brevity, we shall use X to denote the Banach space $L^2_{1/w}(0,1)$ with the inner product

$$\langle u, v \rangle_w = \int_0^1 u\bar{v}w^{-1}dx.$$

In view of assumption (7.19) we have $H^1_0(0,1) \subset X$ algebraically and continuously, since

(7.21) $|u(x)| \leq \sqrt{x}\|u\|_{H^1_0}, \quad |u(x)| \leq \sqrt{1-x}\|u\|_{H^1_0}.$

Let us introduce the operator A by means of

(7.22) $\begin{cases} \mathcal{D}(A) = \{u \in H^1_0(0,1);\ Au \in X\}, \\ \quad Au = \alpha u'' + \beta u'. \end{cases}$

Notice that $Au = w(\eta u')'$, $u \in \mathcal{D}(A)$, when w, η are given by (7.20).

LEMMA 1. *Under assumptions (7.17),(7.19), $-A$ is self adjoint, positive definite and A^{-1} is compact.*

PROOF. We have

(7.23) $\langle -Au, v \rangle_w = \int_0^1 \eta u' \bar{v}' dx, \quad u, v \in \mathcal{D}(A).$

Indeed

$$\langle -Au, v \rangle_w = -[\eta u' \bar{v}]_{x=0}^{x=1} + \int_0^1 \eta u' \bar{v}' dx.$$

Last integral converges from $u,\ v \in H^1_0(0,1)$ and (7.19). Hence there exist the limits

$$\lim_{x \to 0+} \eta(x)u'(x)\overline{v(x)} \quad \text{and} \quad \lim_{x \to 1-} \eta(x)u'(x)\overline{v(x)}.$$

If λ_0 denotes the first limit, we show that $\lambda_0 = 0$. Otherwise, in virtue of (7.19) and (7.21), we have

$$|\eta(x)u'(x)|^2 \geq \tfrac{|\lambda|^2}{2}|v(x)|^{-2} \geq \tfrac{|\lambda|^2}{2}\|v\|_{H^1_0}^{-2}x^{-1}$$

on a suitable interval $(0,\delta)$, so that

$$\int_0^\delta |u'|^2 dx \geq C \int_0^\delta x^{-1}dx = \infty,$$

contradicting $u \in \mathcal{D}(A)$. The same argument applies to the second limit. Hence (7.23) is established.

A second integration by parts shows that A is symmetric. Self adjointness follows from the fact that $I - A$ is onto X, as in the proof of Theorem 7.1.

Equality (7.23) yields that $A^{-1} \in \mathcal{L}(X, H_0^1(0,1))$ is compact, too. Indeed, let $\{f_n\}$ be a bounded sequence in X and let $u_n = A^{-1}f_n$. Since $\{u_n\}$ is bounded in $H_0^1(0,1)$ and by (7.17),(7.19),

$$\frac{|u_n|^2}{w} \leq C \frac{|u_n|^2}{x} \frac{x}{\alpha} \leq C \frac{x}{\alpha} \|u_n\|_{H_0^1}^2, \quad 0 < x < \tfrac{1}{2},$$

$$\frac{|u_n|^2}{w} \leq C \frac{1-x}{\alpha} \|u_n\|_{H_0^1}^2, \quad \tfrac{1}{2} < x < 1,$$

we have

$$\int_0^1 \frac{|u_n|^2}{w} dx \leq C \int_0^1 \frac{x(1-x)}{\alpha} dx \|u_n\|_{H_0^1}^2 \leq C'.$$

On the other hand, for all $\varepsilon > 0$ there exists $\delta(\varepsilon) > 0$ such that

$$\int_0^t \frac{|u_n|^2}{w} dx + \int_{1-t}^1 \frac{|u_n|^2}{w} dx \leq \varepsilon \quad \text{for } 0 < t \leq \delta(\varepsilon),$$

while on a subsequence $\frac{u_n}{\sqrt{w}} \to \frac{u}{\sqrt{w}}$ in $L^2(t, 1-t)$. Hence, $\{u_n\}$ is compact in X, as claimed.

In particular, the Lemma implies that there are $\{\lambda_n\} \to \infty$, $\lambda_n > 0$ for all natural number n, and $\{\varphi_n\}$ forming an orthonormal complete system in X such that $A\varphi_n = -\lambda_n \varphi_n$, that is

(7.24)
$$\begin{cases} \alpha\varphi_n'' + \beta\varphi_n' = -\lambda_n\varphi_n, & \text{a.e. in } (0,1), \\ \varphi_n(0) = \varphi_n(1) = 0. \end{cases}$$

Moreover, $\varphi_n \in H_0^1(0,1) \subset \mathcal{C}([0,1])$ and

$$\lambda_n = \int_0^1 \eta|\varphi_n|^2 dx, \quad n \in \mathbb{N}.$$

Since $\alpha\varphi'' \in L^2(0,1)$, we conclude by (7.24) that $\varphi_n \in \mathcal{C}^1(0,1)$ and $\alpha\varphi_n'' \in \mathcal{C}(0,1)$. Therefore we can assume $\varphi_n \in \mathcal{C}^2(0,1) \cap \mathcal{C}_0([0,1])$.

Before treating the operator A_0, we want to extend the preceding Theorems 7.4 and 7.5 to $\beta \not\equiv 0$. Therefore, we introduce the operator A_1 by

$$\begin{cases} \mathcal{D}(A_1) = \{u \in H_0^1(0,1); \ A_1 u \in H_0^1(0,1)\}, \\ A_1 u = \alpha u'' + \beta u'. \end{cases}$$

We then have

LEMMA 2. *Under assumptions (7.17),(7.19), $-A_1$ is self adjoint, positive definite in $H_0^1(0,1)$ and thus A_1 generates an analytic semigroup in $H_0^1(0,1)$.*

PROOF. We show that A_1 is symmetric. Since $A_1 u = w(\eta u')'$ for all $u \in \mathcal{D}(A_1)$ and, in view of (7.19),

$$\langle u, v \rangle_1 = \int_0^1 \eta u' \overline{v}' dx$$

is an inner product on $H_0^1(0,1)$ equivalent to the usual one in this space, we deduce that for all $u, v \in \mathcal{D}(A_1)$

$$\langle A_1 u, v \rangle_1 = \int_0^1 (w(x)(\eta(x)u'(x))')' \eta(x) \overline{v}'(x) dx$$

$$= [\alpha(x)(\eta(x)u'(x))' \overline{v}'(x)]_{x=0}^{x=1} - \int_0^1 w(x)(\eta(x)u'(x))'(\eta(x)\overline{v}'(x))' dx.$$

Notice that the last integral converges for $u, v \in \mathcal{D}(A_1)$. We prove that the boundary term vanishes. We already know that the limits

$$\lim_{x \to 0+} \alpha(x)(\eta(x)u'(x))' \overline{v}'(x), \quad \lim_{x \to 1-} \alpha(x)(\eta(x)u'(x))' \overline{v}'(x)$$

exist. If $\lambda_0 \neq 0$ is the first one, then in a neighbourhood $(0,\delta)$ of 0 there is a positive constant C such that

$$|v'(x)| \geq \tfrac{|\lambda_0|}{2} |\eta(x)(A_1 u)(x)|^{-1} \geq C x^{-\frac{1}{2}}, \quad x \in (0,\delta),$$

so that $\int_0^1 |v'|^2 dx$ cannot converge. It follows that $\lambda_0 = 0$. Therefore we conclude that for all $u, v \in \mathcal{D}(A_1)$ we have

$$\langle A_1 u, v \rangle_1 = - \int_0^1 w(x)(\eta(x)u'(x))'(\eta(x)\overline{v}'(x))' dx$$

and any $u \in \mathcal{D}(A_1)$ belongs to the space V given by

$$V = \{u \in H_0^1(0,1); \int_0^1 w(x)|(\eta(x)u'(x))'|^2 dx < \infty\},$$

endowed with the natural norm.

Now V coincides with the completion of $C_0^\infty(0,1)$ with respect to the norm

$$\|u\|_V^2 = \int_0^1 \eta(x)|u'(x)|^2 dx + \int_0^1 w(x)|(\eta(x)u'(x))'|^2 dx.$$

For all $u \in V$ we have:

$$\|u'\|_{L^2}^2 \leq C \int_0^1 \eta(x)|u'(x)|^2 dx \leq C\|u'\|_{L^2}^2,$$

and

$$\int_0^1 \eta(x)|u'(x)|^2 dx = [u(x)(\eta(x)\overline{u}'(x))]_{x=0}^{x=1}$$

$$- \int_0^1 \frac{u(x)}{\sqrt{\alpha(x)}} \sqrt{\alpha(x)}(\eta(x)\overline{u}'(x))' dx.$$

The last integral converges and is majorized by

$$\left(\int_0^1 \frac{\eta(x)}{\alpha(x)}|u(x)|^2 dx \right)^{\frac{1}{2}} \left(\int_0^1 w(x)|(\eta(x)u'(x))'|^2 dx \right)^{\frac{1}{2}}.$$

Moreover,

$$\int_0^1 \frac{\eta(x)}{\alpha(x)}|u(x)|^2 dx \leq C \int_0^1 \frac{|u(x)|^2}{\alpha(x)} dx$$

$$\leq C \left(\int_0^{\frac{1}{2}} \frac{x}{\alpha(x)} dx + \int_{\frac{1}{2}}^1 \frac{1-x}{\alpha(x)} dx \right) \|u\|_{H_0^1}^2,$$

so that $\int_0^1 \frac{\eta}{\alpha}|u|^2 dx < \infty$ in view of (7.19). Thus the limits in the brackets exist. We show that they vanish. It suffices to show that this happens for

$$\lim_{x \to 0+} \eta(x)\overline{u'(x)}u(x).$$

If $\eta(x)\overline{u'(x)}u(x) \to \mu \neq 0$, from (7.21) we deduce that on $(0, \delta)$ the bound

$$\int_0^\delta |\eta(x)u'(x)|^2 dx \geq C|\mu|^2 \int_0^\delta x^{-1} dx = \infty$$

holds, in contrast to $u \in H_0^1(0,1)$.

The sesquilinear form

$$b(u,v) = \int_0^1 \eta u'\overline{v}' dx + \int_0^1 w(\eta u')'(\eta \overline{v}')' dx$$

is continuous on $V \times V$ and coercive, for $\|u\|_V^2 = b(u,u)$. Hence we can again exploit [167; Theorems 2.22 and 2.23; pp. 28-29]) to deduce that the operator $\tilde{B}$ associated to $b(u,v)$ is an isomorphism from V onto its dual V' and the part B of $\tilde{B}$ into $H_0^1(0,1)$ is positive definite and selfadjoint. Since $\mathcal{D}(B) = \{u \in H_0^1(0,1); \tilde{B}u \in H_0^1(0,1)\}$ and for all $u, v \in \mathcal{D}(A_1)$

$$\langle (I - A_1)u, v \rangle_1 = b(u,v),$$

we deduce that B coincides with $I - A_1$, and $I - A_1$ is onto $H_0^1(0,1)$.

THEOREM 7.11. *Let us assume (7.17),(7.19) and, in addition, $\beta \in C^1([0,1])$. Then the operator $(A, \mathcal{D}_1(A))$, where*

$$\begin{cases} \mathcal{D}_1(A) = \{u \in H^1(0,1); \text{ there exists } u'' \text{ in the sense of distributions} \\ \qquad\qquad\qquad\qquad\qquad\qquad \text{with } \alpha u'' + \beta u' \in H_0^1(0,1)\}, \\ \quad Au = \alpha u'' + \beta u', \end{cases}$$

generates an analytic semigroup in $H^1(0,1)$.

PROOF. Let g be given in $H^1(0,1)$. We must show that, if $\Re\lambda$ is sufficiently large, then the equation

(7.25) $$\lambda u - Au = g$$

has a (necessarily unique) solution u such that

$$\|u\|_{H^1} \le C(1 + |\lambda|)^{-1}\|g\|_{H^1}.$$

Let us introduce the new function

$$f_\lambda(x) = g(x) - \left((1-x)g(0) + xg(1) + \frac{\beta(x)}{\lambda}[g(0) - g(1)] \right),$$

$$0 < x < 1, \quad \Re\lambda \text{ large.}$$

In view of our assumptions, the Lemma above applies to solve uniquely in $H_0^1(0,1)$ the equation

(7.26) $$\lambda v - A_1 v = f_\lambda,$$

with the estimate

$$\|v\|_{H_0^1} \le C'(1 + |\lambda|)^{-1}\|f_\lambda\|_{H_0^1}.$$

Now (7.26) reads equivalently

$$\lambda \left\{ v + \frac{(1-x)g(0) + xg(1)}{\lambda} \right\} - \alpha \left\{ v + \frac{(1-x)g(0) + xg(1)}{\lambda} \right\}''$$
$$- \beta \left\{ v + \frac{(1-x)g(0) + xg(1)}{\lambda} \right\}' = g,$$

with $\alpha v'' + \beta v' + \beta \frac{g(1)-g(0)}{\lambda} \in H_0^1(0,1)$. Therefore,

$$u = v + \lambda^{-1}\{(1-x)g(0) + xg(1)\} \in H^1(0,1),$$
$$\alpha u'' + \beta u' \in H_0^1(0,1)$$

so that $u \in \mathcal{D}_1(A)$ satisfies (7.25) and

$$\|u\|_{H^1} \le C_1(1 + |\lambda|)^{-1}\|f_\lambda\|_{H_0^1} + C_2|\lambda|^{-1}\|g\|_{H^1} \le C_3(1 + |\lambda|)^{-1}\|g\|_{H^1}.$$

This finishes the proof.

Next we come back to the operator A_0 defined by (7.18). Our first result is an analogous of the famous result by Ph. Clément and C. A. Timmermans [53].

PROPOSITION. *Let us assume (7.17),(7.19) and (7.27), where*

(7.27)
$$\frac{\beta^2}{\alpha} \in L^\infty(0,1).$$

Then A_0 generates a C_0-contraction semigroup in $C_0([0,1])$.

PROOF. We can let $f \in C_0([0,1])$ real valued. Let $\lambda > 0$ fixed and let $\{f_\varepsilon\} \subset X = L^2_{1/w}(0,1)$, $w(x) = \alpha(x)e^{-\int_0^x \frac{\beta}{\alpha}dt}$, as in (7.20), be such that $f_\varepsilon \to f$ in $C_0([0,1])$ as $\varepsilon \to 0$. We can also choose $f_\varepsilon \in C_0([0,1])$ with $\mathrm{supp} f_\varepsilon \subset [\varepsilon, 1 - \varepsilon]$. Since, for all $\lambda > 0$, $(\lambda - A)^{-1} \in \mathcal{L}(X)$, as given in (7.22), we deduce that the problem

$$\lambda u_\varepsilon - \alpha u_\varepsilon'' - \beta u_\varepsilon' = f_\varepsilon \quad a.e. \ x \in (0,1),$$
$$u_\varepsilon(0) = u_\varepsilon(1) = 0,$$

has a unique solution $u_\varepsilon \in H^1_0(0,1) \subset X$: see Lemma 1.

Moreover, we have

(7.28)
$$\|u_\varepsilon\|_{C_0([0,1])} \le \lambda^{-1}\|f_\varepsilon\|_{C_0([0,1])}, \quad \lambda > \lambda_0 > 0,$$
(7.29)
$$\|u_\varepsilon - u_{\varepsilon'}\|_{C_0([0,1])} \le \lambda^{-1}\|f_\varepsilon - f_{\varepsilon'}\|_{C_0([0,1])}, \quad \varepsilon, \varepsilon' > 0, \ \lambda > \lambda_0 > 0,$$

where λ_0 is suitably taken.

It suffices to show (7.28). Let $M_\varepsilon = \lambda^{-1}\|f_\varepsilon\|_{C_0([0,1])}$. Then,

$$\frac{\lambda}{\alpha}(u_\varepsilon - M_\varepsilon) - (u_\varepsilon - M_\varepsilon)'' - \frac{\beta}{\alpha}(u_\varepsilon - M_\varepsilon)' = \frac{1}{\alpha}(f_\varepsilon - \|f_\varepsilon\|_{C_0}).$$

If we multiply the latter by $(u_\varepsilon - M_\varepsilon)^+$, integrate on $(0,1)$ and apply Stampacchia's lemma, we easily have

(7.30)
$$\int_0^1 |(u_\varepsilon - M_\varepsilon)_x^+|^2 dx - \int_0^1 \frac{\beta}{\alpha}(u_\varepsilon - M_\varepsilon)_x^+(u_\varepsilon - M_\varepsilon)^+ dx$$
$$+ \int_0^1 \frac{\lambda}{\alpha}((u_\varepsilon - M_\varepsilon)^+)^2 dx \le 0, \quad \varepsilon > 0.$$

Since Schwarz inequality gives

$$\left| \int_0^1 \frac{\beta}{\alpha}(u_\varepsilon - M_\varepsilon)_x^+(u_\varepsilon - M_\varepsilon)^+ dx \right|$$
$$\le \frac{1}{2}\int_0^1 \frac{\beta^2}{\alpha^2}|(u_\varepsilon - M_\varepsilon)^+|^2 dx + \frac{1}{2}\int_0^1 |(u_\varepsilon - M_\varepsilon)_x^+|^2 dx,$$

(7.30) implies that

$$\frac{1}{2}\int_0^1 |(u_\varepsilon - M_\varepsilon)_x^+|^2 dx + \int_0^1 \left(\frac{\lambda}{\alpha} - \frac{\beta^2}{2\alpha^2}\right)|(u_\varepsilon - M_\varepsilon)^+|^2 dx \le 0.$$

Hence,

$$\frac{1}{2}\int_0^1 |(u_\varepsilon - M_\varepsilon)_x^+|^2 dx + \int_0^1 \frac{1}{\alpha}\left(\lambda - \frac{\beta^2}{2\alpha}\right)|(u_\varepsilon - M_\varepsilon)^+|^2 dx \le 0.$$

Taking $\lambda > \frac{1}{2}\left\|\frac{\beta^2}{\alpha}\right\|_{L^\infty(0,1)}$, we conclude that $(u_\varepsilon - M_\varepsilon)^+ = 0$ and thus

$$u_\varepsilon(x) \le \lambda^{-1}\|f_\varepsilon\|_{C_0([0,1])}, \quad \text{for all } \varepsilon > 0, \ x \in [0,1].$$

A similar argument gives

$$u_\varepsilon(x) \ge -\lambda^{-1}\|f_\varepsilon\|_{C_0([0,1])}, \quad \text{for all } \varepsilon > 0, \ x \in [0,1],$$

and $\lambda > \frac{1}{2}\|\frac{\beta^2}{\alpha}\|_{L^\infty(0,1)}$. Hence (7.28) is proven.

We continue the proof, observing that by (7.29), $u_\varepsilon \to u$ in $\mathcal{C}([0,1])$ as $\varepsilon \to 0$. Moreover, since $\{u_\varepsilon\}$ is bounded in $H_0^1(0,1)$, we infer that $u_\varepsilon \to u$ weakly in $H_0^1(0,1)$ and u is a solution to $\lambda u - A_0 u = f$.

By (7.28) we see that $\|(\lambda - A_0)^{-1}f\|_{C_0([0,1])} \le \lambda^{-1}\|f\|_{C_0([0,1])}$ for $\lambda > \frac{1}{2}\|\frac{\beta^2}{\alpha}\|_{L^\infty(0,1)}$. Then it is well known that $-A_0$ is m-accretive and the previous inequality holds for all positive λ.

Our next purpose is to prove that the semigroup generated by A_0 possesses substantial regularity. In order to obtain this we need both something more than (7.19) and assumption concerning the asymptotic behaviour of the eigenvalues λ_n.

The first assumption reads

(7.31) $\dfrac{\beta}{\alpha} \in L^1(0,1), \quad \dfrac{\sqrt{x(1-x)}}{\alpha} \in L^1(0,1).$

Concerning the second assumption, we suppose that the eigenvalues λ_n in (7.25) satisfy

(7.32) $C_1(n^N + 1) \le \lambda_n \le C_2(n^N + 1), \quad n = 0,1,2,\ldots.$

for some $C_1, C_2 > 0$ and an integer $N \ge 1$.

THEOREM 7.12. *Under assumptions (7.17), (7.27) and (7.31), the operator A_0 generates a C_0-contraction differentiable semigroup in $C_0([0,1])$ which is real analytic on $(0,\infty)$.*

PROOF. Notice that we do not assume (7.32). But we begin our proof by supposing that (7.32) holds, too.

Let $S(t)$ be the semigroup generated on $C_0([0,1])$ and $X(= L^2_{1/w}(0,1))$ by A_0 and A, respectively. Since the second one is an extension of the first one, we use the same notation for both. Since $\{\varphi_n\}$ in (7.24) forms an orthonormal complete system in X, for all $u_0 \in C_0([0,1]) \cap X$, we have

$$(7.33) \qquad (S(t)u_0)(x) = \sum_{n=0}^{\infty} u_0^n e^{-\lambda_n t} \varphi_n(x), \quad t > 0, \, 0 \le x \le 1,$$

where $u_0^n = \int_0^1 \frac{u_0 \varphi_n}{w} dx = \langle u_0, \varphi_n \rangle_w$. This is a well known result in Hilbert spaces and the representation (7.33) is precisely the one given by G. Fichera [103].

We then deduce that

$$(7.34) \qquad \|S(t)u_0\|_{C_0([0,1])} \le C \sum_{n=0}^{\infty} \lambda_n e^{-\lambda_n t} \|u_0\|_{C_0([0,1])}, \quad t \ge 0,$$

because from (7.25), $\|\varphi_n\|^2_{H_0^1} \le C\lambda_n$ and therefore

$$|\varphi_n(x)| \le C_1 \sqrt{x(1-x)} \lambda_n^{\frac{1}{2}}, \quad 0 \le x \le 1,$$

$$|u_0^n| \le \int_0^1 \frac{|\varphi_n|}{w} dx \|u_0\|_{C_0([0,1])}$$

$$\le C_1 \lambda_n^{\frac{1}{2}} \int_0^1 \frac{\sqrt{x(1-x)}}{w} dx \|u_0\|_{C_0} \le C_2 \lambda_n^{\frac{1}{2}} \|u_0\|_{C_0([0,1])},$$

since (7.31) is assumed. Notice that

$$\frac{\sqrt{x(1-x)}}{w(x)} = \frac{\sqrt{x(1-x)}}{\alpha(x)} \eta(x) \le C' \frac{\sqrt{x(1-x)}}{\alpha(x)}.$$

This yields the estimate (7.34), as claimed. By density, we deduce that $S(t)u_0$ has the representation (7.33) for all $t > 0$ and any $u_0 \in C_0([0,1])$. Moreover, for all $u_0 \in C_0([0,1])$, $S(t)u_0 \in C^{\infty}(\mathbb{R}^+; C_0([0,1]))$ and

$$\left(\frac{d^k}{dt^k} S(t)u_0 \right)(x) = \sum_{n=0}^{\infty} u_0^n (-\lambda_n)^k e^{-\lambda_n t} \varphi_n(x),$$

for all $t > 0$, $0 \leq x \leq 1$. On the other hand, assumption (7.32) guarantees that there exist C_3, $C_4 > 0$ such that for $k \in \mathbb{N}_0$

$$\left\| \frac{d^k}{dt^k} S(t)u_0 \right\|_{C_0([0,1])} \leq C_3 \sum_{n=0}^{\infty} (n+1)^{N(k+1)} e^{-ct(n+1)^N} \|u_0\|_{C_0([0,1])}$$

$$\leq C_4 \sum_{j=1}^{\infty} j^{k+1} e^{-\gamma j} \|u_0\|_{C_0([0,1])}$$

for some $\gamma\, (= ct) > 0$.

Since

$$\sum_{j=1}^{\infty} j^{k+1} e^{-\gamma j} = \frac{d^k}{d\gamma^k} \left(\frac{d}{d\gamma} \frac{1}{e^\gamma - 1} \right),$$

and the function $f(\gamma) = \frac{d}{d\gamma} \frac{1}{e^\gamma - 1}$ is analytic on $(0, \infty)$, we get

$$\left\| \frac{d^k}{dt^k} S(t)u_0 \right\|_{C_0([0,1])} \leq M^k k!$$

on each compact interval $[a, b]$ in $t \geq \delta$, where $\delta > 0$. Since δ is arbitrary positive, we conclude by H. Tanabe [167; p. 163] that $S(t)u_0$ is analytic on $(0, \infty)$, as claimed.

It remains to show that assumption (7.32) is not necessary. First of all, our hypotheses imply $\frac{1}{\sqrt{\alpha}} \in L^1(0, 1)$ since

$$\int_0^1 \frac{dx}{\sqrt{\alpha(x)}} \leq \left(\int_0^1 \frac{dx}{\sqrt{x(1-x)}} \right)^{\frac{1}{2}} \left(\int_0^1 \frac{\sqrt{x(1-x)}}{\alpha(x)} \right)^{\frac{1}{2}} < \infty.$$

Then we are allowed to apply Theorem 5.2 of H. D. Niessen and A. Zettl [143; p. 566], according to which

$$\frac{\lambda_n}{n^2 \pi} \to \int_0^1 \frac{dx}{\sqrt{\alpha(x)}} < \infty.$$

Hence $\{\lambda_n\}$ has the property (7.23). The proof is complete.

REMARK. In proving Theorem 7.12, we have in fact shown that there exist positive constants N_k, $k \in \mathbb{N}_0$, such that

$$\left\| \frac{d^k}{dt^k} S(t)u_0 \right\|_{C_0([0,1])} \leq \frac{N_k}{t^{k+2}}, \quad t > 0, \ k = 0, 1, 2, \ldots .$$

Although real analyticity of the semigroup $S(t)$ implies the analyticity of the solution u to the Cauchy problem $u'(t) = A_0 u(t)$ on $(0, \infty)$, $u(0) =$

$u_0 \in C_0([0,1])$, see A. Friedman [104; pp. 206-216], this property must be distinguished from complex analyticity of holomorphic semigroups.

We are now in a position to establish a regularity property of the semigroup generated in $C([0,1])$ by the operator $Au = \alpha u'' + \beta u'$ with Ventcel's boundary conditions. Precisely, let us define an operator W by

$$\begin{cases} \mathcal{D}(W) = \{u \in C([0,1]) \cap C^2((0,1)); \; \alpha(x)u'' + \beta(x)u' \to 0 \text{ as } x \to 0, 1\}, \\ Wu = \alpha u'' + \beta u', \end{cases}$$

where α, β satisfy (7.17). We have

THEOREM 7.13. *Under assumptions (7.17),(7.27) and (7.31), the operator W generates a real analytic contraction semigroup in $C([0,1])$.*

PROOF. We already know from Ph. Clément and C. A. Timmermans [53] that W generates a C_0 contraction semigroup in $C([0,1])$. Before proving analyticity of the semigroup, we first prove that e^{tW} is infinitely differentiable. To this end, in view of Lemma 4.5 of A. Pazy [145; p. 53], we only need to show that e^{tW} is differentiable. We recall that by Theorem 7.12 the semigroup generated by A_0 (in $C_0([0,1])$) is differentiable, so that (see [145: Theorem 4.7, p. 54]) there exist real constants a, b, c with $b, c > 0$, such that $\lambda - A_0$ has a bounded inverse for $\Re\lambda \geq a - b\log|\Im\lambda|$, and $\|(\lambda - A_0)^{-1}\|_{\mathcal{L}(C_0([0,1]))} \leq C|\Im\lambda|$, for such λ's satisfying $\Re\lambda \leq 0$. Our argument is similar to the one for Theorem 7.11. Let $\Xi = \{\lambda \in \mathbb{C}; 0 \geq \Re\lambda \geq a - b\log|\Im\lambda|\}$. Given $f \in C([0,1])$, $\lambda \in \Xi$, $|\lambda|$ large, let

$$f_\lambda(x) = f(x) - \{(1-x)f(0) + xf(1) + \frac{\beta(x)}{\lambda}(f(0) - f(1))\} \in C_0([0,1]).$$

Then $(\lambda - A_0)v = f_\lambda$ has a unique solution $v \in \mathcal{D}(A_0)$. Define

$$u(x) = v(x) + \lambda^{-1}\{(1-x)f(0) + xf(1)\}$$

and observe that $C([0,1])$ can be identified as a direct sum of $C_0([0,1])$ and of polynomials with degree ≤ 1.

On the other hand $(\lambda - A_0)v = f_\lambda$ reads

$$\lambda u(x) - \alpha(x)u''(x) - \beta(x)u'(x) = f(x), \quad 0 \leq x \leq 1,$$

with

$$\begin{aligned} \alpha(x)u''(x) + \beta(x)u'(x) &= \lambda u(x) - f(x) \\ &= \lambda v(x) + (1-x)f(0) + xf(1) - f(x) \to 0 \text{ as } x \to 0, 1. \end{aligned}$$

By uniqueness of the solution, we deduce that necessarily $(\lambda - W)^{-1}f = u = v + \lambda^{-1}\{(1-x)f(0) + xf(1)\} = (\lambda - A_0)^{-1}f_\lambda + \lambda^{-1}P_1f$, where P_1f

is the polynomial $P_1 f(x) = (1 - x)f(0) + xf(1)$. Since A_0 generates a differentiable semigroup in $\mathcal{C}_0([0, 1])$, we see from this expression of $(\lambda - W)^{-1}$ that it satisfies just the estimates of $(\lambda - A_0)^{-1}$ as an operator from $\mathcal{C}_0([0, 1])$ into itself.

From A. Pazy [145; pp. 54-55] we also know that

$$e^{tW} = \frac{1}{2\pi i} \int_\Gamma e^{zt}(z - W)^{-1} dz, \quad t > 0,$$

where Γ is the path composed of three parts: Γ_1 is given by $\Re e z = 2a - b\log(-\Im m z)$ for $-\infty < \Im m z < -L = -e^{\frac{2a-\omega}{b}}$, Γ_2 is $\Re e z = \omega$, $-L \leq \Im m z \leq L$, and Γ_3 is $\Re e z = 2a - b\log(\Im m z)$, for $L < \Im m z < \infty$, ω being a fixed positive number. Γ is oriented so that $\Im m z$ increases along Γ. Using notation $S(t)$ for the semigroup generated by A_0 in $\mathcal{C}_0([0, 1])$, we have

$$e^{tW} f = S(t)(1 - P_1)f + \int_0^t S(\tau)\beta(f(1) - f(0))d\tau + P_1 f$$

and hence, by Theorem 7.12, we conclude that e^{tW} itself is real analytic on $(0, \infty)$.

Thus the proof is complete.

7.3 GENERATION THEOREMS FOR
$\alpha u'' + \beta u'$: OTHER APPROACHES

We begin this paragraph by proving that under suitable assumptions on the functions $\alpha, \beta \in \mathcal{C}([0, 1])$ we can extend the results in Theorems 7.8 and 7.9 to the general $Au = \alpha u'' + \beta u'$, this time taking into account that $Au = \alpha W(W^{-1}u')'$, where

(7.35) $W(x) = e^{-\int_{1/2}^x \frac{\beta}{\alpha} dt}, \quad 0 < x < 1.$

Precisely, we assume that $\alpha(\cdot)$ satisfies (7.1) on $[0, 1]$ and

(7.36) $\alpha W \in \mathcal{C}([0, 1]), \quad \min \alpha W > 0, \quad \alpha \in \mathcal{C}^1([0, 1]),$

so that there exist two positive constants C_1, C_2 such that

$$C_1 \leq \alpha(x)W(x) \leq C_2, \quad x \in [0, 1].$$

Since $\frac{d}{dx}(W(x)^{-1}) = \frac{\beta(x)}{\alpha(x)W(x)}$, we know that $W(x)^{-1} = Y(x) \in \mathcal{C}^1([0, 1])$. If $Z(x) = \alpha(x)W(x)$, $0 \leq x \leq 1$, then the differential operator A is represented by means of

$$Au = \alpha W(Yu')' = Z(Yu')'.$$

We point out that $W(x)$ is just the basic function entering in the result of Ph. Clément and C. A. Timmermans [53] referred after Remark 7.1.

We shall consider the differential operator A with domain $\mathcal{D}_p(A)$, $1 < p \leq \infty$, where

$$\mathcal{D}_p(A) = \{u \in L^p(0,1); \ u \text{ is locally absolutely continuous in } (0,1)$$
$$\text{and } Yu' \in W_0^{1,p}(0,1)\},$$

if $1 < p < \infty$, and

$$\mathcal{D}_\infty(A) = \{u \in \mathcal{C}([0,1]) \cap \mathcal{C}^1((0,1)); \ Yu' \in \mathcal{C}^1([0,1])\},$$

if $p = \infty$. We observe that if $u \in \mathcal{D}_\infty(A)$, then $\lim_{x\to 0,1} Y(x)u'(x) = 0$. Indeed, if $\lim_{x\to 0} Y(x)u'(x) = \lambda \neq 0$, then necessarily one of the two limits $\lim_{x\to 0} Y(x)\Re eu'(x)$ and $\lim_{x\to 0} Y(x)\Im mu'(x)$ is non zero. Suppose that this happens for the limit of $Y(x)\Re eu'$. Let $0 < \varepsilon < x$, and suppose that $\Re e\lambda = \lambda_0 > 0$. An analogous argument works for $\lambda_0 < 0$. Since $\Re eu'(x)Y(x) \geq \frac{\lambda_0}{2}$ for $0 < x$ small, we have for $0 < \varepsilon < x$,

$$\Re eu(x) - \Re eu(\varepsilon) = \int_\varepsilon^x Y(t)^{-1}[\Re eu'(t)Y(t)]dt$$
$$\geq \frac{\lambda_0}{2} \int_\varepsilon^x Y(t)^{-1}dt \geq \frac{\lambda_0}{2} \int_\varepsilon^x \frac{dt}{\alpha(t)} \to \infty \quad \text{as } \varepsilon \to 0+,$$

in view of $\alpha \in \mathcal{C}^1([0,1])$. This contradicts the continuity of u on $[0,1]$. In view of assumption (7.36) we conclude that

$$\mathcal{D}_\infty(A) = \{u \in \mathcal{C}([0,1]) \cap \mathcal{C}^1(0,1); \ Au \in \mathcal{C}([0,1])\}$$

and therefore it coincides with the one studied by C. A. Timmermans in [171]. According to Theorem 3 of C. A. Timmermans [171; p. 311], $(A, \mathcal{D}_\infty(A))$ generates a $\mathcal{C}_0$-semigroup on $\mathcal{C}([0,1])$ if and only if $\int_{\frac{1}{2}}^0 R(x)dx = \int_{\frac{1}{2}}^1 R(x)dx = \infty$, where $R(x) = W(x)\int_{\frac{1}{2}}^x \frac{dt}{Z(t)}$. Now, if $\frac{1}{2} < x < 1$, in view of (7.36), $C_2^{-1}(x - \frac{1}{2}) \leq \int_{\frac{1}{2}}^x \frac{dt}{Z(t)} \leq C_1^{-1}(x - \frac{1}{2})$, so that

$$C_2^{-1} \int_{\frac{1}{2}}^x (x - \tfrac{1}{2})W(x)dx \leq \int_{\frac{1}{2}}^1 R(x)dx \leq C_1^{-1} \int_{\frac{1}{2}}^x (x - \tfrac{1}{2})W(x)dx.$$

Take $\delta > 0$ small. Then $\int_{1-\delta}^1 W(x)(x-\frac{1}{2})dx \geq c \int_{1-\delta}^1 W(x)dx$ for a constant $c > 0$. But $\int_{1-\delta}^1 W(x)dx \geq c' \int_{1-\delta}^1 \alpha(x)^{-1}dx = \infty$, since $\alpha \in \mathcal{C}^1([0,1])$. Therefore $\int_{\frac{1}{2}}^1 R(x)dx = \infty$.

Analogously, on $(0, \frac{1}{2})$, $-R(x) = W(x) \int_x^{\frac{1}{2}} Z(t)^{-1} dt$ gives

$$C_2^{-1} \int_0^{\frac{1}{2}} W(x)(\tfrac{1}{2} - x) dx \leq - \int_0^{\frac{1}{2}} R(x) dx \leq C_1^{-1} \int_0^{\frac{1}{2}} W(x)(\tfrac{1}{2} - x) dx$$

and thus $\int_0^{\frac{1}{2}} R(x) dx = -\infty$ follows from

$$\int_0^\delta W(x)(\tfrac{1}{2} - x) dx \geq c \int_0^\delta W(x) dx \geq c' \int_0^\delta \alpha(x)^{-1} dx = \infty.$$

This also implies that for all $\lambda > 0$, $(\lambda - A)(\mathcal{D}_p(A)) \supset (\lambda - A)(\mathcal{D}_\infty(A))$ is dense in $L^p(0,1)$, since $\mathcal{D}_\infty(A) \subset \mathcal{D}_p(A)$.

We finish these preliminaries by observing that $\mathcal{D}_p(A)$ can be equivalently described by

$\mathcal{D}_p(A) = \{u \in L^p(0,1);\ u$ is locally absolutely continuous on $(0,1),$
$\qquad \alpha u' \in L^p(0,1),\ \alpha u'' + \beta u' \in L^p(0,1),\ \alpha(x)u'(x) \to 0$ as $x \to 0, 1\}.$

Notice that we used hypothesis (7.36) to write

$$Y(x)u'(x) = Z(x)^{-1}\alpha(x)u'(x)$$

and to observe that $\alpha(x)u'(x) = Z(x)(Y(x)u'(x))$, where $Z'(x)$ is well defined on $(0,1)$. The element $\alpha u''$ must be read $(\alpha u')' - \frac{\alpha'}{\alpha}(\alpha u')$.

For sake of brevity, sometimes we shall write A_p for $(A, \mathcal{D}_p(A))$.

We have

THEOREM 7.14. *Let us suppose that (7.1) holds, $\beta \in \mathcal{C}([0,1])$ and (7.36) is verified. Then A_p generates an analytic semigroup $T_p(t)$ in $L^p(0,1)$, $1 < p < \infty$.*

PROOF. Let $\Re\lambda > 0$ and $u \in \mathcal{D}_p(A)$. If $(\lambda - A_p)u = f \in L^p(0,1)$, then

$$(7.37) \quad \Re\lambda \int_0^1 Z(x)^{-1}|u(x)|^p dx - \Re[Y(x)u'\overline{u}|u|^{p-2}]_{x=0}^{x=1}$$

$$+ \int_0^1 Y(x)|u'(x)||u(x)|^{p-2} dx$$

$$+ (p-2)\int_0^1 Y(x)(\Re eu'\overline{u})^2|u(x)|^{p-4} dx$$

$$= \Re e \int_0^1 Z(x)^{-1}f(x)\overline{u}(x)|u(x)|^{p-2} dx,$$

$$(7.38) \quad \Im m\lambda \int_0^1 Z(x)^{-1}|u(x)|^p dx - \Im m[Y(x)u'\overline{u}|u|^{p-2}]_{x=0}^{x=1}$$

$$= -(p-2)\int_0^1 Y(x)(\Im mu'\overline{u})(\Re eu'\overline{u})|u(x)|^{p-4} dx$$

$$+ \Im m\int_0^1 Z(x)^{-1}f(x)\overline{u}(x)|u(x)|^{p-2} dx.$$

Notice that $\|u\|_p' = (\int_0^1 Z(x)^{-1}|u(x)|^p dx)^{1/p}$ is a norm equivalent to $\|u\|_p$. Now (7.37) reads equivalently

$$\Re e\lambda(\|u\|_p')^p - \Re e[Y(x)u'\overline{u}|u|^{p-2}]_{x=0}^{x=1} + (p-1)\int_0^1 Y(x)(\Re eu'\overline{u})^2|u|^{p-4} dx$$

$$+ \int_0^1 Y(x)(\Im mu'\overline{u})^2|u|^{p-4} dx = \Re e\int_0^1 Z(x)^{-1}f(x)\overline{u}(x)|u(x)|^{p-2} dx$$

and thus the limits inside the brackets exist. If

$$\lambda_0 = \lim_{x\to 0} Y(x)(\Re e\overline{u}(x)u'(x))|u(x)|^{p-2} \neq 0,$$

from $Y(x)u'(x) = \int_0^x \frac{d}{dt}(Y(t)u'(t))dt$ it follows that

$$|Y(x)u'(x)| \leq x^{\frac{1}{p'}}\left(\int_0^x \left|\frac{d}{dt}(Y(t)u'(t))\right|^p dt\right)^{\frac{1}{p}} \leq Cx^{\frac{1}{p'}}.$$

Therefore

$$Y(x)|u'(x)\overline{u}(x)| \leq Cx^{\frac{1}{p'}}|u(x)|$$

or else

$$Y(x)|\Re e(u'(x)\overline{u}(x))| \leq Cx^{\frac{1}{p'}}|u(x)|.$$

Hence, on $(0,\delta)$ we have:

$$|u(x)|^{p-2} \geq \frac{|\lambda_0|}{2}(Y(x)|\Re e(u'(x)\overline{u}(x))|)^{-1} \geq c'x^{-\frac{1}{p'}}|u(x)|^{-1},$$

$x^{\frac{1}{p'}}|u(x)|^{p-1} \geq c'$, that gives $|u(x)|^p x \geq c'' > 0$, contradicting $u \in L^p(0,1)$. We have then seen that for all $u \in \mathcal{D}_p(A)$, $\Re e[Y(x)u'(x)\overline{u}(x)|u(x)|^{p-2}]_{x=0}^{x=1} = 0$.

This then yields immediately

$$\Re e\lambda\|u\|_p' \leq \|f\|_p' = \|(\lambda - A_p)u\|_p'.$$

So that, taking into account that the range of $\lambda - A_p$ is dense in $L^p(0,1)$ and A_p is a closed operator, as it is easily shown, we conclude that A_p

generates a $\mathcal{C}_0$-contraction semigroup in $(L^p(0,1), \|\cdot\|'_p)$. Since $\|\cdot\|'_p$ and $\|\cdot\|_p$ are equivalent norms, A_p generates a uniformly bounded $\mathcal{C}_0$-semigroup $T_p(t)$ on $L^p(0,1)$.

We also deduce

$$(7.39) \quad (p-1)\int_0^1 Y(x)(\Re e u'(x)\overline{u}(x))^2|u(x)|^{p-4}dx \le \|f\|'_p(\|u\|'_p)^{p-1},$$

$$\int_0^1 Y(x)(\Im m(u'(x)\overline{u(x)}))^2|u(x)|^{p-4}dx \le \|f\|'_p(\|u\|'_p)^{p-1}.$$

Then (7.39) gives that $\Im m[Y(x)u'(x)\overline{u}(x)|u(x)|^{p-2}]_{x=0}^{x=1}$ in (7.38) vanishes and

$$|\Im m\lambda|(\|u\|'_p)^p$$

$$\le |p-2|\int_0^1 Y(x)\Im m(u'(x)\overline{u(x)})|\Re e(u'(x)\overline{u(x)})||u(x)|^{p-4}dx$$

$$+ \|f\|'_p(\|u\|'_p)^{p-1} \le |p-2|\left(\int_0^1 Y(x)|\Im m(u'(x)\overline{u(x)})|^2|u(x)|^{p-4}dx\right)^{\frac{1}{2}}$$

$$\times \left(\int_0^1 Y(x)|\Re e(u'(x)\overline{u(x)})|^2|u(x)|^{p-4}dx\right)^{\frac{1}{2}} + \|f\|'_p(\|u\|'_p)^{p-1}$$

$$\le \frac{|p-2|}{\sqrt{p-1}}\|f\|'_p(\|u\|'_p)^{p-1} + \|f\|'_p(\|u\|'_p)^{p-1}.$$

And thus there exists $C_p > 0$ such that

$$(1+|\lambda|)\|u\|_p \le C_p\|(\lambda - A_p)u\|_p,$$

for all $u \in \mathcal{D}(A_p)$, $\Re e\lambda > 0$.

This proves the assertion.

REMARK. The estimates (7.37) yield, letting $p \to \infty$, that

$$\Re e\lambda\|u\|_\infty \le \|(\lambda - A_\infty)u\|_\infty,$$

but we can not conclude an analogous estimate for $|\Im m\lambda|\|u\|_\infty$ since $C_p \sim 1 + \frac{|p-2|}{\sqrt{p-1}}$. See the proof of Theorem 7.14.

We now extend Theorem 7.9 to the operator $Bu = \alpha u'' + \beta u'$, where α satisfies (7.1), $\alpha \in \mathcal{C}^1([0,1])$, $\beta \in \mathcal{C}([0,1])$, $\frac{\beta}{\alpha} \in L^1(0,1)$, (and this guarantees that $W \in \mathcal{C}([0,1])$, see (7.35)), with domain

$$\mathcal{D}_p(B) = \{u \in W^{1,p}(0,1) \cap W^{2,p}_{loc}(0,1); Bu \in W^{1,p}_0(0,1)\}.$$

THEOREM 7.15. *Let $1 < p < \infty$, $\alpha \in C^1([0,1])$, $\alpha(0) = \alpha(1) = 0$, $\alpha(x) > 0$ on $(0,1)$, $\beta \in C([0,1])$. If $\frac{\beta}{\alpha} \in L^1(0,1)$, then $(B, \mathcal{D}_p(B))$ generates an analytic semigroup in $W^{1,p}(0,1)$.*

PROOF. Let $\beta_1(x) = \alpha'(x) - \beta(x)$. Let $F \in W^{1,p}(0,1)$ with $F' = f \in L^p(0,1)$. Theorem 7.14 applies to the operator $Av = e^{-\int_{1/2}^x \frac{\beta_1}{\alpha} dt} \times \alpha(e^{\int_{1/2}^x \frac{\beta_1}{\alpha} dt} v')'$. In particular, for all $\lambda \in \mathbb{C}$, $\Re\lambda > 0$, there exists $v \in \mathcal{D}_p(A)$ such that

$$(7.40) \qquad \lambda v - e^{-\int_{1/2}^x \frac{\beta_1}{\alpha} dt} \alpha(e^{\int_{1/2}^x \frac{\beta_1}{\alpha} dt} v')' = F'\alpha e^{-\int_{1/2}^x \frac{\beta_1}{\alpha} dt},$$

$$(7.41) \qquad e^{\int_{1/2}^x \frac{\beta_1(t)}{\alpha(t)} dt} v' \in W_0^{1,p}(0,1),$$

so that $e^{\int_{1/2}^x \frac{\beta_1}{\alpha} dt} v' \to 0$ as $x \to 0, 1$.

We observe, in fact, that $\alpha(x) e^{-\int_{1/2}^x \frac{\beta_1(t)}{\alpha(t)} dt} = \alpha(\frac{1}{2}) W(x)^{-1}$ is continuous on $[0,1]$ and $\inf \alpha e^{-\int_{1/2}^x \frac{\beta_1}{\alpha} dt} > 0$. Dividing both the numbers of (7.40) by $\alpha e^{-\int_{1/2}^x \frac{\beta_1}{\alpha} dt}$, integrating the obtained equation on $(0,x)$ and taking into account (7.40), we get

$$\lambda \left[\int_0^x \frac{v(t)}{\alpha(t)} e^{\int_{1/2}^t \frac{\beta_1(y)}{\alpha(y)} dy} dt + \frac{F(0)}{\lambda} \right] - e^{\int_{1/2}^x \frac{\beta_1(y)}{\alpha(y)} dy} v'(x) = F(x).$$

Let $\int_0^x \frac{v(t)}{\alpha(t)} e^{\int_{1/2}^t \frac{\beta_1(y)}{\alpha(y)} dy} dt + \frac{F(0)}{\lambda} = u(x)$. Then $\lambda u - e^{\int_{1/2}^x \frac{\beta_1(y)}{\alpha(y)} dy} v'(x) = F(x)$ and $e^{\int_{1/2}^x \frac{\beta_1}{\alpha} dt} v'(x) \to 0$ as $x \to 0, 1$ by (7.41). On the other hand, $u'(x) = \frac{v(x)}{\alpha(x)} e^{\int_{1/2}^x \frac{\beta_1}{\alpha} dt}$ yields that

$$v(x) = \alpha(x) e^{-\int_{1/2}^x \frac{\beta_1}{\alpha} dt} u'(x),$$

$$v'(x) = e^{-\int_{1/2}^x \frac{\beta_1(t)}{\alpha(t)} dt} [(\alpha'(x) - \beta_1(x)) u'(x) + \alpha(x) u''(x)],$$

or else

$$e^{\int_{1/2}^x \frac{\beta_1(t)}{\alpha(t)} dt} v'(x) = \alpha(x) u''(x) + \beta(x) u'(x) = Bu(x) \to 0 \text{ as } x \to 0, 1.$$

Moreover,

$$\|u\|_{W^{1,p}} = \left(\int_0^1 |u(x)|^p dx \right)^{\frac{1}{p}} + \left(\int_0^1 |u'(x)|^p dx \right)^{\frac{1}{p}}$$

$$\leq C\|v\|_p + \frac{|F(0)|}{|\lambda|} + C'\frac{\|F'\|_p}{|\lambda|}.$$

But, $|F(0)| \leq \|F\|_\infty \leq C\|F\|_{W^{1,p}}$, where we denoted by $\|F\|_\infty$ the norm in $\mathcal{C}([0,1])$. Hence,

$$\|u\|_{W^{1,p}} \leq C|\lambda|^{-1}\|F\|_{W^{1,p}}.$$

This completes the proof.

As it was remarked previously, under assumption (7.36) the operator $(A, \mathcal{D}_\infty(A))$ generates a $\mathcal{C}_0$-contraction semigroup in $\mathcal{C}([0,1])$. Our next aim is to prove that under an additional assumption the semigroup $T(t)$ generated by $A(= A_\infty)$ is in fact differentiable. To this end, we shall extend an argument from M. Campiti, G. Metafune and D. Pallara [53: Theorem 3.3].

If $W(x)$ is given as in (7.35), let us assume (7.36) and

$$(7.42) \qquad\qquad N(x) = \int_x^{\frac{1}{2}} W(t)dt \in L^p(0,1),$$

where $1 < p < \infty$. Then we have

THEOREM 7.16. *Under assumptions (7.36), (7.42), the semigroup $T(t)$ is differentiable on $\mathcal{C}([0,1])$.*

PROOF. As in M. Campiti, G. Metafune and D. Pallara [53], we show that $\mathcal{D}_{p'}(A)$, $\frac{1}{p'} + \frac{1}{p} = 1$, endowed with the graph norm, is continuously embedded into $\mathcal{C}([0,1])$. Let $u \in \mathcal{D}_{p'}(A)$ and let $Au = f \in L^{p'}(0,1)$. Since

$$e^{\int_{1/2}^x \frac{\beta(t)}{\alpha(t)}dt} u'(x) = Y(x)u'(x) = \int_0^x Z(y)^{-1}f(y)dy \equiv g(x) \in W_0^{1,p'}(0,1),$$

we have $u'(x) = W(x)g(x)$, and for $0 < x < \frac{1}{2}$

$$u(x) - u(\tfrac{1}{2}) = \int_{\frac{1}{2}}^x W(y)g(y)dy$$

$$= \int_0^x \int_x^{\frac{1}{2}} W(y)Z(s)^{-1}f(s)dyds + \int_x^{\frac{1}{2}} \int_s^{\frac{1}{2}} W(y)Z(s)^{-1}f(s)dyds$$

$$= \int_x^{\frac{1}{2}} W(y)dy \int_0^x Z(s)^{-1}f(s)ds + \int_x^{\frac{1}{2}} Z(s)^{-1}f(s)\left(\int_s^{\frac{1}{2}} W(y)dy\right)ds$$

$$= N(x)\int_0^x Z(s)^{-1}f(s)ds + \int_x^{\frac{1}{2}} N(s)Z(s)^{-1}f(s)ds.$$

In view of assumption (7.42), $N \in L^p(0,\frac{1}{2})$ and decreases on $(0,\frac{1}{2})$. Moreover,

$$\left|N(x)\int_0^x Z(y)^{-1}f(y)dy\right| \leq CN(x)x^{\frac{1}{p}}\|f\|_{p'} \to 0 \text{ as } x \to 0.$$

Hence, there exists the limit

$$\lim_{x \to 0} \{u(x) - u(\tfrac{1}{2})\} = \int_0^{\frac{1}{2}} N(s)Z(s)^{-1}f(s)ds.$$

Therefore u is continuous at $x = 0$, with

$$\sup_{0 \leq x \leq \frac{1}{2}} |u(x) - u(\tfrac{1}{2})| \leq C\|f\|_{p'} \sup_{0 \leq x \leq \frac{1}{2}} x^{\frac{1}{p}} N(x) + C\|N\|_p \|f\|_{p'} = C_1\|f\|_{p'}.$$

The same argument assures that u is continuous at $x = 1$ and

$$\sup_{\frac{1}{2} \leq x \leq 1} |u(x) - u(\tfrac{1}{2})| \leq C\|f\|_{p'} \{\|N\|_p + \sup_{\frac{1}{2} \leq x \leq 1} (1-x)^{\frac{1}{p}} N(x)\}.$$

It follows that $u \in C([0,1])$ and

$$\sup_{0 \leq x \leq 1} |u(x) - u(\tfrac{1}{2})| \leq C\|Au\|_{p'}.$$

Then from the continuity property of u we know that there exist $x_0, x_1 \in [0,1]$ such that $\Re eu(x_0) = \int_0^1 \Re eu(x)dx$, $\Im mu(x_1) = \int_0^1 \Im mu(x)dx$. Since $|u(x_0) - u(\tfrac{1}{2})| \leq C\|Au\|_{p'}$, $|u(x_1) - u(\tfrac{1}{2})| \leq C\|Au\|_{p'}$, we deduce that

$$|\Re eu(x_0) - \Re eu(\tfrac{1}{2})| \leq C\|Au\|_{p'}, \quad |\Im mu(x_1) - \Im mu(\tfrac{1}{2})| \leq C\|Au\|_{p'},$$

or else

$$|\Re eu(x_0) + i\Im mu(x_1) - u(\tfrac{1}{2})| \leq 2C\|Au\|_{p'},$$

i.e. $|\int_0^1 u(x)dx - u(\tfrac{1}{2})| \leq 2C\|Au\|_{p'}$. Hence,

$$|u(x)| \leq |u(x) - u(\tfrac{1}{2})| + |u(\tfrac{1}{2})|$$

$$\leq C\|Au\|_{p'} + \left| u(\tfrac{1}{2}) - \int_0^1 u(x)dx + \int_0^1 u(x)dx \right|$$

$$\leq C'\|Au\|_{p'} + \int_0^1 |u(x)|dx \leq C''(\|Au\|_{p'} + \|u\|_{p'}) = C''\|u\|_{\mathcal{D}(A_{p'})}.$$

Recall also that every A_p in $L^p(0,1)$, $1 < p < \infty$, is an extension of $A_\infty = A$, with $T_p(t)u = T_\infty(t)u = T(t)u$ for all $u \in C([0,1])$.

Finally, we arrive at the differentiability property of $T(t)$. Since $T(t)$ is differentiable in $L^p(0,1)$ for all $p \in (1, \infty)$ from Theorem 7.14, there exists $C_t \geq 0$ such that

$$\|AT(t)f\|_{\mathcal{D}_p(A)} \leq C_t\|f\|_\infty, \quad f \in C([0,1]).$$

In virtue of the continuous imbedding $\mathcal{D}_{p'}(A) \subset C([0,1])$, we deduce that there is a constant $C_t' \geq 0$ such that

$$\|AT(t)f\|_\infty \leq C_t'\|f\|_\infty, \quad f \in C([0,1]), \ t > 0,$$

thus proving that $T(t)$ is a differentiable semigroup.

Before enunciating the next result, we set

$$C_0^1([0,1]) = \{u \in C^1([0,1]); \ u(0) = u(1) = 0\}.$$

THEOREM 7.17. *Let us assume that* $\alpha \in \mathcal{C}^1([0,1])$, $\alpha(0) = \alpha(1) = 0$, $\alpha(x) > 0$ *for* $x \in (0,1)$, $\beta \in \mathcal{C}([0,1])$, $\frac{\beta}{\alpha} \in L^1(0,1)$ *and*

$$(7.43) \qquad x \to \int_x^{\frac{1}{2}} Z(t)^{-1} dt \in L^p(0,1)$$

for some $p > 1$. *Then the semigroup* $S(t)$ *generated by* $Bu = \alpha u'' + \beta u'$ *in* $\mathcal{C}^1([0,1])$ *with Ventcel's boundary conditions:*

$$\mathcal{D}(B) = \{u \in \mathcal{C}^1([0,1]) \cap \mathcal{C}^3(0,1); \ \alpha u'' + \beta u' \in \mathcal{C}_0^1([0,1])\}$$

is differentiable.

PROOF. First of all, fix $\lambda > 0$ and consider the operator A with Neumann boundary conditions

$$\begin{cases} \mathcal{D}(A) = \{u \in \mathcal{C}([0,1]) \cap \mathcal{C}^2(0,1); \ Au \in \mathcal{C}([0,1])\}, \\ Av = \alpha v'' + \beta_1 v' = e^{-\int_{1/2}^x \frac{\beta_1}{\alpha} dt} \alpha \left(e^{\int_{1/2}^x \frac{\beta_1(t)}{\alpha(t)} dt} v'(x) \right)', \end{cases}$$

where $\beta_1(x) = \alpha'(x) - \beta(x)$. We know that every $v \in \mathcal{D}(A)$ satisfies in fact $e^{\int_{1/2}^x \frac{\beta_1}{\alpha} dt} v'(x) \to 0$ as $x \to 0, 1$. Since $\alpha(x) e^{-\int_{1/2}^x \frac{\beta_1(t)}{\alpha(t)} dt} = \alpha(\frac{1}{2}) W(x)^{-1}$, Theorem 7.16 applies so that for all $f \in \mathcal{C}^1([0,1])$, the problem

$$\lambda v - Av = \alpha e^{-\int_{1/2}^x \frac{\beta_1(t)}{\alpha(t)} dt} f'$$

has a unique solution v with $e^{\int_{1/2}^x \frac{\beta_1(t)}{\alpha(t)} dt} v'(x) \to 0$ as $x \to 0, 1$. We then deduce that

$$u(x) = \int_0^x \alpha(t)^{-1} e^{\int_{1/2}^t \frac{\beta_1(s)}{\alpha(s)} ds} v(t) dt + \frac{f(0)}{\lambda}$$

satisfies

$$\lambda u - \alpha u'' - \beta u' = f \quad \text{on } (0,1),$$

and $e^{\int_{1/2}^x \frac{\beta_1(t)}{\alpha(t)} dt} v'(x) = \alpha(x) u''(x) + \beta(x) u'(x) \to 0$ as $x \to 0, 1$. Hence, $u \in \mathcal{D}(B)$ with $(\lambda - B)u = f$.

On the other hand, let us introduce on $\mathcal{C}^1([0,1])$ the norm

$$\|u\|_1 = \max \left\{ |u(0)|, \max_{0 \le x \le 1} \left| \alpha(x) e^{-\int_{1/2}^x \frac{\beta_1(t)}{\alpha(t)} dt} u'(x) \right| \right\}.$$

Then $|u(0)| = \frac{|f(0)|}{\lambda}$, $\alpha(x) e^{-\int_{1/2}^x \frac{\beta_1(s)}{\alpha(s)} ds} u'(x) = v(x)$ and since

$$\|v\|_\infty \le \lambda^{-1} \left\| e^{-\int_{1/2}^x \frac{\beta_1(t)}{\alpha(t)} dt} \alpha f \right\|_\infty,$$

we conclude that

$$\|u\|_1 \le \lambda^{-1} \max\left\{|f(0)|, \left\|e^{-\int_{1/2}^x \frac{\beta_1(t)}{\alpha}(t)dt}\alpha f'\right\|_\infty\right\} = \lambda^{-1}\|f\|_1.$$

It follows that $(B, \mathcal{D}(B))$ generates a $\mathcal{C}_0$-contraction semigroup in $\mathcal{C}^1([0,1])$.

It remains to prove that $S(t)$ is differentiable. Now, hypothesis (7.43) implies that $N(x) = \int_x^{\frac{1}{2}} e^{-\int_{1/2}^t \frac{\beta_1(s)}{\alpha(s)}ds}dt = \alpha(\frac{1}{2})\int_x^{\frac{1}{2}} Z(t)^{-1}dt \in L^p(0,1)$ and thus by Theorem 7.16 the semigroup $T(t)$ generated by A is differentiable in $\mathcal{C}([0,1])$.

Moreover, $(\lambda - B)u = f$ if and only if $u(x) = \int_0^x \alpha(t)^{-1}e^{\int_{1/2}^t \frac{\beta_1(s)}{\alpha(s)}ds}v(t)dt$ $+\lambda^{-1}f(0)$, where $(\lambda - A)v = \alpha e^{-\int_{1/2}^x \frac{\beta_1(s)}{\alpha(s)}ds}f'$. Since $T(t)$ is differentiable by Theorem 4.7 of A. Pazy [145; p. 54], there exist $a, \omega \in \mathbb{R}$, $b, c > 0$ such that

$$\|v\|_\infty \le C|\Im m\lambda|\|f'\|_\infty$$

for all $\lambda \in \Sigma = \{\lambda; \Re e\lambda \ge a - b\log|\Im m\lambda|\}$, $\Re e\lambda \le \omega$. The relation between u and v allows us to recognize that a corresponding estimate

$$\|u\|_{\mathcal{C}^1([0,1])} \le C|\Im m\lambda|\|f\|_{\mathcal{C}^1([0,1])}$$

holds for all $f \in \mathcal{C}^1([0,1])$ and any $\lambda \in \Sigma$, $\Re e\lambda \le \omega$.

A new application of A. Pazy [145; p. 54] yields the result.

We now treat the analyticity of the semigroup $T(t)$ in $\mathcal{C}([0,1])$. Let us suppose that $\sqrt{\alpha} \in \mathcal{C}^1([0,1])$, $\alpha(0) = \alpha(1) = 0$, $\alpha(x) > 0$ on $(0,1)$, $\beta \in \mathcal{C}([0,1])$ satisfy (7.36), which is equivalent to assume that $\beta = \alpha' + \alpha\omega$, where $\alpha\omega \in \mathcal{C}([0,1])$, $\omega \in \mathcal{L}(0,1)$. Observe that, if $\omega \in \mathcal{R}([0,1])$, that is, ω is Riemann integrable on $[0,1]$, then (7.36) implies that $\frac{\beta}{\sqrt{\alpha}} = \frac{\alpha'}{\sqrt{\alpha}} + \sqrt{\alpha}\omega \in \mathcal{C}([0,1])$.

Now $\sqrt{\alpha} \in \mathcal{C}^1([0,1])$ yields that $\sqrt{\alpha}$ is R-admissible according to the following definition.

DEFINITION. A function $m \in \mathcal{C}([0,1])$, $m(x) > 0$ on $(0,1)$, $m(0) = m(1) = 0$ is R-admissible if

$$\int_0^{\frac{1}{2}} \frac{dx}{m(x)} = \int_{\frac{1}{2}}^1 \frac{dx}{m(x)} = \infty.$$

Let $\mathcal{C}([-\infty, \infty])$ denote the Banach space of all continuous complex valued functions f on $\mathbb{R}$ such that $f(x)$ converges as $x \to \infty$ and $x \to \infty$, endowed with the supremum norm.

The result below is essential for the next theorem and further statements.

LEMMA. *Let* $m(\cdot)$ *be an R-admissible function in* $\mathcal{C}^1([0,1])$. *Then* $\varphi(x)$ $= \int_{\frac{1}{2}}^{x} \frac{dy}{m(y)}$ *induces the map* $V : \mathcal{C}([-\infty, \infty]) \to \mathcal{C}([0,1])$ *such that*

$$Vf(x) = f(\varphi(x)), \quad x \in [0,1],$$

which is an isometry and

$$V^{-1}(\mathcal{C}([0,1])) = \mathcal{C}([-\infty, \infty]),$$
$$V^{-1}(\mathcal{D}(C_\infty)) = \{v \in \mathcal{C}([-\infty, \infty]); \ v' \in \mathcal{C}([-\infty, \infty])\},$$

where $C_\infty u = mu'$ *with*

$$\mathcal{D}(C_\infty) = \{u \in \mathcal{C}([0,1]) \cap \mathcal{C}^1((0,1)); \text{ there exist } \lim_{x \to 0,1} m(x)u'(x) \in \mathbb{C}\}.$$

Notice that, if $v' \in \mathcal{C}([-\infty, \infty])$, *then* $v'(\pm\infty) = 0$, *so that for all* $u \in \mathcal{D}(C_\infty)$, $\lim_{x \to 0,1} m(x)u'(x) = 0$.

PROOF. Since $\varphi'(x) = m(x)^{-1}$, $m(x)(Vf)'(x) = f'(\varphi(x))$, and hence the assertions follow.

THEOREM 7.18. *Let us assume (7.36) and let* $\frac{\beta}{\sqrt{\alpha}} \in \mathcal{C}([0,1])$ *and* $\sqrt{\alpha} \in \mathcal{C}^1([0,1])$. *Then* $(A, \mathcal{D}(A_\infty))$ *generates an analytic semigroup in* $\mathcal{C}([0,1])$.

PROOF. Let $a(x) = e^{\int_{1/2}^{x} \frac{\beta}{\alpha} dt}$ and $m(x) = a(x)^{\frac{1}{2}}$. Then, $m \in \mathcal{C}^1([0,1])$ from $\frac{\beta m}{\alpha} = \frac{\beta}{\sqrt{\alpha}}(\alpha w)^{-\frac{1}{2}}$. We also introduce the operator C_∞ defined by $C_\infty u = mu'$, where $\mathcal{D}(C_\infty)$ is given in the Lemma. Again the Lemma yields that C_∞ generates a C_0-group on $\mathcal{C}([0,1])$ which is similar to the translation group on $\mathcal{C}([-\infty, \infty])$.

By Corollary 1.13 of [138; A-II], $(C_\infty^2, \mathcal{D}(C_\infty^2))$ generates a bounded analytic semigroup of angle $\frac{\pi}{2}$. Moreover, since $\frac{1}{\sqrt{a}}$ and $\frac{1}{a}$ are not summable in a neighbourhood of 0 and of 1, $\mathcal{D}(C_\infty^2) = \mathcal{D}(A_\infty) = \mathcal{D}_\infty(A)$. Since

$$C_\infty^2 u = m(mu')' = m^2 u'' + mm'u' = e^{\int_{1/2}^{x} \frac{\beta}{\alpha} dt} u'' + \frac{1}{2} e^{\int_{1/2}^{x} \frac{\beta}{\alpha} dt} \frac{\beta}{\alpha} u'$$

$$= (a(x)u'(x))' - \frac{1}{2}\frac{\beta}{\alpha}a(x)u'(x) = Ku - \frac{1}{2}\frac{\beta}{\sqrt{\alpha}}\left(\frac{a(x)}{\alpha}\right)^{\frac{1}{2}} C_\infty u,$$

where $\mathcal{D}(K) = \mathcal{D}(A_\infty)$, $Ku = (a(x)u'(x))'$, then $(K, \mathcal{D}(K))$ generates an analytic semigroup on $\mathcal{C}([0,1])$, for C_∞ has C_∞^2-bound equal to 0 in view of the estimate

$$\|C_\infty u\|_\infty^2 \le 4M^2 \|C_\infty u\|_\infty \|u\|_\infty, \quad u \in \mathcal{D}(C_\infty^2),$$

see A. C. McBride [135: Theorem 2.25, p. 47]. Notice that C_∞ generates a semigroup of type $(M, 0)$ for a certain $M > 0$, and this is the M appearing in the above inequality.

Now, our operator $A(= A_\infty)$ coincides with $A = HK$, where H is the multiplication operator by $p(x) = \alpha(x)e^{-\int_{1/2}^x \frac{\beta}{\alpha} dt}$ in $C([0, 1])$.

Since K generates a bounded analytic semigroup, we know by K.-J. Engel and R. Nagel [76] that $e^{\pm i\theta}K$ generates a bounded C_0-semigroup in $C([0, 1])$ for some $\theta > 0$ (and conversely). On the other hand, since $p(x) > 0$ on $[0, 1]$ and continuous, $e^{\pm i\theta}HK$ generates a bounded C_0-semigroup in $C([0, 1])$, see J. R. Dorroh [70; pp. 36-37]. A new application of K.-J. Engel and R. Nagel [76] implies that $HK = A = A_\infty$ generates a bounded analytic semigroup in $C([0, 1])$.

REMARK. If $\beta = \alpha' + k\alpha$, where $k \in C([0, 1])$, then $\frac{\beta}{\sqrt{\alpha}} = \frac{\alpha'}{\sqrt{\alpha}} + k\sqrt{\alpha} \in C([0, 1])$.

The following variant of Theorem 7.18 holds.

THEOREM 7.19. *Let assumption (7.1) be verified with* $\sqrt{\alpha} \in C^1([0, 1])$, $\beta \in C([0, 1])$ *and* $\frac{\beta^2}{\alpha} \in C([0, 1])$. *Then* $(A, \mathcal{D}(A))$ *generates an analytic semigroup in* $C([0, 1])$.

PROOF. It only needs to observe that

$$Au = \sqrt{\alpha}(\sqrt{\alpha}u')' + \frac{2\beta - \alpha'}{2\sqrt{\alpha}}(\sqrt{\alpha}u') = C_\infty^2 u + \frac{2\beta - \alpha'}{2\sqrt{\alpha}}C_\infty u.$$

THEOREM 7.20. *Let* $\beta, \sqrt{\alpha} \in C^1([0, 1])$, $\alpha(0) = \alpha(1) = 0$, $\alpha(x) > 0$ *on* $(0, 1)$. *If* $\frac{\beta}{\alpha} \in C([0, 1])$, *then the operator* B *given by* $Bu = \alpha u'' + \beta u'$,

$$\mathcal{D}(B) = \{u \in C^1([0, 1]) \cap C^3(0, 1); \ Bu \in C_0^1([0, 1])\}$$

generates an analytic semigroup in $C^1([0, 1])$.

PROOF. Let $\beta_1(x) = \alpha'(x) - \beta(x)$ and define, as in the proof of Theorem 7.17, an operator A by

$$\begin{cases} \mathcal{D}(A) = \{v \in C([0, 1]) \cap C^2(0, 1); \ \alpha v'' + \beta_1 v' \in C([0, 1])\} \\ Av = \alpha v'' + \beta_1 v' = \alpha e^{-\int_{1/2}^x \frac{\beta_1}{\alpha} dt}(e^{\int_{1/2}^x \frac{\beta_1}{\alpha} dt}v')'. \end{cases}$$

Then we have: $Av = e^{\int_{1/2}^x \frac{\beta(t)}{\alpha(t)} dt}(\alpha(x)e^{-\int_{1/2}^x \frac{\beta(t)}{\alpha(t)} dt}v')'$. In addition, the assumption $\frac{\beta}{\alpha} \in C([0, 1])$ guarantees that $\eta(x) = e^{\int_{1/2}^x \frac{\beta_1(t)}{\alpha(t)} dt} \in C^1([0, 1])$. Since $\beta_1 = \alpha' - \frac{\beta}{\alpha}\alpha \in C([0, 1])$, the Remark above implies that A generates an analytic semigroup in $C([0, 1])$. Therefore, for all $f \in C^1([0, 1])$, the problem

$$\lambda v - Av = \alpha(\tfrac{1}{2})e^{\int_{1/2}^x \frac{\beta(t)}{\alpha(t)} dt}f'$$

has a unique solution v with $\eta(x)v'(x) \to 0$ as $x \to 0, 1$. Then it is readily seen that

$$u(x) = \frac{f(0)}{\lambda} + \int_0^x \alpha(t)^{-1} e^{\int_{1/2}^t \frac{\beta_1(s)}{\alpha(s)} ds} v(t) dt$$

$$= \frac{f(0)}{\lambda} + \alpha(\tfrac{1}{2})^{-1} \int_0^x e^{-\int_{1/2}^t \frac{\beta(s)}{\alpha(s)} ds} v(t) dt$$

satisfies

$$(\lambda - B)u = \lambda u - \alpha u'' - \beta u' = f \quad \text{on } (0, 1),$$

and $\eta(x)v'(x) = \alpha(x)u''(x) + \beta(x)u'(x) \to 0$ as $x \to 0, 1$. The estimate that ensures the analyticity of the semigroup is deduced by introducing the new norm $\| \cdot \|_1$ on $\mathcal{C}^1([0,1])$ as in the proof of Theorem 7.17.

This completes the proof.

Our final aim is to show that under the hypotheses $\sqrt{\alpha} \in \mathcal{C}^1([0,1])$ and $\frac{\beta}{\sqrt{\alpha}} \in \mathcal{C}([0,1])$, the operator $(W, \mathcal{D}(W))$, $Wu = \alpha u'' + \beta u'$, $\mathcal{D}(W) = \{u \in \mathcal{C}([0,1]) \cap \mathcal{C}^2((0,1)); Wu(x) \to 0$ as $x \to 0, 1\}$, i.e. $\alpha u'' + \beta u'$ with Ventcel's boundary conditions, generates an analytic semigroup in $\mathcal{C}([0,1])$, as well.

As a preliminary step, we introduce in the space $\mathcal{C}_0([0,1])$ of all $u \in \mathcal{C}([0,1])$ vanishing at $x = 0, 1$, the operator V given by $Vu = \alpha u'' + \beta u'$, $u \in \mathcal{D}(V) = \{u \in \mathcal{C}_0([0,1]) \cap \mathcal{C}^2((0,1)); Vu \in \mathcal{C}_0([0,1])\}$, and we show that V generates an analytic semigroup in $\mathcal{C}_0([0,1])$.

THEOREM 7.21. *Let* $\alpha, \beta \in \mathcal{C}([0,1])$, $\alpha > 0$ *on* $(0,1)$, $\alpha(0) = \alpha(1) = 0$. *If* $\sqrt{\alpha} \in \mathcal{C}^1([0,1])$ *and* $\frac{\beta}{\sqrt{\alpha}} \in \mathcal{C}([0,1])$, *then* V *generates an analytic semigroup in* $\mathcal{C}_0([0,1])$.

PROOF. Given $\Re\lambda > 0$, $\lambda \in \mathbb{C}$, and $f \in \mathcal{C}_0([0,1])$, the equation

$$(\lambda - V)u = f$$

is equivalent to the problem, in $\mathcal{C}_0(\overline{\mathbb{R}}) = \mathcal{C}_0([-\infty, \infty]) = \mathcal{C}_0(\mathbb{R})$,

$$(\lambda - V')v = g,$$

where $\mathcal{D}(V') = \{v \in \mathcal{C}_0(\mathbb{R}) \cap \mathcal{C}^2(\mathbb{R}); V'v \in \mathcal{C}_0(\mathbb{R})\}$,

$$(V'v)(t) = v''(t) - \eta(t)v'(t),$$

$$\eta(t) = \frac{\alpha'(\varphi(t)) - 2\beta(\varphi(t))}{2\sqrt{\alpha(\varphi(t))}}, \quad t \in \mathbb{R}, \quad g(t) = f(\varphi(t)),$$

$\varphi(t)$, $t \in \overline{\mathbb{R}}$, solves

$$\varphi'(t) = \sqrt{\alpha(\varphi(t))},$$

and is strictly increasing from $\overline{\mathbb{R}}$ onto $[0,1]$.

Notice that
$$\int_{\frac{1}{2}}^{x} \frac{dy}{\sqrt{\alpha(y)}} = \psi(x), \quad x \in [0,1],$$

satisfies $\psi'(x) = \dfrac{1}{\sqrt{\alpha(x)}} > 0$ on $(0,1)$ and then $\varphi = \psi^{-1}$ has just the defined properties.

Let us observe also that $\eta(t)$ is bounded and uniformly continuous on $\mathbb{R}$ since there exist the limits

$$\lim_{t \to \pm\infty} \eta(t) \in \mathbb{R}.$$

Then it is known by Corollary 3.1.9 of A. Lunardi [131; p. 81], that $(V',$ $\mathcal{D}(V'))$ generates an analytic semigroup in $\mathcal{C}_0(\mathbb{R})$. Therefore, $u(x) = v(\psi(x))$ is the desired solution to $(\lambda - V)u = f$ and

$$\|u\|_\infty = \sup_{\mathbb{R}} |v(t)| \le C|\lambda|^{-1} \sup_{\mathbb{R}} |g(t)| = C|\lambda|^{-1}\|f\|_\infty.$$

This completes the proof.

THEOREM 7.22. *Let* $\sqrt{\alpha} \in \mathcal{C}^1([0,1])$, $\alpha(0) = \alpha(1) = 0$, $\alpha > 0$ *on* $(0,1)$, $\beta \in \mathcal{C}_0([0,1])$. *If* $\frac{\beta}{\sqrt{\alpha}} \in \mathcal{C}([0,1])$, *then* W *generates an analytic semigroup in* $\mathcal{C}([0,1])$.

PROOF. Notice that if $\alpha(x) = a^2x^2 + o(x^2)$, $\beta(x) = \sigma a x + o(x)$, then $\frac{\beta(x)}{\alpha(x)} \sim \frac{\sigma}{a}\frac{1}{x}$ as $x \to 0$ and thus $W(x)^{-1} = e^{\int_{1/2}^{x} \frac{\beta(t)}{\alpha(t)} dt} \sim Cx^{\frac{\sigma}{a}}$. Hence,

$$\int_0^{\frac{1}{2}} W(x) \int_{\frac{1}{2}}^{x} \alpha(t)^{-1} W(t)^{-1} dt dx \sim \int_0^{\frac{1}{2}} x^{-\frac{\sigma}{a}} \left(\int_{\frac{1}{2}}^{x} t^{\frac{\sigma}{a}-2} dt \right) dx.$$

If $\frac{\sigma}{a} \ne 1$,

$$\int_0^{\frac{1}{2}} x^{-\frac{\sigma}{a}} [\int_{\frac{1}{2}}^{x} t^{\frac{\sigma}{a}-2} dt] dx = \int_0^{\frac{1}{2}} \frac{a}{\sigma - a}(x^{-1} - 2^{1-\frac{\sigma}{a}} x^{-\frac{\sigma}{a}}) dx$$

is divergent. If $\frac{\sigma}{a} = 1$, then it reduces to $\int_0^{\frac{1}{2}} x^{-1}(\log x + \log 2) dx = \infty$. Then the result of Ph. Clément and C. A. Timmermans [53; p. 381] applies and W generates a $\mathcal{C}_0$-contraction semigroup $S(t)$ in $\mathcal{C}([0,1])$.

We can repeat the same argument for $\alpha(x) = a^2x^{2m} + o(x^{2m})$, $m > 1$. It remains to prove that $S(t)$ is an analytic semigroup. To this end, we observe that for all $\lambda \in \mathbb{C}$, $\Re\lambda > 0$, and every $f \in \mathcal{C}([0,1])$, the problem

(7.44) $\lambda v - Vv = f_\lambda \quad$ on $(0,1),$

where

$$f_\lambda(x) = f(x) - (1-x)f(0) - xf(1) - \lambda^{-1}\beta(x)[f(0) - f(1)] \in \mathcal{C}_0([0,1]),$$

has a unique solution $u \in \mathcal{D}(V) \subset \mathcal{C}_0([0,1])$, with

$$\|v\|_\infty \leq C(1 + |\lambda|)^{-1}\|f_\lambda\|_\infty,$$

where V is precisely the operator entering in Theorem 7.21.

Define $u(x) = v(x) + \frac{1}{\lambda}((1-x)f(0) + xf(1))$, so that $u \in \mathcal{C}([0,1]) \cap \mathcal{C}^2((0,1))$, $u'(x) = v'(x) + \frac{1}{\lambda}(f(1) - f(0))$.

Now, (7.44) is written into the form

$$\lambda u - \alpha u'' - \beta u' = f$$

for

$$\lambda[v(x) + \lambda^{-1}\{(1-x)f(0) + xf(1)\}] - \alpha(x)[v(x) + \lambda^{-1}\{(1-x)f(0) + xf(1)\}]''$$
$$- \beta(x)[v(x) + \lambda^{-1}\{(1-x)f(0) + xf(1)\}]' = f(x).$$

On the other hand, $\alpha(x)v''(x) + \beta(x)v'(x) + \lambda^{-1}\beta(x)\{f(1) - f(0)\} \to 0$ as $x \to 0, 1$. It follows that u is the (unique) solution to $(\lambda - W)u = f$. Since

$$\|u\|_\infty \leq \|v\|_\infty + \frac{2}{|\lambda|}\|f\|_\infty \leq \frac{C+2}{|\lambda|}\|f\|_\infty, \quad f \in \mathcal{C}([0,1]),$$

the proof is complete.

Notice that Theorems 7.19 and 7.22 substantially affirm that under the indicated assumptions no boundary conditions reduce to Ventcel boundary conditions and conversely.

EXAMPLE. Take $\alpha(x) = x^j(1-x)^j$, $j \geq 2$, and $\beta(x) = x^k(1-x)^k\gamma(x)$, where $\gamma \in \mathcal{C}([0,1])$ and $k \geq \frac{j}{2}$.

Then all the assumptions in Theorem 7.19 (no boundary conditions or Neumann boundary conditions) and 7.22 are verified.

If $j \geq 2$, $k \geq j$, we conclude from Theorem 7.20 that $\alpha u'' + \beta u'$ with Ventcel's boundary conditions generates an analytic semigroup in $\mathcal{C}^1([0,1])$.

If $j = k = 1$, $\gamma(x) = $ const., we have that the operator $Bu = x(1-x)u'' + \gamma x(1-x)u'$ describing influence of selection in genetics, (see P. Mandl [133; p. 48]), with Ventcel's boundary conditions generates a differentiable semigroup in $\mathcal{C}^1([0,1])$, since condition (7.43) is verified by $Z(x) = x(1-x)e^{-\gamma(x-\frac{1}{2})}$.

If $\alpha(x) = x^2(1-x)^2$, $\beta(x) \equiv 0$, then $\alpha(x)u'' + \beta(x)u'$ is the operator related to changes in gene frequency under random fluctuation of selective advantages, P. Mandl [133; p. 49]. Our last Theorem 7.22 affirms that under Ventcel's boundary conditions the operator generates an analytic semigroup in $\mathcal{C}([0,1])$.

7.4 DIFFERENTIAL OPERATORS ON $(0, \infty)$

The key idea in treating the second order differential operator $Au = \alpha u'' + \beta u'$ on the interval $(0, \infty)$ should be to operate a change of variable $x = \varphi(t)$, $t \in [0, 1]$, like $\varphi(t) = \tan(\frac{\pi}{2}t)$, to handle the corresponding operator

$$Bv = \frac{\alpha(\varphi(t))}{\varphi'(t)^2} v'' + \frac{\beta(\varphi(t))\varphi'(t)^2 - \alpha(\varphi(t))\varphi''(t)}{\varphi'(t)^3} v', \quad 0 < t < 1,$$

where $v(t) = u(\varphi(t))$, by applying the results above and then to go back to the operator A.

Here we confine ourselves to a direct approach and show that the technique of using spaces with weight works as well to treat the linear Kompaneets equation

$$\frac{\partial u}{\partial t} = \frac{1}{\beta(x)} \frac{\partial}{\partial x} \left[\alpha(x) \left(\frac{\partial u}{\partial x} + k(x)u \right) \right], \quad t > 0, \ x > 0,$$

with initial condition

$$u(0, x) = u_0(x), \quad x > 0,$$

and Neumann boundary conditions

$$\lim_{x \to 0, \infty} \alpha(x) \left[\frac{\partial u}{\partial x}(t, x) + k(x)u(t, x) \right] = 0, \quad t > 0.$$

For this equation, in the general nonlinear case, we refer to J. A. Goldstein [110], while for the linear case we indicate K. Wang [174].

The following analyticity result extends K. Wang's theorem [174; p. 568] in that weaker regularity will be assumed regarding α, β.

Our assumptions read

$$(7.45) \quad \begin{cases} \alpha \in \mathcal{C}((0, \infty)), \ \beta \in L^\infty_{loc}((0, \infty)), \ \alpha(x) > 0, \ \beta(x) > 0 \\ \qquad\qquad\qquad\qquad\qquad\qquad\qquad\qquad \text{for all } x > 0, \\ \alpha(x) = O(x^j) \ \text{as } x \to 0 \text{ for some } j \geq 1; \end{cases}$$

$$(7.46) \qquad\qquad\qquad k \in \mathcal{C}([0, \infty)),$$

$$(7.47) \qquad\qquad\qquad \int_0^\infty \beta(t)e^{-\int_1^t k(s)ds} dt < \infty,$$

(7.48) $\displaystyle\inf_{0<x<\infty}\beta(x)^{-1}e^{\int_1^x k(t)dt}>0,\qquad \inf_{0<x<\infty}\alpha(x)^{-1}e^{\int_1^x k(t)dt}>0.$

We observe that $Ku=\frac{1}{\beta}[\alpha(u'+ku)]'$ is formally expressed by means of

$$Ku=\frac{1}{\beta}\left(\frac{\alpha}{\gamma}(\gamma u)'\right)',$$

where $'$ denotes the derivative and $\gamma(x)=e^{\int_1^x k(s)ds}$, $x>0$.

The space L^2 with weight that we use is given by

$$X=\{u\colon(0,\infty)\to\mathbb{C};\ u\ \text{measurable},\ \int_0^\infty\beta(x)\gamma(x)|u(x)|^2dx<\infty\},$$

endowed with the inner product

$$\langle u,v\rangle=\int_0^\infty\beta(x)\gamma(x)u(x)\overline{v(x)}dx,\quad u,\,v\in X.$$

$(X,\langle\cdot,\cdot\rangle)$ is a Hilbert space. Let us introduce the Hilbert space V as

$$V=\left\{u\in\mathcal{C}^1(0,\infty)\cap X;\ \int_0^\infty\frac{\alpha(x)}{\gamma(x)}|(\gamma u)'(x)|^2dx<\infty,\right.$$
$$\left.\int_0^\infty\frac{\gamma(x)}{\beta(x)}\left|\left(\frac{\alpha(x)}{\gamma(x)}(\gamma u)'(x)\right)'\right|^2dx<\infty\right\},$$

with the inner product

$$\langle u,v\rangle_V=\langle u,v\rangle+\int_0^\infty\frac{\alpha(x)}{\gamma(x)}(\gamma u)'(x)(\gamma\bar v)'(x)dx$$
$$+\int_0^\infty\frac{\gamma(x)}{\beta(x)}\left(\frac{\alpha(x)}{\gamma(x)}(\gamma u)'(x)\right)'\left(\frac{\alpha(x)}{\gamma(x)}(\gamma\bar v)'(x)\right)'dx,\quad u,\,v\in V.$$

In other words, $u\in\mathcal{C}^1(0,\infty)$ belongs to V if and only if

$$\int_0^\infty\frac{\beta(x)}{\gamma(x)}|(\gamma u)(x)|^2dx<\infty,$$
$$\int_0^\infty\frac{\gamma(x)}{\alpha(x)}\left|\frac{\alpha(x)}{\gamma(x)}(\gamma u)'(x)\right|^2dx<\infty,$$
$$\int_0^\infty\frac{\gamma(x)}{\beta(x)}\left|\left(\frac{\alpha}{\gamma}(\gamma u)'\right)'(x)\right|^2dx<\infty.$$

In virtue of assumption (7.48), $u \in V$ satisfies

$$\frac{\alpha}{\gamma}(\gamma u)' \in H^1(0, \infty),$$

and thus $\lim_{x \to \infty} \frac{\alpha(x)}{\gamma(x)}(\gamma u)'(x)$ exists and equals zero.

In addition, there exists

$$\lim_{x \to 0+} \frac{\alpha(x)}{\gamma(x)}(\gamma u)'(x) = \lambda \in \mathbb{C}.$$

But if $\lambda \neq 0$, then necessarily

$$\int_0^1 \frac{\gamma(x)}{\alpha(x)}\,dx < \infty,$$

contradicting (7.45).

One then deduces that every $u \in V$ satisfies the boundary conditions

$$\lim_{x \to 0, \infty} \frac{\alpha(x)}{\gamma(x)}(\gamma u)'(x) = 0.$$

We verify that the operator $(A, \mathcal{D}(A))$, where $\mathcal{D}(A) = V$ and $Au = Ku$, $u \in \mathcal{D}(A)$, is symmetric with respect to the inner product $\langle \cdot, \cdot \rangle$ and non positive. Let $u, v \in V$. Then

$$\langle Au, v \rangle = \int_0^\infty \gamma(x)\left(\frac{\alpha}{\gamma}(\gamma u)'\right)'(x)\overline{v(x)}dx$$

$$= \left[\frac{\alpha(x)}{\gamma(x)}(\gamma u)'(x)(\gamma \overline{v})(x)\right]_{x=0}^{x=\infty} - \int_0^\infty \frac{\alpha(x)}{\gamma(x)}(\gamma u)'(x)(\gamma \overline{v})'(x)dx.$$

Since both the above integrals converge, the limits

$$\lim_{x \to 0, \infty} \frac{\alpha(x)}{\gamma(x)}(\gamma u)'(x)(\gamma \overline{v}(x))$$

exist. We show that they vanish.

Let $\lim_{x \to 0} \frac{\alpha(x)}{\gamma(x)}(\gamma u)'(x)(\gamma(x)\overline{v(x)}) = \mu \neq 0$; since $u \in V$, we have

$$\left|\frac{\alpha(x)}{\gamma(x)}(\gamma u)'(x)\right| = \left|\int_0^x \left(\frac{\alpha}{\gamma}(\gamma u)'\right)'(t)dt\right|$$

$$\leq \left(\int_0^x \frac{\beta(t)}{\gamma(t)}dt\right)^{\frac{1}{2}}\|u\|_V \leq C\left(\int_0^x \frac{\beta(t)}{\gamma(t)}dt\right)^{\frac{1}{2}}.$$

Then there is a positive constant C_1 such that

$$|\gamma(x)v(x)|^2 \geq C_1 \left(\int_0^x \frac{\beta(t)}{\gamma(t)} dt \right)^{-1},$$

where $x \in (0, \delta)$, $\delta > 0$ small. This implies that

$$\int_0^\delta \frac{\beta(x)}{\gamma(x)} |\gamma(x)v(x)|^2 dx \geq C_1 \int_0^\delta \frac{\beta(x)}{\gamma(x)} \left(\int_0^x \frac{\beta(t)}{\gamma(t)} dt \right)^{-1} dx$$

$$= C_1 \left[\log \int_0^x \frac{\beta(t)}{\gamma(t)} dt \right]_{x=0}^{x=\delta} = \infty,$$

contradicting $v \in V$.

Analogously, if $\lim_{x \to \infty} \frac{\alpha(x)}{\gamma(x)} (\gamma u)'(x)(\gamma(x)\overline{v(x)}) = \mu \neq 0$, then, from

$$\left| \frac{\alpha(x)}{\gamma(x)} (\gamma u)'(x) \right| = \left| -\int_x^\infty \left(\frac{\alpha}{\gamma} (\gamma u)' \right)'(t) dt \right| \leq \left(\int_0^\infty \frac{\beta(t)}{\gamma(t)} dt \right)^{\frac{1}{2}} \|u\|_V,$$

Assumption (7.47) gives

$$\frac{\beta(x)}{\gamma(x)} |\gamma(x)v(x)|^2 \geq C_1 \frac{\beta(x)}{\gamma(x)} \left(\int_x^\infty \frac{\beta(t)}{\gamma(t)} dt \right)^{-1}.$$

Thus, for a suitable $\delta \geq 1$,

$$\int_\delta^\infty \frac{\beta(x)}{\gamma(x)} |\gamma(x)v(x)|^2 dx \geq C_1 \int_\delta^\infty \frac{f'(x)}{f(x)} dx = -C_1 [\log f(x)]_{x=\delta}^{x=\infty},$$

where $f(x) = \int_x^\infty \frac{\beta(t)}{\gamma(t)} dt \to 0$ as $x \to \infty$. Since $u \in V$, a contradiction is again obtained.

A second integration by parts yields

$$\langle Au, v \rangle = \langle Av, u \rangle, \quad \text{for } u, v \in V,$$

and $\langle Au, u \rangle \leq 0$ for all $u \in V$.

Let us consider the sesquilinear form $b(u, v)$ on $V_1 \times V_1$, where V_1 is the Hilbert space

$$V_1 = \{ f \in X; \int_0^\infty \frac{\alpha(x)}{\gamma(x)} |(\gamma f)'(x)|^2 dx < \infty \},$$

equipped with the inner product

$$b(u, v) = \langle u, v \rangle_1 = \langle u, v \rangle + \int_0^\infty \frac{\alpha(x)}{\gamma(x)} (\gamma u)'(x)(\gamma \overline{v})'(x) dx.$$

For all $u \in \mathcal{D}(A) = V$ we have

$$\langle (I - A)u, u \rangle = b(u, u) = \|u\|_1^2,$$

and thus Lax-Milgram theorem implies that

$$X \subset V_1' \subset \mathcal{R}(I - A).$$

Therefore $(A, \mathcal{D}(A))$ is self adjoint and A generates an analytic semigroup in X.

REMARK. Take $k(x) \equiv k_0 > 0$, $\alpha(x) = x^j$, $\beta(x) = x^s$, with $j \geq 1$ and $s > 0$. Then all assumptions (7.45)-(7.48) are satisfied.

The preceding discussion yields then the following statement.

THEOREM 7.23. *Under assumptions (7.45)-(7.48), the operator* $Au = \frac{1}{\beta}(\alpha(u' + ku))'$, $u \in V$ *(and hence u satisfies the Neumann boundary conditions $\alpha(u' + ku) \to 0$ as $x \to 0, \infty$), generates an analytic semigroup in* $L^2_{\beta\gamma}(0, \infty)$.

BIBLIOGRAPHICAL REMARKS

CHAPTER 0. The theory of linear operators between Banach spaces is carefully described in several monographs. We confine ourselves to quote T. Kato [117], A. E. Taylor [169], K. Yosida [187].

Interpolation spaces have received much attention even in very recent years, both for their theoretical relevance and very useful applications. For this topic we used in a special way H. Amann [5]; see also P. L. Butzer and H. Berens [39], J. L. Lions and E. Magenes [129], J. L. Lions and J. Peetre [130], A. Lunardi [131] and H. Triebel [172]. For fractional powers of linear operators and interpolation we again refer to H. Triebel [172], but also to H. Tanabe [167,168].

Semigroup theory and its applications to abstract Cauchy problem are treated in several valuable monographs like H. O. Fattorini [78], A. Friedman [104], J. A. Goldstein [109], S. G. Krein [121], A. C. McBride [135], A. Pazy [145], H. Tanabe [167,168], K. Yosida [187].

CHAPTER I. Multivalued operators are chiefly used for in literature in order to treat nonlinear problems; the corresponding theory is well organized and developed for monotone (accretive) ones. Basic reference texts for this subject are the monographs by V. Barbu [17,19], H. Brezis [36]; see also R. W. Carroll [47], R. Martin [134] and Ph. Bénilan's thesis [26].

Fractional powers of non negative multivalued linear operators are treated by El. H. Alaarabiou [3]. Here we introduce fractional powers of graphs that are not necessarily non negative, following A. Favini and A. Yagi [97]. Theorem 1.12 is an extension to multivalued operators of previous results due to C. Wild [180] relative to holomorphic semigroups of growth $\alpha < 1$. We remark that, if applied to positive operators in the usual sense, it reduces to Proposition 0.8 with $p = \infty$, for which one can also see the far-reaching research of J. L. Lions and J. Peetre [130].

The results on dual operators given in Theorems 1.14 and 1.15 are new.

CHAPTER II. The generation results here described are the multivalued counter part of the theory, due to G. Da Prato and E. Sinestrari [66], for the linear operators with a not necessarily dense domain. Theorem 2.7 should be compared with the famous theorem of M. G. Crandall and T. Liggett

[59]. The main part of Section 2.4 is an extension of results due to A. Yagi [185].

The initial value problem (D-E.1) is also studied by M. Povoas in [146], but she considers only solvability in the space $L^2(0, T; X)$, where X is a Hilbert space. For semigroup distributions we refer to J. L. Lions [127]. The monograph [128] by J. L. Lions is really a remarkable and enlightening text on weak solutions to operator equations.

Other general approaches to degenerate differential equations can be found in N. Sauer [149,150], V. P. Skripnik [158], G. A. Sviridyuk [162,163]. The nonlinear Volterra equation in Hilbert spaces

$$Mu(t) - \int_0^t a(t-s)Lu(s)ds = f(t),$$

reducing to $\frac{d}{dt}Mu - Lu = f'$ when $a(t) \equiv 1$, is treated in V. Barbu [18].

CHAPTER III. For representation of semigroups with weak singularity, we refer the reader to S. G. Krein [121] and K. Taira [166]. The Theorems in Sections 3.3-3.5 improve and refine some previous results of the authors in [96,97] and extend regularity of solutions to $u'(t) = Au(t) + f(t)$ for sectorial operators A; see P. Acquistapace and B. Terreni [1,2], G. Da Prato and P. Grisvard [62,63,64,65], P. Grisvard [113,114], A. Lunardi [131], E. Sinestrari [157].

In Section 3.6, Theorem 3.29 is due to A. Favini and M. Fuhrman [86] and gives a version for graphs of Trotter-Kato Theorem (see Theorem 2.17 of T. Kato's monograph [117; p. 503]). Theorem 3.31 assuring convergence of the approximate solutions u_n to the limit solution u, is new in showing that the derivative $\frac{du_n}{dt}$ converges to $\frac{du}{dt}$ in the $C^{\omega'}([0, T]; X)$ norm.

Example 3.2 of an equation in Hilbert space is new, while Examples 3.3-3.5, devoted to equations of elliptic-parabolic type, are based upon A. Favini and A. Yagi [96,97]. The estimates appearing in Example 3.6 are a direct extension of arguments in P. Pazy [145] and in P. E. Sobolevskii [159]. The treatment of degenerate parabolic equations as given in Examples 3.8-3.9 is new. Differential equations of Sobolev type are widely treated in literatue: see G. A. Sviridyuk [163] and also R. E. Showalter [156] for the nonlinear case. Example 3.10 is closely related to a problem considered by J. Lagnese [122], who used a completely different method. The idea to view the (linear) Stokes equation as a multivalued linear equation is outlined in A. Favini and A. Yagi [97]. The stability results in Examples 3.12-3.13 are simple corollaries of Sections 3.6, but more general situations are investigated in V. Barbu and A. Favini [20] and A. Favini and M. Fuhrman [86], where most applications are inspired by the concrete partial differential equations appearing in the monograph [11] by H. Attouch.

CHAPTER IV. The main part of this chapter improves the results in A. Favini and A. Yagi [97] and provides, among other things, extensions to the multivalued case, of well known theorems in abstract differential equations. See P. Acquistapace and B. Terreni [1,2], T. Kato and H. Tanabe [118], H. Tanabe [167,168], and, above all, A. Yagi [182,183,184].

The main result for maximal regularity of solutions, as given in Theorem 4.12, is new. Applications to $\frac{d}{dt}M(t)u(t) = L(t)u(t) + f(t)$ in Section 4.3 are either derived from A. Favini and A. Yagi [96,97] or new, as well as the contents of Section 4.4, where $M(t)$ is the multiplication by $t^\ell (t \geq 0)$, $\ell > 0$. Related, but quite different, statements on this last subject were obtained by P. E. Sobolevskii [160] and by A. Friedman and Z. Schuss [105].

Singular, linear or nonlinear, Cauchy problems are also handled from a completely different point of view, by M. S. Baouendi and C. Baiocchi [14], M. S. Baouendi and C. Goulaouic [16], M. L. Bernardi [28,29], and M. L. Bernardi and G. A. Pozzi [30,31].

CHAPTER V. Sections 5.1-5.2 follow from A. Favini [85] and extend the operational method of P. Grisvard [113,114]. A similar technique was also delivered by Yu. Dubinskii [72]. It is an abstract version of the Laplace transform method. This tool was very often applied to (abstract) differential equations, see R. Beals [23,24] and J. Chazarain [49], but it was modified in order to handle degenerate equations by A. Favini in [83,85], and, above all, by A. Rutkas in [147,148].

As long as the examples are concerned, we remark that most of them can be either improved or refined. (For instance, Example 5.7 has been already carefully generalized by G. Coppoletta in [58]). Our aim here in fact is to indicate the wide range of possible applications. Nevertheless, we observe that the treatment of degenerate parabolic equations in $\mathcal{C}(\overline{\Omega})$ and in $L^1(\Omega)$ as given in Examples 5.10-5.11, respectively, is new. One could handle, by similar arguments, the corresponding equation in $L^1(\mathbb{R}^n)$. For interesting related nonlinear problems, see Ph. Bénilan, H. Brezis and M. G. Crandall [26]. For Example 5.12, we refer also to L. Boccardo and T. Gallouët [33] and to T. Gallouët and J. -M. Morel [107].

The abstract Gevrey case in Section 5.4 is new and it finds its motivation from some control problems for hyperbolic partial differential equations as described in A. Favini and R. Triggiani [95]. See also I. Lasiecka and R. Triggiani [125]. It should be observed that under condition $\mathcal{D}(B)$ comparable to $\mathcal{D}(A^\alpha)$, where A, B are two self adjoint operators, $A > 0$, the second order differential equation in Hilbert space H, $u'' + Bu' + Au = f$ can be transformed into a first order equation in the product space $\mathcal{D}(A^{\frac{1}{2}}) \times H$ in which the operator matrix $\begin{pmatrix} 0 & I \\ -A & -B \end{pmatrix}$ generates either an analytic semigroup (for $\frac{1}{2} \leq \alpha \leq 1$) or just a Gevrey semigroup (for $0 < \alpha < \frac{1}{2}$). For more details, see S. Chen and R. Triggiani [50,51]. Very interesting stabilization results relative to the last equation are described by J. Lagnese in his monograph [123]. Application of abstract potential operators to degenerate differential equations is due to A. Favini [82]. The case when $\lambda = 0$ is an isolated singular point of $(\lambda L + M)^{-1}$ has been extensively studied. One starts with the finite-dimensional frameworks, as in F. R. Gantmacher [108] and continues, in more recent years, with the researches by S. L. Campbell [40,41], S. L. Campbell, C. D. Meyer Jr. and N. J. Rose [42], J. D. Cobb [54], E. Griepentrog and R. Marz [112], M. Z. Nashed and Y. Zhao [140], S. Zubova and K. Tchernyshov [188]. Related literature can also be found in the monograph on generalized inverses of linear transformations by S. L. Campbell and C. D. Meyer Jr. [44]. Control problems for finite dimensional systems that are either degenerate by themselves or have a singular cost functional are treated in S. L. Campbell [40,41,43], J. D. Cobb [55], L. Dai [60]. The last book contains a wide bibliography on the subject until 1989, too.

For an interesting minimum problem for a Sobolev equation in Hilbert space, see G. A. Sviridyuk and A. A. Efremov [164]. Numerical treatment of differential-algebraic equations is developped in K. E. Brenan, S. L. Campbell and L. R. Petzold [34]. The results in Section 5.6 were mainly established by A. Favini [83,84]. For the representation of the Banach space X as the direct sum $X = \mathcal{K}(B^m) \oplus \mathcal{R}(B^m)$, see also A. E. Taylor [169]. Basic theory of Fredholm operators is described in M. Schechter [151,152]. Example 5.14 should be compared to some results by G. A. Sviridyuk [163].

The content of Section 5.7 on equations of higher order in time are new. Usually one seeks to reduce them to first order equations in suitable product spaces, but it is difficult to find general conditions on the coefficients guaranteeing that the best estimates as in Chapter II holds. Partial results for the non degenerate case and second order equations are given in H. O. Fattorini [78], K. -J. Engel [75], K. -J. Engel and R. Nagel [76].

Integrated semigroups received much attention in last years for their application to concrete partial differential equations, like Schrödinger equations. See, e. g. F. Neubrabder [141]. Section 5.8 is based on W. Arendt and A. Favini's paper [8]. The second approach described therein uses Laplace transform method from W. Arendt [9] and must be compared to F. Neubrander and A. Rhandi [142]. Convergence theorems for integrated semigroups were obtained by S. Busenberg and B. Wu [38].

CHAPTER VI. Theorems 6.1-6.5 are due to A. Favini and A. Yagi [98] and extend some results by S. G. Krein [121], S. Yu. Yakubov [186] and A. Favini and E. Obrecht [90] to degenerate equations. Theorems 6.6 and 6.7 are new and constitute a version for implicit higher order in time differential equations of previous results due to M. K. Balaev [13] and A. Favini and H. Tanabe [91]. Theorems 6.8-6.9 on second order equations with time dependent operator coefficients are new, too. Also the major part of Examples 6.1-6.17 in Section 6.2 appear here for the first time. Here we use, in particular, some results by R. deLaubenfels [67] on powers of generators and by R. Seeley [154] on complex interpolation spaces.

The main results in Section 6.3 are due to P. Colli and A. Favini [56]. See also P. Colli and A. Favini [57] for a different method and related references. The equations of hyperbolic-parabolic type in Example 6.18 are very much studied, above all when operator B is nonlinear, after the paper [27] by A. Bensoussan, J. L. Lions and G. C. Papanicolau. We quote O. A. de Lima [68], A. B. Maciel [132], L. A. Medeiros [136]. Examples 6.19-6.22 are new as well as the theory and its applications in Section 6.4.

The abstract elliptic-hyperbolic equations discussed in Section 6.6 are widely treated in literature. See R. W. Carroll and R. E. Showalter [48], R. W. Carroll [47]. The main results here are due to A. Favini [79,80]. A similar technique (for the hyperbolic case) was developed by J. A. Donaldson [69].

CHAPTER VII. Degenerate parabolic equations, regularity of their solutions together with their important applications have a very extensive literature. Concerning their connection to approximation theory, we quote F. Altomare and M. Campiti [4]. An approach to the one dimensional case by means of multiplicative perturbation of generators in L^p spaces with weight is given in J. R. Dorroh and A. Holderrieth [71].

Singular ordinary differential operators are carefully investigated by classical methods in the monographs [74] and [139] by J. Elschner and M. A. Naimark, respectively.

For spectral theory of ordinary differential operators, connected in particular to regular Sturm-Liouville problem, see J. Weidmann [176]. For singular Sturm-Liouville problems we recall also H. D. Niessen and A. Zettl [143], M. M. H. Pang [144], M. G. Krein [120], W. N. Everitt, M. K. Kwong and A. Zettl [77]. All quoted references deal mainly with one dimensional space domains. For results in one-several dimensions, we refer to M. S. Baouendi and C. Goulaouic [15], to the monograph [166] by K. Taira, and to V. Vespri [173].

Our theorems 7.1-7.2 were estabished in V. Barbu, A. Favini and S. Romanelli [22], whereas Theorems 7.3-7.4 are new. Related results are found in M. Campiti and G. Metafune [42]. Theorems 7.5-7.7 were proven substantially in A. Favini, J. Goldstein and S. Romanelli [89]. A result similar to Theorem 7.8 is proven in M. Campiti, G. Metafune, D. Pallara [46] by using a slightly different technique. Theorem 7.9 is new.

The generation theorems in Section 7.2 are due to the results of V. Barbu and A. Favini [21]. Also here in some particular but important cases complex analyticity of the semigroup $T(t)$ in $\mathcal{C}([0,1])$ was proven by M. Campiti and G. Metafune [45], G. Metafune [137], extending the arguments by S. Angenent [7] relative to Neumann boundary conditions. Ventcel's boundary conditions appear in a natural way when treating important probabilistic phenomena. See, e. g. the pioneering works by A. D. Ventcel [177,178].

Section 7.3 is new. Some extensions of it are contained in the forth-coming paper [89] by A. Favini, J. Goldstein and S. Romanelli. Theorems 7.21-7.22 were proven in A. Favini and S. Romanelli [99]. Further genera-tion results in L^p spaces with weight and in L^p are described in A. Favini, J. Goldstein and S. Romanelli [88]. Second order differential operators on $(0, \infty)$ are considered in A. Favini and S. Romanelli [100]. The interesting paper [10] by W. Arendt describes some spectral properties of certain op-erators of this kind. Theorem 7.23 is due to A. Favini, J. Goldstein and S. Romanelli [88].

At last, we recall that an important generation result of C_0-semigroup in $C([0, \infty])$ was obtained by H. Brezis, W. Rosenkranz and B. Singer in [35]. Their technique of using special functions, like Bessel's functions has been adapted by S. Angenent [7], G. Metafune [137], M. Campiti and G. Metafune [45] in treating their interesting problems.

BIBLIOGRAPHY

[1]   P. Acquistapace and B. Terreni, *Some existence and regularity results for abstract non autonomous parabolic equations*, J. Math. Anal. Appl. **99** (1984), 9–64.

[2]   P. Acquistapace and B. Terreni, *A unified approach for abstract linear non-autonomous parabolic equations*, Rend. Sem. Mat. Univ. Padova **78** (1987), 47–107.

[3]   El. H. Alaarabiou, *Calcul fonctionnel et puissance fractionnaire d'opérateurs linéaires multivoque non négatifs*, C. R. A. S. Paris **313** (1991), 163–166.

[4]   F. Altomare and M. Campiti, *Korovkin-type Approximation Theory and its Applications*, De Gruyter, Berlin, 1994.

[5]   H. Amann, *Linear and Quasilinear Parabolic Problems, Vol.1: Abstract Linear Theory*, Birkhäuser, Basel, 1995.

[6]   K. T. Andrews, K. L. Kuttler and M. Shillor, *Second order evolution equations with dynamic boundary conditions*, J. Math. Anal. Appl. **197** (1996), 781–795.

[7]   S. Angenent, *Local existence and regularity for a class of degenerate parabolic equations*, Math. Ann. **280** (1988), 465–482.

[8]   W. Arendt and A. Favini, *Integrated solutions to implicit differential equations*, Rend. Sem. Mat. Univ. Polit. Torino **51** (1993), 513–529.

[9]   W. Arendt, *Vector Laplace transforms and Cauchy problems*, Israel J. Math. **59** (1987), 327–352.

[10]   W. Arendt, *Gaussian estimates and interpolation of the spectrum in L^p*, Diff. Int. Eqs. **7** (1994), 1153–1168.

[11]   H. Attouch, *Variational Convergence for Functions and Operators*, Pitman, London, 1984.

[12]   C. Baiocchi and M. S. Baouendi, *Singular evolution equations*, J. Funct. Anal. **25** (1977), 103–120.

[13] M. K. Balaev, *Higher order parabolic type evolution equations*, Dokl. Akad. Nauk. Azerbaidjan **41** (1988), 7–10. (russian)

[14] M. S. Baouendi and C. Baiocchi, *Equations d'evolution a coefficients singuliers*, Semin. Goulaouic -Schwartz 1972-73, Exposé **XXVIII** (1973).

[15] M. S. Baouendi and C. Goulaouic, *Régularité et théorie spectrale pour une classe d'opérateurs elliptiques dégénérés*, Archiv Rational Mech. Anal. **34** (1969), 361–379.

[16] M. S. Baouendi and C. Goulaouic, *Singular nonlinear Cauchy problems*, J. Diff. Eqs. **22** (1976), 268–291.

[17] V. Barbu, *Nonlinear Semigroups and Differential Equations in Banach Spaces*, Noordhoff, Leyden, 1976.

[18] V. Barbu, *Existence for nonlinear Volterra equations in Hilbert Spaces*, SIAM J. Math. Anal. **10** (1979), 552–569.

[19] V. Barbu, *Analysis and Control of Infinite Dimensional Systems*, Academic Press, New York, Boston, 1992.

[20] V. Barbu and A. Favini, *Convergence of solutions of implicit differential equations*, Diff. Int. Eqs. **7** (1994), 665–688.

[21] V. Barbu and A. Favini, *The analytic semigroup generated by a second order degenerate differential operator in* $C[0,1]$, 3rd International Conference on Functional Analysis and Approximation Theory, Maratea, 1996, Suppl. Rend. Circolo Matem. Palermo **52** (1998), 23–42.

[22] V. Barbu, A. Favini and S. Romanelli, *Degenerate evolution equations and regularity of their associated semigroups*, Funkc. Ekv. **39** (1996), 421–448.

[23] R. Beals, *On the abstract Cauchy problem*, J. Funct. Anal. **10** (1972), 281–299.

[24] R. Beals, *Laplace transform method for evolution equations*, "Boundary Value Problems for Linear Evolution and Partial Differential Equations" (H. G. Garnir, ed.), Reidel, Dordrecht, 1977, pp. 1–26.

[25] Ph. Bénilan, *Equations d'evolution dans une espace de Banach quelconque et applications*, Thèse d'Etat, Orsay (1972).

[26] Ph. Bénilan, H. Brezis and M. G. Crandall, *Semilinear elliptic equations in* $L^1(\mathbb{R}^n)$, Ann. Sc. Norm. Sup. Pisa **33** (1975), 523–555.

[27] A. Bensoussan, J. L. Lions and G. C. Papanicolau, *Perturbation and "augmentation" des conditions initiales*, "Singular Perturbations and

Boundary Layer Theory", Lecture Notes in Math., vol. 594, Springer, Berlin, 1977, pp. 10–29.

[28] M. L. Bernardi, *Su alcune equazioni d'evoluzione singolari*, Boll. U. M. I. **5, 13-B** (1976), 498–517.

[29] M. L. Bernardi, *On some singular nonlinear evolution equations*, "Differential Equations in Banach Spaces", Lecture Notes in Math. (A. Favini and E. Obrecht, eds.), vol. 1223, Springer-Verlag, Berlin, Heidelberg, New York, 1986, pp. 12–24.

[30] M. L. Bernardi and G. A. Pozzi, *On some singular or degenerate parabolic variational inequalities*, Houston J. Math. **15** (1989), 163–192.

[31] M. L. Bernardi and G. A. Pozzi, *On a class of singular nonlinear parabolic variational inequalities*, Ann. Mat. Pura Appl. (IV) **159** (1991), 117–131.

[32] A. V. Bitsadze, *Equations of the Mixed Type*, Pergamon Press, Oxford, 1964.

[33] L. Boccardo and T. Gallouët, *Non-linear elliptic and parabolic equations involving measure data*, J. Funct. Anal. **87** (1989), 149–169.

[34] K. E. Brenan, S. L. Campbell and L. R. Petzold, *Numerical Solution of Initial-Value Problems in Differential-Algebraic Equations*, North-Holland, New York, Amsterdam, London.

[35] H. Brezis, W. Rosenkrantz and B. Singer, *On a degenerate elliptic-parabolic equation occuring in the theory of probability*, Comm. Pure Appl. Math. **24** (1971), 395–416.

[36] H. Brezis, *Operateurs Maximaux Monotones et Semi-groupes de Contractions dans les Espaces de Hilbert*, North-Holland, New York, Amsterdam, London, 1973.

[37] H. Brezis and W. Strauss, *Semilinear elliptic equations in L^1*, J. Math. Soc. Japan **25** (1973), 565–590.

[38] S. Busenberg and B. Wu, *Convergence theorems for integrated semigroups*, Diff. Int. Eqs. **5** (1992), 509–520.

[39] P. L. Butzer and H. Berens, *Semi-groups of Operators and Approximations, Die Grundlehren der math. Wissenshaften in Einzeldarstellungen*, vol. 145, Springer, Berlin, Heidelberg, 1967.

[40] S. L. Campbell, *Singular Systems of Differential Equations*, Pitman, New York, 1980.

[41] S. L. Campbell, *Singular Systems of Differential Equations II*, Pitman, London, 1982.

[42] S. L. Campbell, C. D. Meyer Jr. and N. J. Rose, *Applications of the Drazin inverse to linear systems of differential equations with singular constant coefficients*, SIAM J. Appl. Math. **31** (1976), 411–425.

[43] S. L. Campbell, *Optimal control of autonomous linear processes with singular matrices in the quadratic cost functional*, SIAM J. Control Optim. **14** (1976), 1092–1106.

[44] S. L. Campbell and C. D. Meyer Jr., *Generalized Inverses of Linear Transformations*, Pitman, London, 1979; Dover, New York, 1991.

[45] M. Campiti and G. Metafune, *Ventcel's boundary conditions and analytic semigroups*, Preprint.

[46] M. Campiti, G. Metafune and D. Pallara, *Degenerate self-adjoint evolution equations on the unit interval*, Semigroup Forum (to appear).

[47] R. W. Carroll, *Abstract Methods in Partial Differential Equations*, Harper and Row, New York, Evanston, London, 1969.

[48] R. W. Carroll and R. E. Showalter, *Singular and Degenerate Cauchy Problems*, Academic Press, New York, San Francisco, London, 1976.

[49] J. Chazarain, *Problèmes de Cauchy abstraits et applications à quelques probl'emes mixtes*, J. Funct. Anal. **7** (1971), 386–446.

[50] S. Chen and R. Triggiani, *Proof of extension of two conjectures on structural damping for elastic systems: The case $\frac{1}{2} \le \alpha \le 1$*, Pacific J. Math. **136** (1989), 15–55.

[51] S. Chen and R. Triggiani, *Gevrey class semigroups arising from elastic systems with gentle dissipation: The case $0 < \alpha < \frac{1}{2}$*, Proceed. A. M. S. **110** (1990), 401–414.

[52] Ph. Clément and J. Prüss, *On second-order differential equations in Hilbert space*, Boll. U. M. I. **B(7)3** (1989), 623–638.

[53] Ph. Clément and C. A. Timmermans, *On $\mathcal{C}_0$-semigroups generated by differential operators satisfying Ventcel's boundary conditions*, Indag. Math. **89** (1986), 379–387.

[54] J. D. Cobb, *On the solution of linear differential equations with singular coefficients*, J. Diff. Eqs. **46** (1982), 310–323.

[55] J. D. Cobb, *Controllability, observability and duality in singular systems*, IEEE Trans. Aut. Control **AC-29** (1984), 1076–1082.

[56] P. Colli and A. Favini, *On some degenerate second order equations of mixed type*, Funkc. Ekv. **38** (1995), 473–489.

[57] P. Colli and A. Favini, *Time discretization of nonlinear Cauchy prob-
      lems applying to mixed hyperbolic-parabolic equations*, Intern. J.
      Math. and Math. Sci. **19** (1996), 481–494.

[58] G. Coppoletta, *Abstract singular evolution equations of "hyperbolic"
      type*, Funct. Anal. **50** (1983), 50–66.

[59] M. G. Crandall and T. Liggett, *Generation of semigroups of non-
      linear transformations in general Banach spaces*, Amer. J. Math. **93**
      (1971), 265–298.

[60] L. Dai, *Singular Control Systems*, Lecture Notes Control and Infor-
      mation Sciences, vol. 118, Springer-Verlag, Berlin, 1989.

[61] G. Da Prato, *Abstract differential equations and extrapolation spaces*,
      "Infinite Dimensional Systems", Lecture Notes in Math., vol. 1076,
      Springer-Verlag, Berlin, Heidelberg, New York, 1983, pp. 53–61.

[62] G. Da Prato and P. Grisvard, *Sommes d'opérateurs linéaires et
      équations différentielles opérationnelles*, J. Math. Pures Appl. **54**
      (1975), 305–387.

[63] G. Da Prato and P. Grisvard, *On an abstract singular Cauchy prob-
      lem*, Comm. Partial. Diff. Eqs. **3** (1978), 1077–1082.

[64] G. Da Prato and P. Grisvard, *Equations d'évolution abstraites non-
      lineaires de type parabolique*, Ann. Mat. Pura Appl. **(IV)120** (1979),
      329–396.

[65] G. Da Prato and P. Grisvard, *Maximal regularity for evolution equa-
      tions by interpolation and extrapolation*, J. Funct. Anal. **58** (1984),
      107–124.

[66] G. Da Prato and E. Sinestrari, *Differential operators with nondense
      domains*, Ann. Sc. Norm. Sup. Pisa Ser. 4 **14** (1987), 285–344.

[67] R. deLaubenfels, *Powers of generators of holomorphic semigroups*,
      Proc. Amer. Math. Soc. **99** (1987), 105–108.

[68] O. A. de Lima, *Existence and uniqueness of solutions for an abstract
      nonlinear hyperbolic-parabolic equation*, Appl. Anal. **24** (1987), 101–
      116.

[69] J. A. Donaldson, *An operational calculus for a class of abstract op-
      erator equations*, J. Math. Anal. Appl. **37** (1972), 167–184.

[70] J. R. Dorroh, *Contraction semi-groups in a function space*, Pacific J.
      Math. **19** (1966), 35–38.

[71] J. R. Dorroh and A. Holderrieth, *Multiplicative perturbation of semi-
      group generators*, Boll. U. M. I. (7) **7-A** (1993), 47–57.

[72] Ju. A. Dubinskii, *On some differential-operator equations of arbitrary order, (russian)*, Matem. Sb. **90(132)** (1973), 3–22; *Engl. transl.*, Math. USSR Sbornik **19** (1973), 1–21.

[73] G. Duvaut and J. L. Lions, *Inequalities in Mechanics and Physics*, Springer-Verlag, Berlin, Heidelberg, New York, 1976.

[74] J. Elschner, *Singular Ordinary Differential Operators and Pseudo-differential Equations*, Lecture Notes in Math., vol. 1128, Springer-Verlag, Berlin, Heidelberg, New York, 1985.

[75] K. -J. Engel, *On dissipative wave equations in Hilbert space*, J. Math. Anal. Appl. **184** (1994), 302–316.

[76] K. -J. Engel and R. Nagel, *One-Parameter Semigroups for Linear Evolution Equations*, Monograph in preparation.

[77] W. N. Everitt, M. K. Kwong and A. Zettl, *Differential equations and quadratic inequalities with a degenerate weight*, J. Math. Anal. Appl. **98** (1984), 378–399.

[78] H. O. Fattorini, *The Cauchy Problem*, Addison-Wesley, London, 1983.

[79] A. Favini, *Su un'equazione astratta di tipo Tricomi*, Rend. Sem. Mat. Univ. Padova **53** (1975), 257–267.

[80] A. Favini, *Su un'equazione astratta di tipo ellittico-iperbolico*, Rend. Sem. Mat. Univ. Padova **55** (1976), 227–242.

[81] A. Favini, *Laplace transform method for a class of degenerate evolution problems*, Rend. Mat. **12** (1979), 511–536.

[82] A. Favini, *Abstract potential operators and spectral methods for a class of degenerate evolution problems*, J. Diff. Eqs. **39** (1981), 212–225.

[83] A. Favini, *Degenerate and singular evolution equations in Banach spaces*, Math. Ann. **273** (1985), 17–44.

[84] A. Favini, *Abstract singular equations and applications*, J. Math. Anal. Appl. **116** (1986), 289–308.

[85] A. Favini, *An operational method for abstract degenerate evolution equations of hyperbolic type*, J. Funct. Anal. **76** (1988), 432–456.

[86] A. Favini and M. Fuhrman, *Approximation results for semigroups generated by multivalued linear operators and applications*, Diff. Int. Eqs. (to appear).

[87] A. Favini, J. A. Goldstein and S. Romanelli, *An analytic semigroup associated to a degenerate evolution equation*, "Stochastic Processes

and Functional Analysis" (J. A. Goldstein, N. E. Gretsky and
J. J. Uhl Jr., eds.), Marcel Dekker, New York, Basel, Hong Kong,
1997, pp. 88–100.

[88] A. Favini, J. A. Goldstein and S. Romanelli, *Analytic semigroups in
      $L_w^p(0,1)$ and in $L^p(0,1)$ generated by some classes of second order
      differential operators*, Taiwanese J. Math. (to appear).

[89] A. Favini, J. A. Goldstein and S. Romanelli, *On some classes of
      differential operators generating analytic semigroups*, forthcoming.

[90] A. Favini and E. Obrecht, *Condition for parabolicity of second order
      abstract differential equations*, Diff. Int. Eqs. **4** (1991), 1005–1022.

[91] A. Favini and H. Tanabe, *On regularity of solutions to n-order differ-
      ential equations of parabolic type in Banach spaces*, Osaka J. Math.
      31 (1994), 225–246.

[92] A. Favini and P. Plazzi, *On some abstract degenerate problems of
      parabolic type - 1: The linear case*, Nonlinear Anal. **12** (1988), 1017–
      1027.

[93] A. Favini and P. Plazzi, *On some abstract degenerate problems of
      parabolic type - 2: The nonlinear case*, Nonlinear Anal. **13** (1989),
      23–31.

[94] A. Favini and P. Plazzi, *On some abstract degenerate problems of
      parabolic type - 3: Applications to linear and nonlinear problems*,
      Osaka J. Math. **27** (1990), 323–359.

[95] A. Favini and R. Triggiani, *Analytic and Gevrey class semigroups
      generated by $-A + iB$ and applications*, "Differential Equations in
      Banach Spaces" (G. Dore, A. Favini, E. Obrecht, A. Venni, eds.),
      Marcel Dekker, New York, 1993, pp. 93–114.

[96] A. Favini and A. Yagi, *Space and time regularity for degenerate evo-
      lution equations*, J. Math. Soc. Japan **44** (1992), 331–350.

[97] A. Favini and A. Yagi, *Multivalued linear operators and degenerate
      evolution equations*, Ann. Mat. Pura Appl. (IV) **163** (1993), 353–384.

[98] A. Favini and A. Yagi, *Absrtact second order differential equations
      with applications*, Funkc. Ekv. **38** (1995), 81–99.

[99] A. Favini and S. Romanelli, *Analytic semigroups on $C([0,1])$ gener-
      ated by some classes of second order differential operators*, Semigroup
      Forum **56** (1998), 367–372.

[100] A. Favini and S. Romanelli, *Second order operators as generators
      of analytic semigroups on $C(0,\infty)$ or on $L_{\alpha-1/2}^p(0,\infty)$*, "Approxima-
      tion and Optimization" (D. D. Stancu, G. Coman, W. W. Breckner,
      P. Blaga, eds.), Transilvania Press, 1997, pp. 93–100.

[101] W. Feller, *Two singular diffusion problems*, Ann. of Math. **54** (1951), 173–182.

[102] W. Feller, *The parabolic differential equations and the associated semigroups of transformations*, Ann. of Math. **55** (1952), 468–519.

[103] G. Fichera, *On a degenerate evolution problem*, "Partial Differential Equations with Real Analysis" (C. H. Begeher, A. Jeffrey, eds.), Res. Notes, vol. 263, Pitman, 1992, pp. 15–42.

[104] A. Friedman, *Partial Differential Equations*, Holt, Rinehart and Winston, New York, 1969.

[105] A. Friedman and Z. Schuss, *Degenerate evolution equations in Hilbert space*, Trans. AMS **161** (1971), 401–427.

[106] K. O. Friedrichs, *Symmetric positive linear differential equations*, Comm. Pure Appl. Math. **11** (1958), 333–418.

[107] T. Gallouët and J. -M. Morel, *On some semilinear problems in L^1*, Boll. U. M. I. **A(6)4** (1985), 123–131.

[108] F. R. Gantmacher, *The Theory of Matrices, Vol. II*, Chelsea, New York, 1964.

[109] J. A. Goldstein, *Semigroups of Linear Operators and Applications*, Oxford Univ. Press, New York, 1985.

[110] J. A. Goldstein, *The Kompaneets equation*, "Differential Equations in Banach Spaces" (G. Dore, A. Favini, E. Obrecht, A. Venni, eds.), Marcel Dekker, New York, 1993, pp. 115–123.

[111] J. A. Goldstein and C. -Y. Lin, *Singular nonlinear parabolic boundary value problems in one space dimension*, J. Diff. Eqs. **68** (1987), 429–443.

[112] E. Griepentrog and R. März, *Differential-Algebraic Equations and Their Numerical Treatment,*, Teubner-Texte zur Mathematik, vol. 88, B. G. Teubner, Leipzig, 1986.

[113] P. Grisvard, *Equations différentielles abstraites*, Ann. Ec. Norm. Sup. Paris **2** (1969), 311–395.

[114] P. Grisvard, *Spazi di tracce e applicazioni*, Rendiconti di Matem. **(4)5** (1972), 657–729.

[115] D. Henry, *Geometric theory of semilinear parabolic equations*, Lecture Notes in Math., vol. 840, Springer-Verlag, Berlin, Heidelberg, New York, 1981.

[116] M. Hieber, A. Holderrieth and F. Neubrander, *Regularized semigroups and systems of linear partial differential equations*, Ann. Sc. Norm. Sup. Pisa **19** (1992), 362–379.

[117] T. Kato, *Perturbation Theory for Linear Operators*, Springer-Verlag, Berlin, Heidelberg, New York, 1966.

[118] T. Kato and H. Tanabe, *On the abstract evolution equation*, Osaka Math. J. **14** (1962), 107–133.

[119] H. Kellermann and M. Hieber, *Integrated semigroups*, J. Func. Anal. **84** (1989), 160–180.

[120] M. G. Krein, *Theory of self-adjoint extensions of semi-bounded Hermitian operators and its applications, I*, Mat. Sb. **20(60)** (1947), 431–495; *II*, Mat. Sb. **21(63)** (1947), 356–404.

[121] S. G. Krein, *Linear Differential Equations in Banach Spaces*, Nauka, Moscow, 1963 (russian); Engl. Transl., Transl. Math. Monogr. vol. 29, A. M. S., Providence, 1972.

[122] J. Lagnese, *Singular differential equations in Hilbert space*, SIAM J. Math. Anal. 4 (1973), 623–637.

[123] J. Lagnese, *Boundary Stabilization of Thin Plates*, SIAM, 1989.

[124] R. K. Lamm and I. G. Rosen, *An approximation theory for the estimation of parameters in degenerate Cauchy problems*, J. Math. Anal. Appl. **162** (1991), 13–48.

[125] I. Lasiecka and R. Triggiani, *Differential and Algebraic Riccati Equations with Applications to Boundary/Point Control Problems: Continuous Theory and Approximation Theory*, Lecture Notes Control. Inform. Sc. **164**, Springer-Verlag, Berlin, Heiderberg, New York, 1991.

[126] P. D. Lax and R. S. Phillips, *Local boundary conditions for dissipative symmetric linear differential operators*, Comm. Pure Appl. Math. **13** (1960), 427–455.

[127] J. L. Lions, *Semi-groupes distributions*, Portug. Math. **19** (1960), 141–164.

[128] J. L. Lions, *Equations Différentielles Operationnelles et Problèmes aux Limites*, Springer-Verlag, Berlin, Heiderberg, New York, 1961.

[129] J. L. Lions and E. Magenes, *Problèmes aux limites non homogènes et applications, Vol. 1*, Dunod, Paris, 1968.

[130] J. L. Lions and J. Peetre, *Sur une classe d'espaces d'interpolation*, Inst. Hautes Etudes Sci. Publ. Math. **19** (1964), 5–68.

[131] A. Lunardi, *Analytic Semigroups and Optimal Regularity in Parabolic Problems*, Birkhäuser, Basel, Boston, Berlin, 1995.

[132] A. B. Maciel, *On hyperbolic-parabolic equations with a continuous nonlinearity*, Nonlinear Anal. **20** (1993), 745–754.

[133] P. Mandl, *Analytical Treatment of one-dimensional Markov Processes*, Die Grundlehr. math. Wissen. Einz. vol. 151, Springer-Verlag, Berlin, Heidelberg, New York, 1968.

[134] R. Martin, *Nonlinear Operators and Differential Equations in Banach Spaces*, J. Wiley & Sons, New York, London, Sidney, Toronto, 1976.

[135] A. C. McBride, *Semigroups of linear operators: an introduction*, Pitman Research Notes Math. Series 156, Longman Scientific and Technical (1987).

[136] L. A. Medeiros, *Nonlinear hyperbolic-parabolic partial differential equations*, Funkc. Ekv. **23** (1980), 151–158.

[137] G. Metafune, *Analyticity for some degenerate one-dimensional evolution equations*, Tübingen Berichte zur Funktionalanalysis **5** (1995/96), 260–284.

[138] R. Nagel (ed.), *One-parameter Semigroups of Positive Operators*, Lecture Notes in Math. vol. 1184, Springer-Verlag, Berlin, Heidelberg, New York, 1986.

[139] M. A. Naimark, *Linear Differential Operators, Part II: Linear Differential Operators in Hilbert Spaces*, Ungar Publ. Co., New York, 1968.

[140] M. Z. Nashed and Y. Zhao, *The Drazin inverse for singular evolution equations and partial differential equations*, "Recent Trends in Differential Equations" (R. R. Agarwal, ed.), World Scientific, 1992, pp. 441–456.

[141] F. Neubrander, *Integrated semigroups and their application to the abstract Cauchy problem*, Pacific J. Math. **135** (1988), 111–155.

[142] F. Neubrander and A. Rhandi, *Degenerate abstract Chauchy problems*, Seminar Notes in Functional Analysis and Partial Differential Equations, Louisiana State University (1992/93).

[143] H. D. Niessen and A. Zettl, *Singular Sturm-Liouville problems: The Friedrichs extension and comparsion of eigenvalues*, Proc. London Math. Soc. **(3)64** (1992), 545–578.

[144] M. M. H. Pang, L^1 *properties of two classes of singular second order elliptic operators*, J. London Math. Soc. **(2)38** (1988), 525–543.

[145] A. Pazy, *Semigroups of Linear Operators and Applications to Partial Differential Equations*, Springer-Verlag, Berlin, Heidelberg, New York, 1983.

[146] M. Povoas, *On some singular hyperbolic evolution equations*, J. Math. Pures et Appl. **60** (1981), 133–192.

[147] A. G. Rutkas, *Cauchy's problem for the equation $Ax'(t) + Bx(t) = f(t)$*, Diff. Uravn. **11** (1975), 1996–2010. (russian)

[148] A. G. Rutkas, *Classification and properties of solutiuons of the equation $Ax'(t) + Bx(t) = f(t)$*, Diff. Uravn. **25** (1989), 1150–1155. (russian)

[149] N. Sauer, *Linear evolution equations in two Banach spaces*, Proc. Roy. Soc. Edinburgh **91 A** (1992), 287–303.

[150] N. Sauer, *Implicit evolution equations and empathy theory*, "Recent Developments in Evolution Equations" (A. C. McBride and G. F. Roach, eds.), Pitman Research Notes in Math. vol. 324, Longman, Harlow, 1995, pp. 32–39.

[151] M. Schechter, *Basic theory of Fredholm operators*, Ann. Sc. Norm Sup. Pisa **21** (1967), 361–380.

[152] M. Schechter, *Principles of Functional Analysis*, Academic Press, New York, 1971.

[153] L. Schwartz, *Analyse Mathématique, Tome I*, Hermann, Paris, 1967.

[154] R. Seeley, *Interpolation in L^p with boundary conditions*, Studia Math. **44** (1972), 47–60.

[155] R. E. Showalter, *Hilbert Space Methods for Partial Differential Equations*, Pitman, London, San Francisco, Melbourne, 1977.

[156] R. E. Showalter, *A nonlinear parabolic-Sobolev equation*, J. Math. Anal. Appl. **50** (1975), 183–190.

[157] E. Sinestrari, *On the abstract Cauchy problem of parabolic type in spaces of continuous functions*, J. Math. Anal. Appl. **107** (1985), 16–66.

[158] V. P. Skripnik, *Degenerate linear systems*, Izvestiya VUZ Matematika **26** (1982), 62–67.

[159] P. E. Sobolevskii, *Equations of parabolic type in Banach space*, Amer. Math. Soc. Transl. Ser. 2 **49** (1966), 1–62.

[160] P. E. Sobolevskii, *On degenerate parabolic operators*, Dokl. Akad. Nauk SSSR **196** (1971); Engl. Transl., Soviet Math. Dokl. **12** (1971), 129–132.

[161] H. B. Stewart, *Generation of analytic semigroups by strongly elliptic operators*, Trans. Amer. Math. Soc. **199** (1974), 141–162.

[162] G. A. Sviridyuk, *On the general theory of operator semigroups*, Uspekhi Mat. Nauk. **49** (1994 no. 4(298)), 47–74.

[163] G. A. Sviridyuk, *Phase portraits of Sobolev-type semilinear equations with a relatively strongly sectorial operator*, Algebra i Analisis **6** (1995), 252–272 (russian); Engl. transl., St. Peterburg Math. J. **6** (1995), 1109–1126.

[164] G. A. Sviridyuk and A. A. Efremov, *Optimal control of Sobolev-type linear equations with relatively p-sectorial operators*, Diff. Uravn. **31** (1995), 1912–1919 (russian); Engl. transl., Diff. Eqs. **31** (1995), 1882–1890.

[165] K. Taira, *Diffusion Processes and Partial Differential Equations*, Academic Press, Boston, San Diego, Tokyo, Toronto, 1988.

[166] K. Taira, *The theory of semigroups with weak singularity and its applications to partial differential equations*, Tsukuba J. Math. **13** (1989), 513–562.

[167] H. Tanabe, *Equations of Evolution*, Pitman, London, 1979.

[168] H. Tanabe, *Functional Analysis Methods for Partial Differential Equations*, Marcel Dekker, New York, 1997.

[169] A. E. Taylor, *Functioal Analysis*, J. Wiley & Sons, New York, 1958.

[170] S. Taylor, *Gevrey class semigroup*, Ph. D. Thesis, Chapter 1, School of Mathematics, University of Minnesota, 1989.

[171] C. A. Timmermans, *On C_0-semigroups in a space of bounded continuous functions in the case of entrance or natural boundary points*, "Approximation and Optimization" (J. A. Gómez Fernández et al., eds.), Springer-Verlag, Berlin, Heidelberg, New York, 1988, pp. 209–216.

[172] H. Triebel, *Interpolation Theory, Function Spaces, Differential Operators*, North Holland, Amsterdam, 1978.

[173] V. Vespri, *Analytic semigroups, degenerate elliptic operators and applications to non-linear Cauchy problems*, Ann. Mat. Pura Appl. **155** (1989), 353–388.

[174] K. Wang, *The linear Kompaneets equation*, J. Math. Anal. Appl. **198** (1996), 552–570.

[175] J. Weidmann, *Lineare Operatoren in Hilberträumen*, B. G. Teubner, Stuttgart, 1976.

[176] J. Weidmann, *Spectral theory of ordinary differential operators*, Lecture Notes in Math. vol. 1258, Springer-Verlag, Berlin, Heidelberg, New York, 1987.

[177] A. D. Wentzell, (*Ventcel'*), *On boundary condotions for multidimensional diffusion processes*, Teoriya Veroyat. i ee Primen (1959), 172–185 (russian); Engl. transl., Theory of Prob. and its Appl. **4** (1959), 164-177.

[178] A. D. Wentzell, (*Ventcel'*), *Semigroups of operators associated with a generalized second order differential operator*, Dokl. Akad. Nauk SSSR **111** (1956), 269–272. (russian)

[179] C. H. Wilcox, *Initial-boundary value problems for linear hyperbolic partial differential equations of the second order*, Archiv Rat. Mech. Anal. **10** (1962), 361–400.

[180] C. Wild, *Semigroupes de croissance $\alpha < 1$ holomorphes*, C. R. Acad. Sc. Paris Série A **285** (1977), 437–440.

[181] X. Xu, *Existence and convergence theorems for doubly nonlinear partial differential equations of elliptic-parabolic type*, J. Math. Anal. Appl. **150** (1990), 205–225.

[182] A. Yagi, *On the abstract evolution equations in Banach spaces*, J. Math. Soc. Japan **28** (1976), 290–303.

[183] A. Yagi, *Parabolic evolution equations in which the coefficients are the generators of infinitely differential semigroups*, Funkc. Ekv. **32** (1989), 107–124.

[184] A. Yagi, *Parabolic evolution equations in which the coefficients are the generators of infinitely differential semigroups, II*, Funkc. Ekv. **33** (1990), 139–150.

[185] A. Yagi, *Generation theorem of semigroup for multivalued linear operators*, Osaka J. Math. **28** (1991), 385–410.

[186] S. Ya. Yakubov, *A nonlocal boundary value problem for a class of Petrovskii well posed equations*, Math. Sb. (N. S.) **118** (1982), 252–261 (russian); Engl. transl., Math. USSR Sb. **46** (1983), 255–265.

[187] K. Yosida, *Functional Analysis*, 6th edition, Springer-Verlag, Berlin, Heidelberg, New York, 1980.

[188] S. Zubova and K. Tchernyshov, *On the linear differential equations with a Fredholm operator at a derivative*, Diff. Uravn. i Prim. **14** (1976), 21–39. (russian)

LIST OF SYMBOLS

INDEX

INDEX

DATE DUE

DEMCO, INC. 38-2931